Life Sciences Research Report LS 53

Held and published on behalf of the
Freie Universität Berlin

Sponsored by:
Deutsche Forschungsgemeinschaft

Twins as a Tool of Behavioral Genetics

Edited by

T.J. BOUCHARD, Jr. and P. PROPPING

Report of the Dahlem Workshop on
What Are the Mechanisms Mediating the Genetic
and Environmental Determinants of Behavior?
Twins as a Tool of Behavioral Genetics
Held in Berlin, 17–22 May 1992

Program Advisory Committee:
T.J. Bouchard, Jr. and P. Propping, Chairpersons
D.A. Hay, C.G.N. Mascie-Taylor, N.L. Pedersen, K.K. Kidd

JOHN WILEY & SONS
Chichester • New York • Brisbane • Toronto • Singapore

Library of Congress Cataloging-in-Publication Data

Dahlem Workshop on What Are the Mechanisms Mediating the Genetic and
  Environmental Determinants of Behavior? Twins as a Tool of
  Behavioral Genetics (1992 : Berlin, Germany)
    Twins as a tool of behavioral genetics : report of the Dahlem
  Workshop on What Are the Mechanisms Mediating the Genetic and
  Environmental Determinants of Behavior? Twins as a Tool of
  Behavioral Genetics, Berlin 1992, May 17–22 / edited by T.J.
  Bouchard, Jr., and P. Propping ; program advisory committee, T.J.
  Bouchard, Jr. . . . [et al.].
        p.   cm. — (Dahlem workshop reports)   (Life sciences research
  report ; 53)
    Includes bibliographical references and indexes.
    ISBN 0-471-94174-3
    1. Behavior genetics—Congresses.   2. Twins—Congresses
  I. Bouchard, T. J. (Thomas J.)   II. Propping, Peter, *1942–*
  III. Title.   IV. Series.   V. Series: Life sciences research report ;
  53.
  QH457.D34   1992
  155.44$'$4—dc20   93–8773
                                                                CIP

British Library Cataloguing in Publication Data

A catalogue record for this book is available from the British Library

ISBN 0-471-94174-3

Dahlem Editorial Staff: J. Lupp, C. Rued-Engel
Typeset in 10/12pt Times from authors' disks by Text Processing Department,
John Wiley & Sons Ltd, Chichester
Printed and bound in Great Britain by Biddles Ltd, Guildford, Surrey

Contents

The Dahlem Konferenzen

The purpose of the Dahlem Konferenzen is to promote an international interdisciplinary exchange of scientific information and ideas as well as to stimulate international cooperation in research. This is achieved by arranging discussion workshops, mainly in the life sciences and environmental sciences, organized according to a model developed and tested by the Dahlem Konferenzen.

Dahlem Konferenzen was founded in 1974 by the Stifterverband für die Deutsche Wissenschaft[1] in collaboration with the Deutsche Forschungsgemeinschaft[2] to promote more effective communication between scientists. It was named after the Dahlem district of Berlin, long a home of the sciences and arts. In January, 1990, it became incorporated into the Freie Universität Berlin. Financial support comes from the Senate of the Land Berlin, the Deutsche Forschungsgemeinschaft, and private foundations.

As scientific research has become increasingly interdisciplinary, a growing need has emerged for specialists in one field to understand the problems and work with the concepts of related fields. New insights can be gained when a problem is approached from the standpoint of another discipline, and because no existing form of scientific meeting provided the forum necessary for such exchanges, Dahlem Konferenzen created a new concept, which has since been tested and refined over the years. Now internationally recognized as the Dahlem Workshop Model, it provides a framework for coherent interdisciplinary discussion of a topic in five working days, and culminates in the draft manuscript for a book.

Dahlem Workshops provide a unique opportunity for scientists to pose questions to colleagues from different disciplines and to solicit alternative opinions on contentious issues. The aim is not to solve problems or to reach a consensus, but to identify gaps in knowledge, to find new ways to approach stubborn issues, and to define priorities for research. This approach is well summed up in the instructions given to participants: please state what you do not know, rather than what is known.

---

[1] The Donors Association for the Promotion of Sciences and Humanities, a foundation created in 1921 in Berlin and supported by German trade and industry to fund basic research in the sciences

[2] German Science Foundation

Workshop topics are proposed by leading scientists and are approved by a scientific board advised by qualified referees. Approximately one year before the workshop, an advisory committee meets to determine the scientific program and select participants, who are invited according to their scientific reputations, with the exception of a number of places reserved for junior German scientists.

The discussions at each workshop are organized around four key questions, each tackled by a group of about twelve participants with a range of expertise. There are no lectures. Instead, prior to the workshop, selected participants write background papers, which review the group's discussion topic and serve as the basis for the discussion. These papers are distributed to all participants before the meeting, with selected participants acting as referees. Based on the discussion during the week, each group prepares a report reflecting the ideas, opinions, and controversies which have emerged, as well as identifies problem areas and directions for future research.

The revised background papers and group reports are published as the Dahlem Workshop Reports. Each volume is edited by the chairperson(s) of the workshop and the Dahlem staff. The Reports provide multidisciplinary surveys by an international group of distinguished scientists, based on discussions of advanced concepts, techniques, and models. The Dahlem Workshop Reports are published in two series: Life Sciences and Environmental Sciences (formerly Physical, Chemical, and Earth Sciences).

Jennifer Altman, Acting Director
Klaus Roth, Director
Dahlem Konferenzen der Freien Universität Berlin
Rothenburgstr. 33, 12165 Berlin, F.R. Germany

List of Participants with Fields of Research

J. ASENDORPF Max-Planck-Institut für Psychologische Forschung, Postfach 44 01 09, Leopoldstr. 24, 80802 München, F.R. Germany

Personality development in the social and cognitive domain, and interrelations between these two domains; behavioral developmental genetics

L.A. BAKER Department of Psychology, S.G. Mudd Building #501,University of Southern California, Los Angeles, CA 90089-1061, U.S.A.

Genetic and environmental models of transmission of human personality, cognitive abilities, mate preferences; genetic and environmental influences in human growth

S.D. BIRYUKOV Institute of Psychology, Academy of Sciences, Yaroslavskaya 13, 129366 Moscow, Russia

Psychology of individual differences, EEG-typology, family genetic study of temperament, cross-cultural study of temperament

D. BISHOP MRC Applied Psychology Unit, 15 Chaucer Rd., Cambridge CB2 2EF, U.K.

Developmental language disorders including a twin study to investigate role of genetic factors

D.I. BOOMSMA Department of Psychonomics, Vrije Universiteit, De Boelelaan 1111, B-105, 1081 HV Amsterdam, Netherlands

Psychophysiology (cardiovascular and EEG research), behavioral genetics (twin studies), genetics of twinning

P. BORKENAU Abteilung für Psychologie, Universität Bielefeld, Postfach 10 01 31, 33501 Bielefeld, F.R. Germany

Personality and appearance, person perception, behavioral genetics

T.J. BOUCHARD, JR. Department of Psychology, University of Minnesota, Elliott Hall, 75 East River Rd., Minneapolis, MN 55455–0344, U.S.A.

Twin research, behavioral genetics, personality, mental abilities, psychological interests

C.R. BRAND Department of Psychology, University of Edinburgh, 7 George Square, Edinburgh EH8 9JZ, U.K.

The nature and origins of individual differences in fluid intelligence (gF) as explored by studies of "inspection time" and IQ

E.M. BRYAN Multiple Births Foundation, Queen Charlotte's and Chelsea Hospital, Goldhawk Rd., London W6 0XG, U.K.

Bereavement in multiple pregnancy including selective reduction; services to families with multiple births including twins clinics

M.S. BUCHSBAUM Brain Imaging Center, University of California, Irvine, CA 92717, U.S.A.

Imaging of brain structure and function

D.M. BUSS Department of Psychology, University of Michigan, 580 Union Drive, Ann Arbor, MI 48109–1346, U.S.A.

Personality psychology; evolutionary psychology; human mating strategies; conflict between sexes; competition; status, prestige, reputation

R. DEROM Centrum Menselijke, Erfelijkheid, Katholieke Universiteit Leuven, Herestraat 49, 3000 Leuven, Belgium

Population genetics, obstetrics and gynecology, didymology, perinatal epidemiology

D. FULKER Institute for Behavioral Genetics, Campus Box 447, University of Colorado, Boulder, CO 80309–0447, U.S.A.

Longitudinal twin research of cognitive abilities

H. GARDNER Project Zero, Harvard University, 323 Longfellow Hall, Appian Way, Cambridge, MA 02138, U.S.A.

Development in normal and gifted children, breakdown of cognitive abilities in brain-damaged adults, assessment of multiple intelligences

H.H. GOLDSMITH Department of Psychology, University of Wisconsin, 1202 W. Johnson St., Madison, WI 53706–1696, U.S.A.

Genetics of personality development; infant and childhood emotionality; psychometrics

I.I. GOTTESMAN Department of Psychology, Gilmer Hall, University of Virginia, Charlottesville, VA 22903, U.S.A.

Psychiatric and behavioral genetics: schizophrenia, alcoholism, personality, and personality disorders; twin strategies

D.A. HAY Department of Psychology, La Trobe University, Bundoora, Victoria 3083, Australia

Behavioral development of twins, effects of twins on the family, twin-family study of the effects of parental psychopathology on their children; attention deficit disorder in twins

A.C. HEATH Department of Psychiatry, Washington University School of Medicine, 4940 Children's Place, St. Louis, MO 63110, U.S.A.

Genetic epidemiology of substance use and other psychiatric disorders

J. HEBEBRAND Klinik für Kinder- und Jugendpsychiatrie, Universität Marburg, Hans-Sachs-Str. 6, 35039 Marburg/Lahn, F.R. Germany

Genetics of Tourette syndrome, X-linked bi-polar illness

J. KAPRIO Department of Public Health, University of Helsinki, P.O. Box 52, Mannerheimintie 96A, SF-00014 Helsinki, Finland

Twin studies of smoking, alcohol use, and obesity; genetic epidemiology of coronary heart disease; twin registry with twin and twin-family studies of other traits and diseases

M. KINSBOURNE 158 Cambridge Street, Winchester, MA 01890, U.S.A.

Developmental neuropsychology and learning disabilities

C. KIRSCHBAUM c/o Prof. D. Felten, Department of Neurobiology and Anatomy, University of Rochester, Box 603, 601 Elmwood Ave., Rochester, NY 14642, U.S.A.

P. KLINE Department of Psychology, University of Exeter, Devon EX4 4QJ, U.K.

Personality test development — the big five; percept genetics

J. KÖRNER Psychiatrische Klinik, Universität Bonn, Sigmund-Freud-Str. 25, 53127 Bonn, F.R. Germany

Psychiatric genetics, genetic association studies in schizophrenia and the affective disorders and linkage studies in the affective disorders, with molecular genetic methods

B. KRACKE Department of Psychology, Justus-Liebig-Universität Giessen, Otto-Behaghel-Str. 10F, 35394 Giessen, F.R. Germany

Relative timing of pubertal maturation and problem behaviors (smoking and drinking) in adolescent females and males

E. KRINGLEN Psychiatriske Klinikk, University of Oslo, PO Box 1072, Vinderen, N-0316 Oslo 3

Twin studies of schizophrenia and affective disorders, Alzheimer's disease

U. LINDENBERGER Max-Planck-Institut für Bildungsforschung, Center for Psychology and Human Development, Lentzeallee 94, 14195 Berlin, F.R. Germany

Intellectual abilities in old and very old age, structure of intelligence, cognitive development and aging

J.C. LOEHLIN Department of Psychology, University of Texas, Austin, TX 78712, U.S.A.

Human behavioral genetics, structural modeling

A. MACDONALD Genetics Section, Institute of Psychiatry, De Crespigny Park, Denmark Hill, London SE5 8AF, U.K.

Neurotic disorders (obsessive compulsive disorder and normal obsessionality); schizophrenia, melanocytic naevi, arthritis, pre-eclampsia, Alzheimer's Disease

K. MCCARTNEY Department of Psychology, University of New Hampshire, Conant Hall, Durham, NH 03824, U.S.A.

Early experience and development (e.g., child care experience and mother-child attachment); genotype-environment correlation

P. MCGUFFIN Department of Psychological Medicine, University of Wales, College of Medicine, Heath Park, Cardiff CF4 4XN, U.K.

Family, twin, and molecular genetic studies of affective disorders in schizophrenia

W. MAIER Department of Psychiatry, University of Mainz, Untere Zahlbacher Str. 8, 55131 Mainz, F.R. Germany

Family and linkage studies in major psychiatric disorders (including anxiety and personality disorders); neuropsychology of schizophrenia, epidemiology of dementia and affective disorders; personality factors in psychiatric disorders

N.G. MARTIN Queensland Institute of Medical Research, 300 Herston Road, Brisbane, Queensland 4029, Australia

Twin and family studies of personality, attitudes, substance use/abuse, and psychiatric symptoms

C.G.N. MASCIE-TAYLOR Department of Biological Anthropology, University of Cambridge, Downing Street, Cambridge CB2 3DZ, U.K.

Genetics, biosocial studies

M. Nöthen Institut für Humangenetik, University of Bonn, Wilhelmstr. 31, 53111 Bonn, F.R. Germany

Linkage and association studies in major psychoses

D.L. Pauls Child Study Center, Yale University School of Medicine, 230 S. Frontage Rd., P.O. Box 3333, New Haven, CT 06510, U.S.A.

Genetics of childhood neuropsychiatric conditions

N.L. Pedersen Division of Epidemiology, Institute of Environmental Medicine, The Karolinska Institute, Box 210, 17177 Stockholm, Sweden

Behavioral genetics, specializing in the study of twins reared apart

P. Propping Institut für Humangenetik, University of Bonn, Wilhelmstr. 31, 53111 Bonn, F.R. Germany

Psychiatric genetics, medical genetics

M. Rietzschel Psychiatrische Universitätsklinik, Sigmund-Freud-Str. 25, 53127 Bonn, F.R. Germany

Psychiatric genetics: genetic association studies in schizophrenia and the affective disorders and linkage studies in the affective disorders, with molecular genetic methods

D.C. Rowe School of Family and Consumer Resources, University of Arizona, Tucson, AZ 85721, U.S.A.

Behavioral genetic studies of adolescent behavior

M. Rutter MRC Child Psychiatry Unit, Institute of Psychiatry, De Crespigny Park, Denmark Hill, London SE5 8AF, U.K.

Twin and family studies of autism; longitudinal twin study of psychopathology in 8–16 year age period; twin study of children with conduct disorder and/or emotional disorder, incorporating a follow-up into adult life; comparison of twins and singletons with respect to both levels and patterns of psychopathology, and also patterns of family interaction; family genetic study of childhood depression, incorporating follow-up into adult life

H. Schepank Psychosomatic Hospital, Zentralinstitut für Seelische Gesundheit, P.O. Box 122 120, Quadrat J5, 68072 Mannheim, F.R. Germany

Twin research: twins with psychoneuroses, character neuroses and psychosomatic disorders; 20-year follow up; epidemiology: field studies of psychogenic disorders with long-term follow-up studies; psychotherapeutic supply, in-patient psychotherapy process research; transcultural psychosomatics

H.H. Stassen Psychiatric University Hospital, Research Department, P.O. Box 68, 8029 Zurich, Switzerland

Psychiatric genetics and twin family studies, particularly in the field of EEG research

N. WALLER Department of Psychology, University of California at Davis, Davis, CA 95616, U.S.A.

Behavioral genetics of personality and attitude

1

Twins: Nature's Twice-told Tale

T.J. BOUCHARD, Jr.[1] and P. PROPPING[2]

[1]Department of Psychology, Institute of Genetics, University of Minnesota,
Minneapolis, MN 55455-0344, U.S.A.
[2]Institut für Humangenetik, Universität Bonn, Wilhelmstr. 31,
53111 Bonn, F.R. Germany

The idea that twins can be used as a natural experiment to subject hypotheses regarding the etiology of behavior to empirical testing is an old one. St. Augustine of Hippo, in his masterpiece *The City of God* (Book V), cited evidence from twins in order to refute astrology and, by implication, support Christianity and his theory of predestination (Augustine of Hippo 415). Francis Galton, however, was the first person to frame the problem in a scientific context. Drawing on a wide variety of sources—including German work in the biology of twinning (Kleinwachter 1871; Spaeth 1860, 1862), statistics on twins gathered for the National Life Assurance Society (Galton 1876), and the new cellular point of view developing in biology (Darlington 1962)—Galton laid the conceptual basis, over a period of many years, for the development of quantitative behavioral genetic methods.

FRANCIS GALTON AND THE HISTORY OF TWINS

Galton's classic paper on twins, "The history of twins as a criterion of the relative powers of nature and nurture" (Galton 1875), is often cited as the first statement of the twin method. In a formal statistical sense it is not; in a conceptual sense it is much more. Galton clearly understood the confounded nature of the data gathered using studies of ordinary families and showed how special types of families—genetically matched (monozygotic or MZ) and genetically unmatched (dizygotic or DZ) twins who both underwent similar environmental experiences—solved the problem.

> The exceedingly close resemblance attributed to twins has been the subject of many novels and plays and most persons have felt a desire to know upon what basis of truth those works of fiction may rest. But twins have many other claims to attention, one of which will be discussed in the present memoir. It is that their *history* [our emphasis] affords a means of distinguishing between the effects of tendencies received at birth, and those that were imposed by the circumstances of their after lives; in other words, between

Twins as a Tool of Behavioral Genetics
Edited by T.J. Bouchard, Jr. and P. Propping © 1993 John Wiley & Sons Ltd.

the effects of nature and of nurture. The objection to statistical evidence in proof of its inheritance has always been: "The persons whom you compare may have lived under similar social conditions and have had similar advantages of education, but such prominent conditions are only a small part of those that determine the future of each man's life. It is to trifling accidental circumstances that the bent of his disposition and his success are mainly due, and these you leave wholly out of account—in fact, they do not admit of being tabulated, and therefore your statistics, however plausible at first sight, are really of very little use" (Galton 1875, p. 391).

It should be emphasized that the main criticism of Galton's previous family studies (Galton 1869), cited in the quote above, is one that behavioral geneticists constantly target at environmentalists today—family studies are uninformative when taken alone, as they completely confound environmental and genetic influences. To solve this problem, Galton turned to the study of "twins who were closely alike in boyhood and youth, and who were educated together for many years" to see if they, in fact, grew subsequently unlike. He also proposed the converse method of studying "the history of twins who were exceedingly unlike in childhood, to learn how far their characters become assimilated under the influence of identical nurtures, in as much as they had the same home, the same teachers, the same associates, and every other respect the same surroundings." He proposed studying both types of twins and formulated an early version of the famous "equal environment assumption" of the ordinary twin method.

Even with a somewhat faulty understanding of the phenomenon of twinning (see below), Galton's early observations were close to the mark.

> There is no escape from the conclusion that nature prevails enormously over nurture when the differences of nurture do not exceed what is commonly to be found among persons of the same rank of society and in the same country. My only fear is that my evidence seems to prove too much, and contrary to all experience that nurture should go for so little. But experience is often fallacious in ascribing great effects to trifling circumstances (Galton 1875, p. 404).

Galton's problem ("my evidence seems to prove too much") remains with us in the behavioral sciences, even though behavioral geneticists continually constrain their generalizations in much the same way Galton did ("when the differences of nurture do not exceed..."). Sandra Scarr recently summarized her views regarding the influence of heredity and environment on behavioral development and, without citing Galton, came to a conclusion remarkably similar to his.

> Ordinary differences between families have little effect on children's development, unless the family is outside of normal developmental range. Good enough, ordinary parents probably have the same effects on their children's development as culturally defined super-parents [Rowe 1990; see also Rowe, this volume]....As Richard Weinberg and I said (Scarr and Weinberg 1978), children's outcomes do not depend on whether parents take children to the ball game or to a museum so much as they depend on genetic transmission, on plentiful opportunities, and on having a good enough environment that supports children's development to become themselves (Scarr 1992).

The plethora of antagonistic responses to her article, which are scheduled to appear in 1993 in the journal of *Child Development*, sound very much like some of the reviews of Galton's work that appeared in *The Times* over one hundred years ago.

Galton's conclusions were based on his rough-and-ready analysis of questionnaire (anecdotal) data gathered from 94 sets of twins, some (35 sets) of whom were classified as alike (MZ) and some (20 sets) of whom were unlike (DZ). Classification was based on reports that they had been physically similar to each other and equally similar in susceptibility to illness from early on in life or that they had been different from each other from early on in life. Galton's method in this early study was exceedingly similar to contemporary studies that diagnose two types of twins via a questionnaire, a method that we now know is quite reliable (Segal 1984a).

Until about 1925 (Lauterbach 1925) there was much confusion regarding the nature of twins. Galton believed there were two types of twins (Corney 1984), as the questionnaire he used to collect data clearly shows. He made many of his fundamental contributions to science not because of his intense interest in quantifying observations, although he was a genius at that (Pearson 1924), or because of his interest in statistical techniques, which he was only good at (Stigler 1986), but rather because he constantly integrated the quantitative and the qualitative. Galton was familiar with the new developments in cell biology (Darlington 1962) and was not misled by the fact that twin differences distributed themselves in a unimodal (although nonnormal) manner. As late as 1905, Thorndike argued against the hypothesis of two types of twins on just such a basis (Thorndike 1905). Interested as he was in "the numbers," Galton had a breadth of perspective and looked across the life histories of twins. Numbers were assessed within the context of theory and empirically testable hypotheses; however, Galton mistakenly believed that placentation was an infallible method of determining zygosity, and it was not until the discovery of blood groups that this issue was fully resolved.

Rende, Plomin, and Vandenberg do not credit Galton with the discovery of the twin method because "Galton did not propose the comparison between identical and fraternal twin resemblance which is the essence of the twin method" (Rende et al. 1990, p. 277). This is a disingenuous argument that seems plausible only because it ranks quantitative arguments above conceptual and qualitative arguments. As shown in the quote above (Galton 1875, p. 404), Galton understood the confounded nature of heredity and environmental influences when traits are studied in ordinary families (he had already presented adoption data as one means of solving this problem, see below) and he clearly proposed the conceptual comparison of the life histories of the two types of twins. We would argue that Galton should also receive credit for inventing developmental behavioral genetics in that he clearly recognized that genes express themselves throughout the life course, an idea that has barely begun to take root in the behavioral sciences (see the numerous calls for longitudinal studies of twins throughout this book). Galton did not carry out a quantitative comparison of the two types of twins because a quantitative method was not available—he had not yet invented it! He demonstrated the phenomenon of "regression to the mean" in 1885

(Galton 1885), but it was not until 1888 that he published his famous paper "Correlations and their measurements, chiefly from anthropometric data" (Galton 1888). In this paper he computed "correlations" between relatives and even introduced the concept of "mid-parent" as an "ideal progenitor." As many of the chapters in this book make clear, there is no single twin method, rather there are a number of ways in which twins can be used informatively, with the comparison of MZ and DZ twins being only one.

Galton, without doubt, was the first scientist to recognize fully that twins can be usefully conceptualized as a tool for addressing a wide variety of interesting questions regarding the genesis of human traits, including behavioral phenotypes. The ways in which twins can be used to help us understand the distal influence of genes on behavior are so diverse that many are not even discussed in this book (see McGue et al. 1993; Segal 1984b, 1993). In line with this focus on the multiplicity of what often appears to be a single method, it should be remembered that Galton also introduced the adoption design as a means of disentangling the role of heredity and environment in the determination of ability (Galton 1869). Given that he proposed adoption as a research method prior to his discussion of twins as a research tool, it is curious that Galton did not invent the twin reared-apart design. A review of his twins papers and correspondence in the University of London archives yields no indication that he mentioned it as a hypothetical design or thought about it systematically. More interestingly, one letter from a correspondent responding to his twin questionnaire explicitly mentions such a pair of twins.

For scientific purposes, however, twins became truly useful after biological scientists finally clarified, in terms of mechanisms, the fundamental distinction between MZ and DZ twins and provided a precise method for determining zygosity. This was not a single event but rather the culmination of a whole series of events which, when taken together, make a convincing case (Dahlberg 1926; Fisher 1919, 1925; Siemens 1924). A careful and thorough documentation of this set of events remains to be written (but cf. Price 1950 and Rende et al. 1990). Curiously, the fundamental cause (set of initiating events) of MZ twinning remains unknown.

THE GOALS OF THIS BOOK

Because of the explosion of research on the influence of heredity on psychological and psychiatric traits in recent years (Plomin 1990) and the particularly strong interest in twins (Hrubec and Robinette 1983), the Dahlem organizing committee believed it timely to initiate a state-of-the-art discussion on the current findings and controversies in this field. The committee (D.A. Hay, K.K. Kidd, C.G.N. Mascie-Taylor, N. Pedersen, P. Propping, and T. Bouchard) met to delineate the overall goal approximately one year before the meeting, as well as to form the discussion groups around the workshop theme. As in all Dahlem conferences, lectures are not given nor are papers presented. Instead, prior to the conference, background papers written specifically around a group's discussion topic are circulated to all participants. The conference week is therefore reserved solely for the exchange of ideas.

This book reflects the structured portion of our workshop, as it contains the background papers and the resulting group discussion reports. Here we would like to delineate the lines of thought behind our approach to the workshop. Then, in the subsequent sections, we will comment on the various chapters in a manner that emphasizes these points and yet has minimum overlap with the superb summaries of the group discussions (see Baker et al., McCartney et al., Macdonald et al., and Kaprio et al., this volume).

As has been decisively established, virtually no dimension of individual differences in behavior is exempt from genetic influence (Bouchard et al. 1990). The demonstration of pervasive and quantitatively significant genetic influence on behavior is important because it directs behavioral scientists toward a domain of explanatory mechanisms that has been consistently ignored for many years. Nevertheless, it is only the beginning of understanding. A full scientific explanation of any phenomenon consists of a theoretical structure that specifies how things come about. It requires the specification of mechanisms and a description of the action of those mechanisms over time and within specifiable boundary conditions. Genetic influences manifest themselves in a bewildering variety of ways (Eaves 1982; Eaves et al. 1989; Lykken et al. 1992; Neale and Cardon 1992). For a scientist, understanding is more convincingly demonstrated by a theory's ability to illuminate the small details than by sketching a larger picture. As with any scientific method, there are numerous facts about the twin method, and how it might be most judiciously applied in each circumstance, that must be known before we can achieve sufficient precision in its application. The purpose of this book is to focus on these details in four substantive research domains: mental abilities, personality, childhood behavioral disorders, and adult psychopathology. Since a superb state-of-the-art presentation of quantitative twin analytic methods was readily available (Neale and Cardon 1992), we chose not to include treatment of this obviously important subject matter. Rather we chose to sample domains where (a) sufficient research had been carried out to illustrate fully the power of twin methods in clarifying fundamental issues, even when many questions in the domains remain controversial (*mental abilities* and *personality*), and (b) where much more limited research had been carried out, but where it was clear that twin research could contribute greatly to the resolution of fundamental theoretical problems (*childhood behavior disorders* and *adult psychopathology*).

g and Special Mental Abilities

The first section of this volume deals with an old problem: the "reality of *g*" and its validity versus special mental abilities. Brand (chapter 2) sets the stage with his provocative conceptual discussion of *g* and of criticisms that have been aimed against it in recent years. He concludes that "IQ is probably real and important, yet few dare say so." This paper generated a very lively and engaging, some might say truculent, discussion at the workshop and is well reflected in the group report (see Baker et al., this volume). This discussion truly represents a fundamental division of opinion in the behavioral sciences regarding the importance of *g*, its biological reality, and its validity relative to special abilities. It is a division that is not easily resolved by simply "resorting to the facts," as can be illustrated by the reaction to a recent review entitled "Intelligence is the best predictor of job performance" (Ree and Earles 1992) from the journal *Current Directions in Psychological Science*. The response to this article was so intense that most of a subsequent issue was devoted to a special section entitled "Controversies" (Jensen 1993; McClelland 1993; Ree and Earles 1993; Schmidt and Hunter 1993; Sternberg and Wagner 1993).

Fulker and Cardon (chapter 3) illustrate what can be discovered with regard to the problem of *g* vs. special mental abilities when powerful quantitative multivariate behavioral genetic methods are applied simultaneously to longitudinal data on twins (Galton's twin life history data), adopted children (Galton's adoption methods), and ordinary siblings (Galton's family study approach). They, like Brand, conclude that the large body of behavioral genetic literature on *g* indicates a "a trait of considerable evolutionary and social significance." They proceed to demonstrate, however, that both general and specific genetic variation exists and that the developmental pattern is complex, at both the genetic and environmental levels. This study is an elegant illustration of the principle that "understanding is in the details." We leave it to the reader to study the chapter, asserting only that the research program on which this work is based sets the standard for all future discussion of this problem.

Mascie-Taylor (chapter 4) provides a concise, but thorough, and illuminating review of the confines that exist in the correlations between a very wide variety of environmental variables and IQ measures based on family data (Galton's problem again). Not surprisingly, for anyone experienced in multivariate analysis, the correlations between what are often purported to be quite different variables are so high that when viewed in terms of independent contributions to explaining variance in the traits of interest (IQ), only a few variables will cover almost all the ground. Again, however, the details are important and some of the specific analyses yield counterintuitive results. This is a domain in which the uninitiated are constantly led down the garden path by authors who treat each variable independently and leave the reader to believe that, when taken together, the joint effect of all environmental variables is sufficient to "explain" IQ, personality, or some other individual difference variable. This chapter clearly illustrates the erroneousness and futility of such an approach.

Boomsma (chapter 5) reviews the current status of the development of cognitive abilities across the life span. The evidence strongly supports the conclusion that heritability of IQ increases with age. The largest increases occur from infancy to childhood; however, this may be because infant tests of cognitive ability are inadequate to the task. Increases in adulthood are more modest, but the data are less compelling because there are so few adult twin studies—there are only four such studies reported in the entire world literature. Yet this conclusion is clearly contrary to the expectation of most psychologists, who would expect environmental factors to cumulate over the life span. Especially striking is the fact that common family environmental influence for IQ is nearly zero in adulthood (McGue et al. 1993). Boomsma makes it clear that while the twin method is powerful, the full explication of the genetic and environmental factors that influence cognitive abilities will require complex multivariate designs, a number of whose features she discusses in detail, which include relatives other than twins.

Personality in the Normal Range

After IQ, twin researchers have studied personality more frequently than any other trait. The proliferation of different inventories using a plethora of trait names has made it very difficult for anyone to summarize the vast literature. The growing consensus around the "big five" personality factors (Costa and McCrae 1992; Goldberg 1992; Hofstee et al. 1992; John 1990) provides a frame of reference for organizing this diverse literature, and Loehlin (1992) has carried out an elegant summary of all the kinship correlations available in the literature organized around the "big five." Here, in chapter 7, he provides us with a succinct summary of his results for extraversion and emotional stability (neuroticism) and embeds the discussion in an encompassing four-level scheme for guiding future research in this domain. To many followers of the genetics and personality literature, it will be surprising to note his solid finding that nonadditive influences play a surprisingly large role in accounting for variance in these dimensions. Consistent with the findings of other investigators, common family environmental influence remains negligible.

Human behavioral geneticists naturally think of genetic influences as highly distal influences mediated by a variety of processes (cf. Table 1 in Eaves 1982). Unlike their colleagues in animal behavioral genetics (or their "other selves" when they play both the role of human behavioral geneticists and animal behavioral geneticists), human behavioral geneticists stray only infrequently into the thicket of speculation regarding the evolutionary adaptiveness of a given trait or character. The new burgeoning field of evolutionary psychology (Barkow et al. 1992) directly challenges this inherent conservatism. These authors argue that the traits of interest to most human behavioral geneticists are nothing more than noise (Tooby and Cosmides 1990). Buss is one of the few psychologists who has taken this challenge to heart and has attempted an integration of the views of evolutionary psychology and human behavioral genetics (Buss 1991; but cf. Bouchard et al. 1993; Crawford et al. 1987; Crawford and Anderson 1989). In this volume (chapter 8), Buss provides us with a frame of reference that allows us to fill with content many of the content-free processes we now apply to traits in a hit-or-miss fashion (G × E interactions and G × E correlations). His illustrative analysis of how jealousy and emotional stability (neuroticism) can be studied from an evolutionary frame of reference will open up a new approach to the study of all major personality traits.

The correlations of MZ twin discordances with discordances in environmental treatments are a powerful means of detecting environmental influences on a trait. As

Schepank points out (chapter 9), even with this simple and powerful design there are idiosyncratic details that one must pay attention to if the design is to be properly implemented. Schepank reports a rare and exceptional set of findings, based on case studies, that provide hypotheses which can be put to more rigorous tests with large samples. This approach could be productively implemented as part of a longitudinal prospective study of twins at high risk for mental disorder.

In the domain of personality, unlike the domain of mental abilities (except for measures taken in infancy), we have very little confidence that measures taken at earlier ages are reasonably comparable to those taken at later ages. This makes Pedersen's evaluation of genetic and environmental continuity and change in person-ality very difficult (see chapter 10). Worse yet, most of the data are cross-sectional as there is a dearth of data based on more than two points in time. As Pedersen points out, all caveats expressed by Boomsma apply to the personality data as well as the mental ability data. We also have powerful designs available but little adequate data. When reading Pedersen's chapter, most of the findings unique to this approach must be considered very tentative and "the challenge is to find a better synergism between behavioral genetics and life-span developmental perspectives."

Childhood Psychopathology

Rutter et al. (chapter 12) point out in impeccable detail the possible confines that may occur in twin studies of child psychopathology and provide a series of textbook examples of twin studies (done in conjunction with other kinships) of childhood disorders that should be carried out. From their chapter, it is very clear that the groundwork, available in the IQ and personality domains for years, remains to be established with respect to the study of childhood disorders. The assumptions we make in the latter domain, with little fear of challenge, must be tested and retested in this newer arena. This chapter provides the necessary highway warning signs but also spells out why this road will be well worth traveling. The authors warn us that "with the dawning era of molecular genetics there is a danger of underplaying the strengths of behavioral genetics and the twin paradigm for genetic research as well as underes-timating how much remains to be done." Any doctoral student looking for a disserta-tion topic in this field will find an embarrassment of riches in this chapter. Reiterating a refrain echoed throughout this book, the authors particularly focus on the value of longitudinal studies.

Apart from the fact that is it unlikely that neither genes nor environment alone will prove sufficient to explain childhood disorders, childhood phenotypes vary in expressivity, as do all phenotypes, and they change with age. This makes the study of such disorders especially difficult. Pauls (chapter 13) argues persuasively, on conceptual and theoretical grounds, that a full understanding of the etiology of these disorders will be very difficult to achieve without detailed longitudinal studies of high risk children carried out in conjunction with new molecular genetic techniques.

Rowe and Rodgers (chapter 14) outline a research program aimed at clarifying the elusive nonshared environmental influences that supposedly underlie childhood behavior problems. They review the MZ twin differences method (also discussed by Schepank, chapter 9) and conclude that since so few results have been replicated, it is difficult to draw conclusions about its empirical worth. They then proceed to implement two new methods: the first depends on a comparison of variances in one- and two-child families; the second is an extension of the DeFries and Fulker multiple regression model developed for behavioral genetic studies. Both methods make use of large samples, although the later study was hindered by lack of knowledge about twin zygosity. In spite of the apparent conceptual elegance of these methods, the authors demonstrate that proving that a finding truly reflects etiological causation is more difficult than one might suppose.

Bryan (chapter 15) discusses the problem of potential biases in twin studies, carefully reviewing recent work on prenatal and perinatal influences on twin children. Both medical and psychological influences are covered. Taken together, this chapter and the methodological chapter by Rutter et al. (chapter 12) create a persuasive case for carefully monitoring potential biasing influences in twin studies of behavioral disorders in childhood. To avoid any misunderstanding, it bears repeating that most likely any bias introduced into the study of normal traits using twins is in the direction of underestimating the influence of heredity. Price's brilliant review of this matter (Price 1950, 1978), while dated, remains well worth reading.

Adult Psychopathology

Virtually every behavioral geneticist has been asked by lay friends or relatives whether schizophrenia, depression, or alcoholism is genetic. When given the complex response such a question requires—there is strong evidence for genetic influence but the results vary from study to study, from one definition of the disorder to another, and the mode of both genetic and environmental transmission still eludes us—they often express dismay and openly wonder why scientists can not "do better." McGuffin et al. (chapter 17) demonstrate that the problem, with regard to schizophrenia and depression, is not one of reliability of diagnosis (high reliability can be achieved using available techniques) but rather one of validity—capturing the construct. The variability of MZ and DZ concordance (particularly for depression) as a function of various definitions of these disorders, while not sufficient to threaten the conclusion that the disorders are heritable, do not augur well for the success of future linkage studies. The large differences in base rates for depression in U.S. vs. British studies are disheartening and puzzling. Nevertheless, as the authors point out, "twin studies are perhaps still the most useful way of defining and refining phenotypes for molecular genetic research into psychiatric disorders."

Even if we are able to establish linkage for a gene involved in a psychiatric disorder, it will be extremely useful to know the pathophysiology of the brain caused by the gene. Brain imaging techniques will, most likely, eventually reveal that pathophysiology. Buchsbaum (chapter 18) takes us on a tour of the results of numerous exploratory studies, most focusing on MZ twins discordant for various disorders in an effort to discover the pathophysiology and its environmental cause. Many suggestive results have been reported, but small sample sizes and the same problems of validity of diagnosis encountered by McGuffin et al. (chapter 17) continue to stifle advances in this exciting new field.

Again, however, both concordant and discordant twins appear to provide the most powerful group of patients for drawing conclusions regarding the conceptual significance of observed brain imaging differences.

In chapter 19 Heath asks whether anything can be learned about the determinants of psychopathology and substance abuse from studies of normal twins. The vast majority of medically oriented investigators, most geneticists, and perhaps most psychologists would answer "no." Yet as Heath demonstrates, with a series of mostly real examples (some are contrived for didactic purposes), the answer is a resounding "yes." The development of all these disorders involves complex linkages between multiple predispositions and the environment, be they genotype–environment correlations, $G \times E$ interactions, or mediation by multivariate personality and/or environmental factors. Even if we "find the genes" for some of these disorders, the behavioral mechanisms by which the disorders "unfold" or are created by the genetics of "self-exposure" will have to be explicated. Heath marshals a great deal of evidence and argues persuasively that we have the necessary tools to carry out highly informative (with respect to prevention and intervention) long-term studies of behavior development with regard to these disorders and that, in the rush "to map rare Mendelian disorders," very important behavioral questions are being neglected.

SUMMARY COMMENTS

It must be stated that none of the authors in this book nor any of the workshop discussants failed to appreciate the fundamental role of the environment in human development. Genes and genomes constitute the blueprints of life (Singer and Berg 1991). They constitute the machinery of gene replication and gene expression. The phenotypic structure of developed organisms reflects both the role of their genes and their environments. Dogs, however, are dogs and mice are mice, and the two are never confused despite the great variability within each of the species. Both species require different but equally adequate environments in order for them to develop, reproduce, and live out their lives. Very detailed questions, such as to what extent individual differences between organisms within species reflect genetic or environmental influences and what mechanisms mediate each effect, can be answered (in theory, but not always easily) with lower organisms by carrying out appropriate breeding experiments and properly analyzing the data (Falconer 1990; Mather and Jinks 1982). The role of the human behavioral geneticist is much more difficult. He/she must depend upon comparisons drawn from experiments of nature (comparisons of identical and fraternal twins, studies of twins reared apart) or social experiments (adoption of children into the homes of unrelated individuals). As the chapters in this book so clearly illustrate, the use of such designs to draw causal inferences about both environmental and genetic influences requires every bit as much knowledge of the details of the phenomenon under study as does, say, experimental work with inbred strains or recombinant crosses.

A Note on Estimation of Heritability

Heritabilities for various traits are reported throughout this book. A criticism often made of work in behavioral genetics is that estimation of heritability is uninformative, because it may vary from population to population and from one type of environment to another, and should therefore never be computed. We disagree with this argument, not because it lacks merit, but because when put in this simple form, it is misleading. First, the argument is general and applies to all descriptive statistics. When the critics of heritability calculations continually reiterate the limits of their own descriptive statistics, they will have earned the right to criticize the reporting of heritabilities. Second, while it is theoretically true that the heritability of a trait can vary depending on the population and range of environments sampled, it is also an empirical question as to what the variation is when a range of populations and environments is actually sampled (Rushton 1989). To state that a heritability coefficient is uninformative and therefore should not be computed precludes our ever obtaining the data necessary to answer an interesting scientific question. Third, the citation of a heritability does say, for the population studied and the range of environments encompassed, that heredity explains a part of the trait variation and environment plus measurement error the remainder. With more complex designs, more detailed statements can be made. Such statements constitute an unambiguous claim to anyone who understands quantitative genetics and they are just as informative as, if not more informative than, the simple correlations between variables widely reported in ordinary psychological journals. As Mascie-Taylor repeatedly points out (chapter 4), the ordinary correlation between a feature of the family environment and a child's IQ is extremely difficult to interpret due to the high correlations between family environmental features and it is totally confounded and uninformative with regard to causality (Bouchard and Segal 1985; Scarr 1992). Environmental researchers seldom mention this fact in their reports. Many supposed environmental variables when treated as phenotypic measures are partly genetic (see Rowe, chapter 14). Human behavioral geneticists are far more careful in specifying the limits of generalizability of their findings than the vast majority of psychologists (cf. Hoffman [1991] and the critique by Bouchard et al. [1993] and Tomlinson-Keasey [1990] and the critique by Bouchard [1993]). For an even more recent commission of this error, see Steinberg et al. (1992). When psychologists decide to carry out their studies on twins and adoptive families, as opposed to biological families, and to study environmental influences uncontaminated by heredity, the accusations of the uninformativeness of heritabilities will ring true.

This criticism has greater generality than even most behavioral geneticists generally presume. Consider divorce: this "event" has been studied extensively as a causal environmental variable with regard to its influence on children's lives (i.e., their mental health, delinquency, etc.), but the only twin study on the topic suggests that the direction of causation is actually the reverse of that which has been assumed (McGue and Lykken 1992). Divorce is heritable! This finding is eminently consistent with the results of longitudinal studies of the effects of divorce on children. These studies show the adjustment difficulties of children of divorce actually predate the time of parental separation. As the authors of this work argue, the direction of effect in such studies is ambiguous (Block et al. 1986). Divorce is probably mediated, in part, by heritable personality traits (Cramer 1993). Consider, as well, the problem of direction of effects of life stress events. Such events are commonly assumed to be causal related to various mental disorders. The events themselves are, however, moderately heritable (Plomin et al. 1990; cf. the replication by Moster 1991).

Envoy

The conference was an enriching and stimulating intellectual adventure for all participants. There are a number of people to thank. First, we would like to express the joint gratitude of all the participants to the twins in our studies and those of others before us who, over many years, have been so generous with their time, energy, and enthusiasm. Without their generous help, most participants at this conference would have had far less knowledge to contribute to our understanding of the problems under consideration.

Second, we would like to thank Silke Bernhard for transmitting her infectious enthusiasm for the Dahlem format to us and the planning committee. We dedicate this book to her in honor of her wide-ranging contribution to the international exchange of scientific ideas. This was the last Dahlem conference of which she was the principle organizer. Thanks also go to Jennifer Altman who, under very difficult circumstances, oversaw the finalization of the conference plans.

Last, but not least, we would like to express our deep thanks and admiration for the skill, enthusiasm, dedication, and hard work of the current Dahlem staff, particularly Julia Lupp and Klaus Roth. Dahlem Konferenzen is in good hands. We wish the staff a long and successful future.

REFERENCES

Augustine of Hippo, St. 415. De civitate Dei (The City of God).

Barkow, J.H., L. Cosmides, and J. Tooby. 1992. The Adapted Mind. Oxford: Oxford Univ. Press.

Block, J.H., J. Block, and P.S. Gjerde. 1986. The personality of children prior to divorce: A prospective study. *Child Devel.* **57**:827–840.

Bouchard, T.J., Jr. 1993. Genetic and environmental influences on adult personality: Evaluating the evidence. In: Basic Issues in Personality, ed. I. Dreary and J. Hetteman. Dordrecht: Kluwer.

Bouchard, T.J., Jr., D.T. Lykken, M. McGue, N.L. Segal, and A. Tellegen. 1990. Sources of human psychological differenes: The Minnesota study of twins reared apart. *Science* **250**:223–228.

Bouchard, T.J., Jr., D.T. Lykken, A. Tellegen, and M. McGue. 1993. Genes, drives, environment, and experience: EPD theory – revised. In: Psychometric and Social Issues Concerning Intellectual Talent, ed. C.P. Benbow and D. Lubinski. Baltimore: Johns Hopkins Univ. Press, submitted.

Bouchard, T.J., Jr., and N.L. Segal. 1985. Environment and IQ. In: Handbook of Intelligence: Theories, Measurements, and Applications, ed. B.J. Wolman. New York: Wiley.

Buss, D.M. 1991. Evolutionary personality psychology. *Ann. Rev. Psychol.* **42**:459–491.

Corney, G. 1984. Sir Francis Galton, 1822–1911. *Acta Gen. Med. Gem.* **33**:13–18.

Costa, P.T., and R.R. McCrae. 1992. NEO PI-R Professional Manual. Odessa, FL: Psychological Assessment Resources, Inc.

Cramer, D. 1993. Personality and divorce. *Pers. Indiv. Diff.* **14**:605–607.

Crawford, C., D. Krebs, and M. Smith, eds. 1987. Sociobiology and Psychology. Hillsdale, NJ: Erlbaum.

Crawford, C.B., and J.L. Anderson. 1989. Sociobiology: An environmentalist discipline? *Am. Psychol.* **44**:1449–1459.

Dahlberg, G. 1926. Twin Births and Twins from a Hereditary Point of View. Stockholm: Bokforlags, A.B.

Darlington, C.D. 1962. Introduction to 1962 edition of "Hereditary Genius: An Inquiry into Its Laws and Consequences" (Meridian Books Edition). New York: The World Book Publ. Co.

Eaves, L.J. 1982. The utility of twins. In: Genetic Basis of the Epilepsies, ed. E. Anderson, W.A. Hauser, J.K. Penry, and C.F. Sing, pp. 249–276. New York: Raven.

Eaves, L.J., H.J. Eysenck, and N.G. Martin. 1989. Genes, Culture, and Personality: An Empirical Approach. New York: Academic.

Falconer, D.S. 1990. Introduction to Quantitative Genetics, 3rd ed. New York: Longman Group Ltd.

Fisher, R. 1919. The genesis of twins. *Genetics* **4**:399–433.

Fisher, R.A. 1925. The resemblance between twins, a statistical examination of Lauterbach's measurements. *Genetics* **10**:569–579.

Galton, F. 1869. Hereditary Genius: An Inquiry into its Laws and Consequences. London: Macmillan.

Galton, F. 1875. The history of twin, as a criterion of the relative powers of nature and nurture. *Fraser's* **Nov**:566–576.

Galton, F. 1876. On twins. *J. Anthro. Inst. GB Ire.* **5**:324–329.

Galton, F. 1885. Regression towards mediocrity in hereditary stature. *J. Anthro. Inst. GB Ire.* **15**:246–263.

Galton, F. 1888. Co-relations and their measurements, chiefly from anthropometric data. *Proc. Roy. Soc. Lond.* **45**:135–145.

Goldberg, L.R. 1992. What the hell took so long? Donald Fiske and the big-five factor structure. In: Advances in Personality Research, Methods, and Theory: A Festschrift honoring Donald W. Fiske, ed. P.E. Shrout and S.T. Fiske. New York: Erlbaum.

Hoffman, L.W. 1991. The influence of the family environment on personality: Evidence for sibling differences. *Psychol. Bull.* **110**:187–203.

Hofstee, W.K.B., B. De Raad, and L.R. Goldberg. 1992. Integration of the big five and circumplex approaches to trait structure. *J. Pers. Soc. Psychol.* **63**:146–163.

Hrubec, Z., and C.D. Robinette. 1983. A study of twins in medical research. *New Eng. J. Med.* **310**:435–441.

Jensen, A.R. 1993. Test validity: g versus "tacit knowledge." *Curr. Dir. Psychol. Sci.* **2**:9–10.

John, O.P. 1990. The "big five" factor taxonomy: Dimensions of personality in the natural language and in questionnaires. In: Handbook of Personality: Theory and Research, ed. L.A. Pervin. New York: Guilford.

Kleinwachter, L. 1871. Die Lehre von den Zwillingen (The Theory of Twins). Prague: Haerpfer.

Lauterbach, C.E. 1925. Studies in twin resemblance. *Genetics* **10**:525–568.

Loehlin, J.C. 1992. Genes and Environment in Personality Development. Newbury Park, CA: Sage.

Lykken, D.T., M. McGue, A. Tellegen, and T.J. Bouchard, Jr. 1992. Emergenesis: Genetic traits that may not run in families. *Am. Psychol.* **47**:1565–1577.

Mather, K., and J.L. Jinks. 1982. Biometrical Genetics: The Study of Continuous Variation, 3rd ed. London: Chapman and Hall.

McClelland, D.C. 1993. Intelligence is not the best predictor of job performance. *Curr. Dir. Psychol. Sci.* **2**:5–6.

McGue, M., T.J. Bouchard, Jr., W.G. Iacono, and D.T. Lykken. 1993. Behavioral genetics of cognitive ability: A life-span perspective. In: Nature, Nurture and Psychology, ed. R. Plomin and G. McClearn. Washington, D.C.: American Psychological Association.

McGue, M., and D.T. Lykken. 1992. Genetic influence on risk of divorce. *Psychol. Sci.* **3**:368–372.

Moster, M. 1991. Stressful Life Events: Genetic and Environmental Components and Their Relationship to Affective Symptomatology. Ph.D. diss., Univ. of Minnesota.

Neale, M.C., and L.R. Cardon, eds. 1992. Methodology for Genetic Studies of Twins and Families. Dordrecht: Kluwer.

Pearson, K. 1924. The Life, Letters, and Labours of Francis Galton, vol. 2. Cambridge: Cambridge Univ. Press.

Plomin, R. 1990. The role of inheritance in behavior. *Science* **248**:183–188.

Plomin, R., P. Lichtenstein, N.L. Pedersen, G.E. McClearn, and J.R. Nesselroade. 1990. Genetic influences on life events during the last half of the life span. *Psychol. Aging* **5**:25–30.

Price, B. 1950. Primary biases in twin studies, a review of prenatal and natal difference-producing factors in monozygotic pairs. *Am. J. Hum. Genet.* **2**:293–352.

Price, B. 1978. Bibliography on prenatal and natal influences in twins. *Acta Gen. Med. Gem.* **27**:97–113.

Ree, J.M., and J.A. Earles. 1992. Intelligence is the best predictor of job performance. *Curr. Dir. Psychol. Sci.* **1**:86–89.

Ree, J.M., and J.A. Earles. 1993. g is to psychology what carbon is to chemistry: A reply to Sternberg and Wagner, McClelland, and Calfee. *Curr. Dir. Psychol. Sci.* **2**:11–12.

Rende, R.D., R. Plomin, and S.G. Vandenberg. 1990. Who discovered the twin method? *Behav. Genet.* **20**:277–285.

Rowe, D.C. 1990. As the twig is bent? The myth of child-rearing influences on personality development. *J. Counsel. Devel.* **6**:606–611.

Rushton, J.P. 1989. The generalizability of genetic estimates. *Pers. Indiv. Diff.* **10**:985–989.

Scarr, S. 1992. Developmental theories for the 1990s: Development and individual differences. *Child Devel.* **63**:1–19.

Scarr, S., and R.A. Weinberg. 1978. The influence of family background on intellectual attainment. *Am. Sociol. Rev.* **43**:674–692.

Schmidt, F.L., and J.E. Hunter. 1993. Tacit knowledge, practical intelligence, general mental ability, and job knowledge. *Curr. Dir. Psychol. Sci.* **2**:8–9.

Segal, N.L. 1984a. Cooperation, competition, and altruism in human twinships: A sociobiological approach. In: Sociobiological Perspectives on Human Development, ed. K.B. MacDonald, pp. 168–206. New York: Springer.

Segal, N.L. 1984b. Zygosity testing: Laboratory and the test investigator's judgment. *Acta Gen. Med. Gem.* **33**:515–521.

Segal, N.L. 1993. Twin, sibling, and adoption methods: Tests of evolutionary hypotheses. *Am. Psychol.*, in press.

Siemens, H. 1924. Die Zwillingspathologies. Berlin: Springer.

Singer, M., and P. Berg. 1991. Genes and Genomes: A Changing Perspective. Mill Valley, CA: Univ. Science Books.

Spaeth, J. 1860. Studien über Zwillingen. *Zeitschrift Wiener Gesellschaft der Ärzte* **16**:225–231, 241–244.

Spaeth, J. 1862. Studies regarding twins (English translation). *Edinburg Med. J.* **7**:841–849.

Steinberg, L. S.D. Lamborn, S.M. Dornbush, and N. Darling. 1992. Impact of parenting practices on adolescent achievement: Authorative parenting, school involvement, and encouragement to succeed. *Child Devel.* **63**:1266–1281.

Sternberg, R.J., and R.K. Wagner. 1993. The g-ocentric view of intelligence and job performance is wrong. *Curr. Dir. Psychol. Sci.* **2**:1–5.

Stigler, S.M. 1986. The History of Statistics: The Measurement of Uncertainty before 1900. Cambridge, MA: Belknap Press of Harvard Univ. Press.

Thorndike, E.L. 1905. Measurement of twins. *J. Phil. Psychol. Sci. Meth.* **2**:547–553.

Tomlinson-Keasey, C., and T.D. Little. 1990. Predicting educational attainment, occupational achievement, intellectual skill and personal adjustment among gifted men and women. *J. Ed. Psychol.* **82**:442–455.

Tooby, J., and L. Cosmides. 1990. On the universality of human nature and the uniqueness of the individual: The role of genetics and adaptation. *J. Pers.* **58**:17–68.

2

Cognitive Abilities: Current Theoretical Issues

C.R. BRAND
Department of Psychology, University of Edinburgh, 7 George Square, Edinburgh EH8 9JZ, U.K.

ABSTRACT

In Britain today, the consensus of opinion in and around psychology and genetics is to ignore or dispute the existence of general intelligence (g). However, the classical tradition of relatively unitarian and hereditarian thought and work on intelligence staged a revival in the 1980s. The "London School"—running from Galton and Spearman, through Burt and P.E. Vernon, to H.J. Eysenck and A. R. Jensen today—has lately scored several new correlational successes that consolidate g's status as a central variable for psychology (Brand et al. 1991).

In this chapter, it is suggested that the conventional piety, which prefers talk of "cognitive abilities" to talk of g, involves undue commitment to four theoretical dogmas of modern psychology. These dogmas, each of which has enjoyed its own period of popularity since 1960, are identified respectively as *multidimensionalism, interactionism, constructivism*, and *nihilism*. These demonstrably strange sets of assertions have all distracted attention from how g itself may yield—or allow identification of—greater psychological differentiation of specific abilities and traits at higher levels of g (Anderson 1992; Brand et al. 1992).

It is concluded that modern, empirical psychogenetic work examining twin correlations separately at different g levels has much (including some interaction effects) to offer to all but the most ideologically entrenched opponents of g and heritability (see, e.g., Pearson 1991). One possible next step is to explore how higher-g twins—despite having similar genes, shared macroenvironments, and g-levels—often differ quite markedly in non-g-related abilities and propensities. The London School was and remains right about the importance of g; beyond g, and beyond the gene-gene interaction (*epistasis*) that contributes so greatly to individuality, Adam Smith and Alfred Adler were perhaps also correct to stress the role of *niche*-selection and competition in human development.

Because intelligence is not the objectively defined explanatory concept it is often assumed to be, it is more an obstacle than an aid to understanding abilities (Howe 1990).

Modern genetics is one of the few sciences that has reduced its expectations. Geneticists nowadays have no interest in what their subject might say about the large and vague

Twins as a Tool of Behavioral Genetics
Edited by T.J. Bouchard, Jr. and P. Propping © 1993 John Wiley & Sons Ltd.

issues that occupied their predecessors....modern geneticists scarcely involve themselves with what their work implies for the future of humanity (S. Jones).[1]

The two experts whose admonitory views these are both hold chairs in British universities today. The first, Michael Howe, who believes "intelligence is an obstacle to understanding abilities," is a professor of educational psychology in Exeter. For the past decade, he has mounted a vigorous media campaign against every major proposition of the London School descendants of Sir Francis Galton's relatively unitarian and hereditarian account of human differences in ability and achievement. The second, Steve Jones, who is glad that "genetics has reduced its expectations," is professor of genetics at the prestigious University College London. He holds the very Chair that Galton helped to establish for the study of eugenics, though Jones himself believes it is impossible to breed selectively for human intelligence.

In Britain, at least in the 1990s, expert, "Establishment" opinion in the social and behavioral sciences is largely indifferent or hostile to the idea that people differ enduringly in intelligence or in any other genetically transmissible and broadly consequential characteristic. The IQs of children in Britain's State schools are not tested by educationalists: in Britain, more is known about children's teeth than about their intelligence. There are currently no psychological studies of representative, unselected twins or adoptees. In Edinburgh—a capital city of half a million souls, sporting the fifth-best university in the United Kingdom—the main scientific journal that reports studies of the heritability of psychological features, *Behaviour Genetics*, is not purchased by anybody, public or private. For the typical student of psychology in Britain today, familiarity with the heritability of intelligence will stop short at the claim that Sir Cyril Burt "faked his data" on the IQs of identical twins reared apart—in 1966, when he was 83. Students will not know of Sandra Scarr's (1992) observation that, at least by adolescence, fraternal twins reared together have diverged considerably—much more than identical twins—in line with their genetic dissimilarity, which leads them to select their own microenvironments (of friends, involvements, and opportunities for distinctive experience). Nor will students know of Plomin's findings—that sharing a family environment with an adoptive sibling for 15 years produces little detectable personal similarity on psychological indices by late adolescence (Plomin and Daniels 1987)—or have been told of Bouchard's finding of strong similarities between identical twins reared apart in largely uncorrelated environments. None of this should come as a surprise: for example, Bouchard et al.'s (1990) major study, though reported in a leading journal, *Science*, received little attention in the British media, while the so-called British Psychological Society (which "tried" Burt, found him guilty, and now rules any re-trial out of order [*Nature*, March 5th, 1992]) provides no scientific news digest for its members. The magazine *Nature* declines

---

[1] 1992, discussing his own *BBC Reith Lectures* (1991), BBCIV (U.K.), February 20, 20.00 hrs.

correspondence that would bring to its readers' attention Bouchard et al.'s (1993) high estimate of IQ's heritability.

Given such a dismal scene of expert indifference—at least in modern Britain—to "heritable IQ," and probably to the nonzero heritability and biological reality of any important trait difference, how is such neglect sustained? The critics of "heritable IQ" (or, more precisely, of the general intelligence factor, g, the first principal component found in all mental ability correlation matrices with reasonable sampling) will not undertake empirical work of their own to see whether Burt was right because they do not believe in, and wish to forget about, heritability. They also will not examine those elementary correlates of IQ that have been discovered by others. Averaged evoked cortical potentials, inspection times (measures of perceptual intake speed, i.e., sensory reaction time), paced serial addition times, nose length, and tongue-rolling are all ignored because self-styled progressives—always protesting the need "to go beyond g"—do not care to believe in the central importance of g to psychology. It matters little to such skeptics that there are ways of refuting their claim that identical twins are alike only because of the similar environments imposed on them artificially by parents and teachers. Plomin and Bergeman's (1991) finding that children's "environments" are themselves partly heritable (presumably because the child's own behavior and personality influence parental child-rearing practices) is rejected by those (such as Hirsch 1990) who deny that any characteristic at all can begin to be quantified for its heritability under nonexperimental conditions. It does not matter to self-styled researchers of diverse "cognitive abilities" that IQ has, across scores of studies since 1976, shown a substantial and surprising correlation, of around $r = -0.50$, with inspection times (even across the restricted range of nonretarded, young adult testees) (e.g., Brand et al. 1991). Skeptics, insofar as they can bring themselves to acknowledge the repeated occurrence of such correlations, assume such r's confirm their own views: that IQ is just one rather narrow aspect of human cognitive abilities and that its correlation with perceptual speed requires complex explanation "in terms of" the development and use of special perceptual "strategies" by higher-IQ subjects (see Brand 1987b: Reply to Pelligrino). Again, champions of diverse "cognitive abilities" feel they cannot be expected to undertake such researches because they long ago rejected IQ and g, just as they reject twin studies and heritability coefficients: their work must necessarily use other techniques and paradigms that allow "old-fashioned" IQ tests and nature/nurture questions to be relegated to the mists of history where they properly belong. Thus, immured from the real world in a supposedly virtuous circle of disdain, the prophets of "cognitive abilities" continue psychology's twentieth-century dream of avoiding the recognition of impartially testable, mental-speed-linked, general, lasting, and important differences between people in intelligence (see Brand et al. 1991).

How, then, can the traditional understandings and the newly supported empirical claims of the London School break through to New Age aspirants of "cognitive abilities" who decline to read Guilford and Jensen? Of course, an optimist would expect that sheer lack of tangible achievement by the avoiders of g and heritability (h^2) would eventually drive out educational, occupational, developmental and experimental psychologists from their late twentieth-century burrows. Alas, in Britain at least, talk of diverse "cognitive abilities" provides the Utopian, egalitarian piety that is so helpful in "university" teaching today, and in begging letters to taxpayer-funded but Establishment-controlled "research" councils. The politicians of Britain's major parties learned from their experiences with Mrs. Thatcher and Sir Keith Joseph that Britain's educational Establishment was so locked in 1960s piety as to be incorrigible, short of privatization. (Admittedly, in response to H.M.G.'s Alvey Report, politicians did fund the new race to discover "artificial intelligence" [AI]; however, the novelty here was essentially slight, for putatively intelligent computers, too, were [at least until *parallel distributed processing* arrived on the scene] considered to involve countless distinct "black box" functions that could be separately engineered. Perhaps because of this neo-traditionalism, achievement in AI is acknowledged to be slight [see Brand et al. 1992], with not even a robot with an all-around IQ of 2 to show for multibillion-dollar expenditures.) So perhaps there is a role for the pessimist who doubts that much change will occur until some hard things are said? Here they are! Some *four main theoretical problems* present themselves repeatedly in the world of "cognitive abilities"; there are arguably four main *points d'appui* at which pressure will bring down the walls of the egalitarian monastery garden; there are four topics at the mention of which even the longest-serving cognitivist will be seen to blush. These theoretical problems may be summarized as concerning four main psychological dogmas of recent times: *multidimensionalism, interactionism, constructivism*, and *nihilism*.

THE LOST CAUSES OF MULTIDIMENSIONALISM

Ever since the great American psychometrician, Louis Thurstone, took on Britain's factor-analytic pioneer, Charles Spearman, in the 1930s, there has been a division between progressive optimists (often American) and hard-bitten realists (more often British, at least in the past) as to how many distinct types or dimensions of intelligence can be identified. In the 1960s, Guilford (entitled to be called the first "cognitive psychologist," surely, for his theoretical analyses of what types of mental ability might be distinguished) raised the number to 150, although even his very popular distinction between "creativity" and intelligence met with limited empirical success. Today things may have calmed down a little; however, well-known American enquirers still insist on distinguishing, at the very least, seven independent dimensions of cognitive variation. These include verbal, spatial, logico-mathematical, musical, bodily/kines-thetic, interpersonal and intrapersonal, according to Gardner (1983), and encoding, decoding, recoding, strategy-forming, meta-encoding, meta-strategic, and many others (and their interaction effects), according to Sternberg (1990).

There are many problems with such ideas, notably when it comes to providing actual tests that are reliable, valid, and demonstrably uncorrelated with each other in usage across full, normal population ranges. Yet lack of empirical success and psychometric vindication present no impediment to continued speculation. (Like-wise, experimental psychologists continue with their own view that the mind involves a large and presently unspecifiable number of "black boxes" [episodic memory, short-term memory, working memory, audiovisual scratchpad, long-term semantic memory, two-and-a-half-dimensional imagers, etc.] despite the boxes having their shapes, sizes, positions, and names changed once every half-decade.) Evidently the causes of multidimensionality might themselves be discovered just around the corner; this would lend further credibility to the modestly egalitarian adage, "We all have our strengths and weaknesses" (across a range of unrelated, separately determined abilities that gives each of us a chance of a few successes, however isolated).

Although multidimensionalist ideology may still retain its early hint of promise, there is a problem. It is not just that even known non-g-related ability differences as reliable as those in hand skill remain of highly debatable provenance despite years of study (Previc 1991). It is not just that 60 years of widespread enthusiasm have failed to deliver a single multidimensional, psychometric ability battery of real, existing, uncorrelated mental tests. Rather, it is that none of the enthusiasts for such a scenario would know how to account, causally, for such a picture if it ever did emerge. Thurstone himself would have had no such problem, for he thought and told his readers plainly that each of his six or so dimensions would have a largely genetic basis. By contrast, most of Thurstone's descendants have a problem. *Where are the hypotheses* as to the distinct causal origins of the putative abilities whose independence has been maintained—however unconvincingly—by Guilford, Sternberg, Gardner, and others?

It may be envisaged by lay supporters of multidimensionality that the "social environment" can furnish dimensions of long-term causal variation to match up with such a multidimensional psychometric picture. Such variables as parental presence in the home, affluence, warmth, standards, sensitivity, supervision, and reward schedules may be thought to provide the necessary variety of influences. Alas, this is far from the case: such variables are correlated in the general population, almost as much as in their exponents' minds, in line with the familiar summary variable of "social class" (alias "social advantage" vs. "disadvantage"), yielding the "good" and frankly "bad" homes of modern social work (e.g., Belsky et al. 1991). Once any one of, e.g., Gardner's cognitive abilities is identified as being caused primarily by "class" differences, multidimensional enthusiasts are going to have to look for equally plausible but entirely uncorrelated explanations of the rest of human cognitive variation. Imagine multidimensionalists admitting frankly to their students that they believe that variation in most human mental abilities has *nothing to do with* coming from a "privileged" vs. "disadvantaged" home! That would be the day when "cognitive abilities" come of scientific age! Yet this will not happen. Multidimensionalist theoreticians do not offer a well-known set of even provisional hypotheses as to multicausation. In any case, the vanishingly slight similarities that are usually found between adoptive siblings suggest that not even "social class" itself has the potency to function as much of a long-term causal variable (e.g., White 1982; Brand 1987a). Of course, multidimensionalists could always retreat to a Thurstonian hereditarianism; however, this would hardly be popular with their publishers.

In mirror image, it must be admitted that genetic accounts of intelligence run into some of their own problems when required to explain how a relatively large number of genes might all contribute to individual differences in a finally unitary *g*. If each gene adds to (or subtracts from) *g*, why might not the individual psychosocial contribution of each gene be specified one day? If so, *g* itself would evaporate into multiple cognitive abilities!

This theoretical problem for the London School can be resolved, however, if we envisage that, once some small number of genes (or even one major gene; see Weiss 1992) has worked developmentally to set up a certain basic level of intellectual functioning (or fluid *g*), further genes make their own distinctive phenotypic contributions to intellectual achievements only insofar as basic resources are available to fuel or support them. The idea that cognitive abilities are markedly more independent of each other in subjects of higher levels of *g* is one that has lately been advanced in several quarters (e.g., Anderson 1992). It explains a problem that has baffled psychometricians for many years. Measures of "cognitive style," "empathy," and, indeed, "creativity" have long seemed largely independent of *g*, so long as testees were quite bright, say of IQ 115 (a condition that obtained in many studies of university students in the past). It is as if, at higher levels of *g*, whether a person excels at chess or crossword puzzles depends on non-*g* factors, even though few such non-*g*-related abilities can be expected (or observed) at lower levels of *g*. The idea that intelligence conspicuously "*differentiates*" at higher levels of *g* even into more distinctive personality features, yielding greater variance in personality for higher IQ subjects, is one that, strangely enough, can be found in the writings and theoretical work of Jungian psychologists whose efforts have not normally coincided with those of psychologists of the London School (see Brand et al. 1993). Appropriately, if differentiation were developmental in origin and traceable to individual choice or environmental control, the possibility that *g* itself might be more heritable across lower IQ ranges has come under investigation (e.g., Bailey and Revelle 1991).

Along such lines, the London School is presently able to venture a prospect of reconciliation in psychology's struggle as to whether intelligence has one or many dimensions. The answer is: both! Intelligence, like a tree, has both a trunk and branches, and looks less linear, and more branching and bush-like, when viewed from above. This answer allows multiple genetic causation to be accommodated. By contrast, the devotees of "cognitive abilities" seem unlikely to be able to find virtually any of the multiple causes for which they are obliged to search. Since their favorite causal variables are drawn mainly from the single covariational package involving "class," "enrichment," and "westernization," theirs is truly a home of lost causes in psychology.

THE UNDETECTABLE INTERACTIONS OF DEVELOPMENTALISM

There was a time—let us say in the 1960s—when the sage psychologist who did not know the causes of crime, racial conflict, or vegetarianism could safely pontificate to what would be an admiring audience that these phenomena were obviously "multidimensional." Faced with such polysyllabic expertise, few had the cheek to ask what any of the proposed dimensions actually *were*. In the days of "the white heat of technology" and early Headstart programs, few suspected that the social scientific emperor had no clothes.

Times change, especially in subjects like psychology, where there has never been agreement as to the set of underlying variables used to explain the chosen surface phenomena. By the 1970s, a number of forces led to the rise of "interactionism" as a new, central dogma in developmental, social, and experimental psychology. Neither side won the historic debates (between behaviorists and ethologists) about whether language or bird song—or sex-role division of labor, nest-building, etc.—were innate or learned; talk of "complex interactions" offered an apparently sophisticated way of saying that nobody really knew the answer (for language), or that it varied by species (for bird-song), or that nature beat nurture hands down but it would be impolitic to say so (for much of the behavioral repertoire of most species). Again, in social psychology, it came to be seen that many interesting effects (falling in love, getting hired) might plausibly reflect, in technical jargon, *Person* × *Situation* interactions in which two people (one of them constituting, in the jargon, the "situation" for the other) achieve a quite unique lift-off that neither achieves with a host of other people (alias "situations"). Finally, experimental psychologists very occasionally found statistical interaction effects (i.e., variables having effects not additively, but as functions, e.g., multiplicative, of each other) in their extreme manipulations of laboratory variables, even though few of these were remembered for long. (Subsequent recall of interaction effects serves more to illustrate experimenters' sensitivities than to document proven explanatory advance. A $300 prize has long been offered by Edinburgh's Structural Psychometrics Group to any student discovering two textbooks by experimental or cognitive psychologists which agree on half-a-dozen interaction effects of importance to their discipline.)

It is no surprise, then, to find a strong demand for interaction effects between nature and nurture—ideally making them "impossible to disentangle"—from critics of g and h^2. Rather, the surprise is to find that some twenty years of such demands have produced not a single memorable finding of such an interaction effect in the long-term causation of individual differences in either cognitive abilities or g. Indeed, it is common for critics to allege that countless interaction effects (which they would expressly hypothesize if only they could think of them) have gone missing through the willful disregard of psychogenetic investigators! Psychogeneticists are said to search their data for effects of genetic factors (making identical twins more similar than fraternal twins) or environmental factors (making for similarity between adoptees and their adoptive siblings) instead of acknowledging that *all* their data might be shot through with *genetic* × *environmental* (G × E) interaction effects that make conventional searches for additive effects of *genes* and *environment* quite inappropriate (e.g., Wahlsten 1990, and co-discussants).

It is hard to know how to understand such critics; however, here are the four main possibilities, and some replies.

1. If interactionists mean, as they sometimes say, "everyone has to have both genes and environment," we can smile politely. The psychogenetic question is "how do *differences* in genes and *differences* in experienced environment relate to eventual individual *differences* in phenotype?"

2. If what is meant is that there is no way of detecting the occurrence of G × E interaction effects, the criticism is simply wrong. If special similarity results from two people having *both* the same genes *and* the same environment to yield crucial G × E interactions with those genes, then identical twins reared together (MZT) should be markedly more similar than identical twins reared apart (MZA)—more than would be expected from the additive effects of environment that yield similarities between unrelated children reared together. Generally, in fact, such discoverable G × E interaction does not seem to have much influence on either abilities or personality: MZT twins are not much more similar than MZA twins, no more than would be expected from the merely additive effects of shared environment.

3. If critics mean that human beings create their own environments—influencing, for example, their own parents' behavior towards them from an early age—it can be claimed that mainstream psychogeneticists have themselves supplied an *embarras de richesses*, indicating with some precision the degree to which the growing child's experience of warmth vs. coldness, supervision and degree of permissiveness is indeed under its own (genetic) control (Plomin and Bergeman 1991).

4. There remains one last possibility as to what "interactionism" may mean. Could phenotypes result largely from G × E interaction effects without there being any additive genetic or environmental effects to be observed at all? Here, once more, we arrive at a theoretical problem for neotraditionalist critics of *g* and h^2. What is being suggested is that human development is replete with effects of perfect interaction such as the following combination of influences on genetic fitness:

Animals with genes for fins survive better in the sea than on land, while animals with genes for legs are quite the opposite (surviving better on land).

In this example, across the specified parameters of G and E, there is no gain to fitness deriving from which G value an animal has, nor from which E value it has. What is important is clearly to have the right G × E combination, for land animals with fins and sea-bound bipeds will not be around long enough for scientists to be able to tell much of a tale.

Perhaps there are countless such effects in human development, all of them missed in the hunt for additive types of causation? Sadly for the critic of conventional heritability estimation, this is unlikely. In nature, important interaction effects (like the above) will yield the evolutionary outcome that particular genes will only occur in particular environments, thus yielding a final state of genetic-environmental covariation (G, E, COV) rather than G × E interaction. Unlike the plant geneticist in the laboratory, nature does not persist with failed experiments just to show there can be affairs, too, parents and teachers, like evolution, abhor interaction effects: they will not persist long in supplying music lessons to children who derive no conspicuous enjoyment or benefit from them. The critic of g and h^2 has lost his $G \times E$ interactions and does not know where or how to find them; however, the hypothetical interactions beloved of theoretical behavioral geneticists just are not going to turn up: they have almost certainly been abolished, by nurture if not by nature, and contribute little or nothing to notable variations in either cognitive abilities or g.

All this is not to deny that *other* types of interaction really *do* create variance in human abilities and propensities. This is an important lesson that psychogeneticists have had to learn through the 1980s, although David Lykken helped prepare the ground with his talk of "emergenesis" (Lykken 1982). The multiplicative interactions between several genes may make MZ twins (having a near-100% genetic similarity, allowing for occasional mutations) much more similar phenotypically than would be expected from genetic influences alone that create intra-familial similarities when around 50% of genetic variations are held in common. In line with the idea that shared gene combinations are important, MZ twins are markedly more similar than DZ twins, especially once environmental factors are allowed for, and especially in the study of personality, interests, and attitudes (Brand 1989; Lykken et al. 1992).

It will come as no surprise to admirers of Sir Cyril Burt's methodological grip that, long before other psychologists, Burt allowed a role for *genetic dominance, epistasis* (i.e., gene-gene interaction), and indeed G, E, COV in the causation of intellectual differences. Burt's own estimate of the narrow heritability ($h^2{}_N$) of g, the degree to which g bred true, was thus on the low side (he estimated $h^2{}_N$ as 0.38), even though he maintained a fairly high estimate ($h^2{}_B = 0.75$) for the overall contribution of genetic factors to human variability in measured IQ. It is perhaps a pity that other broadly hereditarian psychologists were slow to appreciate that $h^2{}_N$ and $h^2{}_B$ can diverge considerably. If $h^2{}_N$ were indeed 0.70, aristocracy would be a workable type of political society; in fact, however, the great-grandchildren of eminent men were thought by Burt to be of mean IQ 104, a considerable regression to average population values. Genius may be of largely genetic origin (especially if genes shape microen-vironments), yet genius cannot be transmitted across several generations, either genetically or environmentally (Waller et al. 1993).

There are, then, plenty of "interactions" to be found in the study of intelligence; however, those who would fulminate about g and h^2 seem to have mislaid the very interactions that are actually on offer. Most talk of "interactionism" is confused or empty, although hereditarians can sometimes make a little sense of it!

THE UNPREDICTED DECONSTRUCTIONS OF CONSTRUCTIVISM

In the 1980s, the vanishingly slight commitments to quantification of multidimensionalists and interactionists were replaced by the outright innumeracy of *constructivism*, a new "movement of thought" in psychology. Spread from Piagetianism (and so ultimately from Kantian constructivism) by Jerome Bruner (see Brand 1988), the idea that we are all constructing ourselves, our relationships, and our experienced world as we develop, picking and choosing "identities" via language as we do so, proved as popular to social psychologists as it did to theologians and to literary critics. Today, the cumbersome *méthode clinique* of Piagetians and the serendipitously targeted "discourse analyses" of identity-seeking social psychologists have largely replaced validated measurement and systematic experimentation in studies of children in many university departments of psychology. Social and linguistic abilities are thought to be built from "proto-conversations" between infants and their innately given "virtual other" in the head; and the analysis of discourse in dieting advertisements, nineteenth-century novels, or politically incorrect lectures has come to count as doing psychology.

Like most briefly fashionable ideas in psychology, "constructivism" has one or two attractive features: it implies that people themselves cause their own futures (as differential psychologists argued for many years against "situationists" and environmentalists), and it stresses—indeed, celebrates—the flexibility and creativity that we all have available to ourselves, thanks to the human resource of language. Of course, as constructivism implies, we do *in some sense* move through time, building up our knowledge and skills (cf. Anderson 1992), developing the argumentative thread of reasoning that may indeed propel us from one major life event to another.

Yet unreconstructed constructivism is seriously flawed as an account of most, if not all, measurable cognitive abilities. First, it cannot easily account for the astonishing skills sometimes seen in grossly motorically handicapped, cerebrally palsied children, who have clearly been unable to go through the stages of exploration of the world (resulting in *assimilation* and *accommodation*) that the developmental psychologist normally envisages. (One such case involves the distinguished young Irish poet, Davoren Hanna [1990], who had no way at all of "interacting with the environment" until he was six years old, and thereafter could only communicate with the greatest difficulty by means of an "alphabet board" onto which he would fall forward from an adult's lap. Piaget used to try to explain away the mental accomplishments of motorically impaired thalidomide victims by saying they still showed "active engagement and problem-solving with the world": no such characterization of Hanna's childhood abilities is remotely plausible—he was deemed profoundly and irrevocably mentally retarded by all the experts who saw him in his early years.)

Second, though constructivism looks plausible superficially when "explaining" the gradual build-up of skills that occurs through a normal childhood, it is quite incapable of explaining why general intelligence shows little increase beyond age 16, and why measures of fluid intelligence actually show decreases in many people beyond age 55 despite welfare-state provision. No theory can or should explain everything, of course. However, since Piaget's constructivist theory of intelligence cannot explain individual differences in g, the failure of g to rise beyond age 16, or g's later decline, one wonders at its one-time wide appeal.

Once more, there is a mirror-image problem for supporters of the London School. Why, exactly, *does* intelligence increase through childhood? Is there a general increase in basic mental speed? Or is it that the improvements through childhood in even inspection time and short-term memory for digits should be thought to reflect the learning of special skills and strategies, as Anderson (1992) prefers? Yet, whichever direction is correct, modern differential psychology arguably has concepts in place that allow such processes to be understood. It may be that mental speed is based on nervous system transmission speed, or it may be that the recent discovery of what is called "working memory" by experimental psychologists is just *g* by another name (Kyllonen and Christal 1990; Salthouse 1993), which explains why the higher-*g* child will do a better job of extracting more from the environment over time. Progress along either line looks much easier than it can be for constructivists, obliged as they are to predict repeated improvements in cognitive abilities that just do not occur. The London School can at least begin to explain advance and recession in intellectual levels; "constructivists" can only hope to celebrate the one in mime and dance while remaining baffled by, and trying to ignore, the other.

THE SELF-DENYING INANITIES OF NIHILISM

Piaget died some ten years ago, so what is the very latest fashion in the study of cognitive abilities? Undoubtedly it can be called *nihilism*: it is the attempt in psychology (a hundred years after Nietzsche's similar effort in philosophy) to show that even *g* itself (or IQ, as critics often prefer to call it) is not a real or important variable, either as a predictor of educational, socioeducational, or moral success, or as an explanation of its own many cognitive correlates.

Ultimately, the psychological nihilist will maintain that *g*—like everything else—is only a form of discourse that misleads its users into errors of reification; however, this is essentially a position of philosophical idealism, applicable in principle across the whole of psychology rather than targeted on *g* in particular. The most impressive targeted endeavor in this direction is James Flynn's (see e.g., Brand 1990) interpretation of the worldwide IQ rise that he detects since 1930, and which he claims was unaccompanied by any notable increase in educability or attainment among the young, as indicating the unimportance of IQ, which he takes to reflect "mere problem-solving ability." But Michael Howe (1990) and Stephen Ceci (1991) have also set out to deny the reality and the causal status of *g*: they stress the calculational achievements of *idiots savants*, the IQ-unrelated (and apparently financially unrewarding) "handicapping skills" of regular gamblers on horse races, and the remote possibilities that IQ-type differences between people arise from the amount of time spent in school or from hypothetical "personality" differences (which themselves strangely elude measurement).

There are numerous ways to reply to such suggestions. Some (Detterman and Spry 1988; Brand 1990) take issue with the empirical claims of these nihilists; others (e.g.,

Lynn 1989) offer interpretations of the nihilists' results that make them compatible with London School accounts of *g*, thus, for example, hypothesizing an influence of improved nutrition on levels of intelligence in the twentieth century to account for Flynn's data. Considerable methodological problems remain (Brand 1990; Blinkhorn 1991); and the persistence of racial differences, as almost certainly requiring explanation in terms of *g* (Jensen 1985), seems unlikely to placate the average nihilist. Flynn (1992) attempts an explanation of the achievements of Orientals under *Pax Americana* in terms of the ancient variables of "hard work" and "achievement motivation"; however, Flynn himself accepts that the black-white difference is a real mystery to present environmentalist thought. It is too early to adjudicate the dispute, which will certainly continue in and around the work of Flynn and Phil Rushton (*vide infra*).

What must be remarked, however, is an astonishing theoretical omission in nihilist offerings so far. If a trait is strikingly valuable to a species, variation in it will be reduced and remaining differences between people will not be strongly heritable: if differences in such features *were* heritable, the lower fitness of those lacking the trait would gradually bring about the trait's nonheritability (by additive genetic means, at least). Conversely, it is when a trait is broadly unimportant to fitness that heritable variation in trait levels can be considerable—as with the variable of height in modern Man.

It is strange, then, that those, like Howe, who contest the importance of IQ should wish simultaneously to contest its heritability and biological basis. One would have thought that a high h^2 for IQ would have served their purpose well! Of course, Flynn *does* take this sensible view; he trips himself up only by insisting that *racial* differences in *g* should not be heritable, which they might as well be if, as Flynn likes to argue, IQ does not really matter much.

A more coherent position is that of Rushton (e.g., 1988), who maintains that *g* differences are substantially heritable but unrelated to long-term fitness. Rushton's view is that high- and low-*g* people may end (presently, as in the past) with similar numbers of great-great-grandchildren, but that they will have done so by the dramatically different routes of breeding, *either* for quality *or* for quantity in their children. (In the rapidly approaching future, increasing use of the Pill and the spread of AIDS mean that quantity-oriented strategies involving casual procreative activities will be evolutionarily less viable without massive assistance from Western medicine.)

Another view (e.g., Brand 1987b) would be that *g* differences are unrelated to differential fitness but are highly relevant to the maintenance of human social structures in which leadership, hierarchy, and a certain amount of deference play a part in generating favorable outcomes for all. According to this account, *g* would be a socially and psychologically important variable; however, since only the range of inter-individual variation would have biological importance (for group fitness), *g* could have a substantial degree of heritability. If this is too subtle (or un-Darwinian), then the differential psychologist who wants to stress the all-around importance of *g* would do well to envisage a relatively low h^2_N estimate—as did IQ's greatest theoretician, Sir Cyril Burt.

The truth is this: IQ is probably real and important, yet few dare to say so. Rather, experts pretend IQ is trivial, yet strangely will not raise their h^2_N estimates accordingly. Nihilism is thus unimaginative and self-defeating in its assault on g so far. Nihilists would do better to stick to discourse analysis: Nietzsche's scholarly biblical criticism was more intellectually persuasive than his full-blown nihilism.

REVIEW AND CONCLUSION

In this chapter, I have identified four main lines of criticism of the "psychometric," "hereditarian," London School view of general intelligence (g). Preferring to dispense with g, self-styled students of "cognitive abilities" have, since 1960 or so (as the tide turned against behaviorism) moved from *multidimensionalism* through *interactionism* and *constructivism* to the latest, ca. 1990, fad of *nihilism* that doubts the importance of measured intelligence. None of these successive dogmas has proved particularly productive; however, each phase of the assault on g (visible as psychology's major scientific achievement once the behaviorists' "laws of learning/conditioning" were laid to rest) has left behind a message for serious students of nature/nurture questions.

1. The search for multidimensionality will probably continue in some quarters. It seems likely, however, that breakthroughs can most obviously be expected if some differentiation—whether apparent or real—of cognitive abilities at higher levels of g is borne in mind (Anderson 1992). The study of high-IQ MZA twins matched for g will perhaps be especially important in resolving such questions: the hope must be at last, by controlling for levels of g, to discover some discrete causes of divergence in non-g-related, "specific" abilities.

2. With regard to "interaction effects," four possibilities beckon: (a) The study of DZA twins matched for the social class of their homes of rearing should help identify (but more likely discount) any likelihood of G × E interaction effects creating important variance that has gone unremarked so far. (b) The study of MZA twins matched for the social class of their homes of rearing (SES) should help pin down the importance of unshared environmental effects resulting from within-family niche-selection and competition. (c) Ordinary comparisons between the similarities of MZ and DZ twins should help firm up the possibility (Brand 1989) that gene–gene interaction makes MZ twins markedly more similar than are DZs (whose genetic similarities must arise mainly from additive, "narrowly heritable," transmissible factors). (d) Studies of separated relatives of all kinds would throw light on whether relatives are more similar when they are not in immediate competition with each other, as has sometimes been observed for the trait of *extraversion* in MZ twins. Using these methods,

London School psychogeneticists might distinguish themselves still further by discovering more interactions than have ever been discovered by official "interactionists"; alternatively, they would crush for good the remaining comprehensible hypotheses about nature/nurture "interaction."

3. The concerns of developmental psychologists (whether accepting or opposing *constructivism*) would seem best served by twin studies that trouble to use modern measures of inspection time and perceptual intake speed (e.g., Zhang 1991; Anderson 1992). It would be of great interest to know the heritability of such parameters—as Tony Vernon once attempted, but without the aid of state-of-the-art devices allowing reliable, tachistoscope-type presentations of target stimuli. In all likelihood, a vindication and deepening of Cattell's (1982) classic distinction between "fluid intelligence" (*gf*) and "crystallized intelligence" (*gc*) will be the outcome; however, Anderson's ideas and work will need to be borne in mind.

4. As to the importance of IQ, no new type of study is as necessary as a change of attitude. The important thing to establish is the impartiality of twin study as a way of arbitrating disputes. All too often it appears that the London School would somehow rejoice in a high h^2_N and that skeptics of g are obliged to argue h^2 estimates lower. In fact, the position is not so simple. Subtleties such as the high h^2_N of differences that are unimportant to individual biological fitness need to be recognized: then it will be possible to persuade more workers of egalitarian or environmentalist leanings to undertake such researches themselves. This would be a major step forward, for the most influential adoption study of all time is undoubtedly, at present, the one carried out by the left-wing nuclear physicist, Maurice Schiff, and his colleagues in Paris. Here, empirical work overcame ideology, even if the researchers were loath to highlight their remarkable finding that the influence of imposed variation in parental social class on child IQ was smaller, if anything, than Eysenck and Jensen themselves had long allowed (Brand 1987a).

The London School will no doubt take some knocks if twin studies flourish (as they have shown every sign of doing in several countries through the 1980s where the educational and psychological establishments are not so frightened of "what is in the genes"). Yet it looks as if g is psychology's great survivor in a century through which psychologists have preferred to trace human behavior to instincts, reflexes, the unconscious mind, the unfalsifiable "strategies" of cognitivism, and the mysteries of multidimensionalism, interactionism, constructivism, and nihilism. Contrary to Michael Howe's view (*vide supra*), the phenomena of g are not going to disappear into a welter of "cognitive abilities," nor will intellectual development turn out to involve impossibly complex interactions yielding never-ending constructions of "mere problem-solving ability," nor will psychogenetics long be ignored by anyone other than myopic professional geneticists, like Prof. Steve Jones, who have old-fashioned, ideological axes to grind.

ACKNOWLEDGEMENTS

I am grateful for the critical and encouraging comments of Bambos Kyriacou, Gail Addis, and my colleagues in the Edinburgh Structural Psychometrics Group.

REFERENCES

Anderson, M. 1992. Intelligence and Development: A Cognitive Theory. Oxford: Blackwell.

Bailey, J.M., and W. Revelle. 1991. Increased heritability for lower IQ levels? *Behav. Genet.* **21(4)**: 397–404.

Belsky, J., L. Steinberg, and P. Draper. 1991. Childhood experience, interpersonal development and reproductive strategy. *Child Devel.* **62(4)**:647–670.

Blinkhorn, S. 1991. A dose of vitamins and a pinch of salt. *Nature* **350(6313)**:13.

Bouchard, T.J., Jr., D.T. Lykken, M. McGue, N.L. Segal, and A. Tellegen. 1990. Sources of human psychological differences: The Minnesota study of twins reared apart. *Science* **250**:223–228.

Bouchard, T.J., Jr., M. McGue, D.T. Lykken, and A. Tellegen. 1993. The Burt business. *Times Lit. Suppl.*, Feb. 19, 1993.

Brand, C.R. 1987a. A touch of (social) class for psychology. *Nature* **325**:767–768.

Brand, C.R. 1987b. The importance of general intelligence. In: Arthur Jensen: Consensus and Controversy, ed. S. Modgil and C. Modgil. Brighton, U.K.: Falmer.

Brand, C.R. 1988. The changing advice of experts on childrearing: The return of individualism. In: Full Circle: Bringing up Children in the Post-Permissive Society, ed. D. Anderson, pp. 16–36. London: Social Affairs Unit.

Brand, C.R. 1989. A fitting endeavor: A review of L.J. Eaves et al. (Genes, Culture and Personality, publ. by Academic.) *Nature* **341**:29.

Brand, C.R. 1990. A "gross" underestimate of "massive" IQ rise? A rejoinder to Flynn. *Irish J. Psychol.* **11(1)**:52–56.

Brand, C.R., P.G. Caryl, I.J. Deary, V. Egan, and H.C. Pagliari. 1991. Is intelligence illusory? *Lancet* **337(8742)**:678–679.

Brand, C.R., V. Egan, and I. Deary. 1992. Personality and general intelligence. In: Personality Psychology in Europe, vol. 4, ed. G. Van Heck, P. Bonainto, I. Deary, and W. Nowack. Lisse, Netherlands: Swets and Zeitlinger.

Brand, C.R., V. Egan, and I.J. Deary. 1993. Intelligence, personality and society: "Constructivist" vs. "essentialist" possibilities. In: Current Topics in Human Intelligence, vol. 4, ed. D.K. Detterman. Norwood, NJ: Ablex.

Cattell, R.B. 1982. The Inheritance of Personality and Ability: Research Methods and Findings. New York: Academic.

Ceci, S. 1991. How much does schooling influence general intelligence and its cognitive components? A re-assessment of the evidence. *Devel. Psychol.* **27(5)**:703–722.

Detterman, D.K., and K.M. Spry. 1988. Is it smart to play the horses? Comment on a day at the races: A study of IQ, expertise and cognitive complexity (by S. Ceci and Liker, 1986.) *J. Exp. Psychol.* **117**:91–100.

Flynn, J.R. 1992. The Achievements of American Orientals. Hillsdale, NJ: Erlbaum Ass.

Gardner, H. 1983. Frames of Mind: The Theory of Multiple Intelligences. New York: Basic.

Hanna, D. 1990. Not Common Speech: The Voice of Davoren Hanna. Dublin: Raven Arts Press.

Hirsch, J. 1990. A nemesis for heritability estimation. *Behav. Brain Sci.* **13(1)**:137–138.

Howe, M.J.A. 1985. Does intelligence exist? *Psychologist* **Nov.**:490–493.

Howe, M.J.A. 1990. Sense and Nonsense about Hothouse Children. Leicester: British Psychological Society.

Jensen, A.R. 1985. The nature of the black-white difference on various psychometric tests: Spearman's hypothesis. *Behav. Brain Sci.* **8**:193–263.

Kyllonen, P.C., and R.E. Christal. 1990. Reasoning ability is little more than working memory capacity?! *Intelligence* **14**:389–433.

Lykken, D.T., M. McGue, A. Tellegen, and T.J. Bouchard, Jr. 1992. Emergenesis: Genetic traits that do not run in families. *Am. Psychol.* **47**:1565–1577.

Lynn, R. 1989. A nutritional theory of the secular increases in intelligence: Positive correlations between height, head size and IQ. *Br. J. Ed. Psychol.* **59**:372–377.

Pearson, R. 1991. Race, Intelligence and Bias in Academe. Washington: Scott-Townsend.

Plomin, R., and C.S. Bergeman. 1991. The nature of nurture. *Behav. Brain Sci.* **14(3)**:373–427.

Plomin, R., and D. Daniels. 1987. Why are children in the same family so different from one another? *Behav. Brain Sci.* **10(1)**:1–60.

Previc, F.H. 1991. A general theory concerning the prenatal origins of cerebral lateralization in humans. *Psychol. Rev.* **98(3)**:299–334.

Rushton, P. 1988. Race differences in behavior: A review and evolutionary analysis. *Pers. Indiv. Diff.* **9(6)**:1021–1031.

Salthouse, T.A. 1993. Influence of working memory on adult age differences in matrix reasoning. *Br. J. Psychol.* **84(2):**171–200.

Scarr, S. 1992. Developmental theories for the 1990s: Development and individual differences. *Child Devel.* **63**:1–19.

Sternberg, R.J. 1990. Metaphors of Mind: Conceptions of the Nature of Intelligence. New York: Cambridge Univ. Press.

Wahlsten, D. 1990. Insensitivity of the analysis of variance to heredity: Environment interaction. *Behav. Brain Sci.* **13(1)**:109–162.

Waller, N.G., T.J. Bouchard, Jr., D.T. Lykken, A. Tellegen, and D.M. Blacker. 1993. Creativity, heritability, familiarity: Which word does not belong? *Psychol. Inq.*, in press.

Weiss, V. 1992. Major genes of general intelligence. *Pers. Indiv. Diff.* **13**:1115–1134.

White, K.R. 1982. The relation between socioeconomic status and academic achievement. *Psychol. Bull.* **91(3)**:461–468.

Zhang, Y. 1991. Inspection time correlation with IQ in Chinese students. *Pers. Indiv. Diff.* **12(3)**:217–219.

3

What Can Twin Studies Tell Us about the Structure and Correlates of Cognitive Abilities?

D.W. FULKER and L.R. CARDON
Institute for Behavioral Genetics, University of Colorado
Boulder, CO 80309–0447, U.S.A.

Twin study is arguably the single most powerful method in human quantitative genetics. It is easy to carry out, since twins are readily available in the general population, it is statistically powerful, it is straightforward to analyze, and its outcome is capable of yielding a variety of insights into the structural properties of traits. When combined with data on parents and offspring, adoptions, and inbreeding, we can validate the basic twin approach and fill out the picture in terms of the biological and social significance of the trait in question. In the case of general cognitive ability, this synthetic approach was outlined by Jinks and Fulker (1970) and later described at some length by Fulker and Eysenck (in Eysenck 1979), employing the wealth of published data available for this trait at that time. More recently published data, in our view, serves only to reinforce the view outlined in those publications.

We have found that the cumulative data suggest a mode of inheritance for general intelligence, of both additive and directionally dominant gene action in the presence of strong assortative mating together with cultural as well as biological transmission from parent to offspring. This picture indicates a trait of considerable evolutionary and social significance. Under this mode of transmission, genetic and environmental influences become correlated both within and between individuals in a complex manner, reinforcing their effects and accentuating individual differences. These processes would tend to preserve social stratification based on general intelligence. The complexity of this picture may be appreciated by examining the path diagram in Figure 3.1, which shows the highly correlated system of variables involved in genetic and environmental transmission that exists just between parents and their offspring for a trait such as IQ.

Twins as a Tool of Behavioral Genetics
Edited by T.J. Bouchard, Jr. and P. Propping © 1993 John Wiley & Sons Ltd.

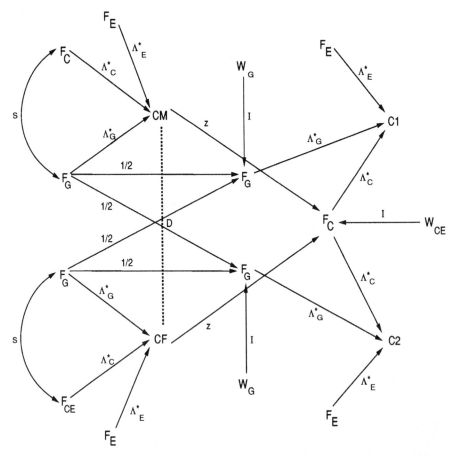

Figure 3.1 Parent–offspring path model.

Furthermore, the presence of directional dominance suggests a trait of some evolutionary as well as social significance, since such gene action is usually associated with fitness traits, such as fertility, fecundity, disease resistance, etc.—traits that have been under strong directional selection. Since general intelligence, of some sort or another, is what distinguishes us from our evolutionary ancestors, the evidence also suggests that what we measure in these studies—IQ—is of some validity, also, in the context of human evolution. Taken as a whole, twin and related studies of general intelligence suggest a trait of considerable sociobiological importance. Such a trait would clearly repay study, in terms of its genetic and environmental relationship to important correlates such as educational achievement, socioeconomic status, social mobility, and differential fertility. It is these issues that are discussed at some length by Fulker and Eysenck (in Eysenck 1979). We mention them again here because we believe they represent the more interesting reasons for carrying out behavioral genetic analysis, and it would be unfortunate to see their importance diminished by the narrow brief offered by the title of this chapter. These features of the genetic and environmental architecture are correlates of general cognitive ability in a broad sense. Some, but not all, will be touched upon again here as they relate to what we take to be the more specific concerns of the present title, namely multivariate analysis.

Twin analyses are able to distinguish between several broad classes of influence that provide preliminary evidence for different kinds of causal mechanisms. The basic biometrical model partitions the causes of individual differences into three classes. First, heritable influences that have their origin, though not necessarily their immediate causes, in DNA. Genetic variation demonstrates the ultimate importance of biological factors. Of course, the manner in which these factors come to express themselves in any particular phenotype may be quite complex and highly contingent on circumstances. Given the complexity of the living organism, gene products may not relate in any simple manner to observed phenotype. Nevertheless, DNA still remains the ultimate cause of variation when we find a trait is heritable. This source of variation is labeled h^2. Next there is environmental variation, which is further partitioned into two classes: shared environmental influences, those shared by individuals reared together, and specific or nonshared environmental influences, those which are unique to the individual. The former source of variation, labeled c^2, refers to systematic environmental influences, often social or parental in origin, while the latter, which we will call e^2, relates more to idiosyncratic experiences, both physical and social. Data on additional kinds of relatives are required to explore the origin of c^2 further, parental data from adoption studies being particularly useful though twins and their parents may suffice. Both kinds of influence can lead to correlated genetic and environmental effects. How to parameterize simple models of this kind and estimate their effects in an optimal manner was outlined by Jinks and Fulker (1970). Distinguishing between these three kinds of influence has the value of placing the explanation of individual differences—theories of individual differences, if you like—in an appropriate context. For example, IQ, which shows all three influences, would seem, therefore, to require explanation at many different levels: biological, social, educational, etc. Indeed, as we have suggested, general intelligence is a richly complex trait at many levels. Schizophrenia, on the other hand, which seems only to provide evidence for genetic and specific environmental influences (Fulker 1973; Gottesman and Shields 1966), may require a different kind of explanatory context. The absence of c^2 suggests that explanations into the causes of schizophrenia, in terms of social or parental influences, may well be of quite limited value, and emphasizes the importance of biological factors. This simple partitioning of variation in the basic twin study is illustrated by means of the path diagram in Figure 3.2.

It is when this form of analysis is extended to the multivariate case that the twin study comes into its own in terms of exploring both the structure and correlates of traits such as cognitive abilities. By entering the traits and correlated variables into such an analysis, we can determine whether the causes of the observed relationships

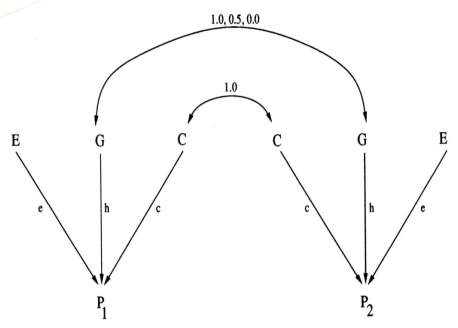

Figure 3.2 Path diagram of sources of variance for pairs of twins, siblings, or unrelated individuals reared together (URT).

among them are genetic or environmental in origin. By entering the same or similar measures at different ages, we can determine whether the developmental process is driven by the genes or the environment. By breaking down general intelligence into its specific components, we can determine whether genetic or environmental influences are responsible for the apparent organization of these specific traits into simpler entities. Finally, if we have suitable data, we can combine these questions and ask what drives the developmental processes that organize specific abilities into more general components at different points in time. We can do all of this, in principle, with twin data alone, although, as we have suggested, additional data make for stronger inference. For this reason, our own studies combine longitudinal data on twins and adopted children.

All twin studies are essentially multivariate. We know of no study in which data on only one single variable have ever been collected. At the present time, the difference is that we now have the statistical methodology and computing power to explore multivariate issues in a reasonably rigorous and searching way.

Various forms of multivariate twin analysis have been suggested and available for some time, e.g., Loehlin and Vandenberg (1968) developed a method that explored the dimensionality of a battery of specific cognitive abilities. Nevertheless, it was not until comparatively recently that rigorous hypothesis testing in this area was possible

(Martin and Eaves 1977; Fulker, Baker, and Bock 1983; Martin, Boomsma and Neale 1989; Vogler and Fulker 1983; Neale and Cardon 1992). This approach stemmed from the sophisticated work of Karl Joreskog (1978) in the field of structural modeling, which enabled the simple univariate approach of Jinks and Fulker (1970) to be extended to the multivariate case. In view of the importance of systematic analysis in this area, we will focus on twin and related studies involving the structural modeling approach.

One central issue in the field of cognition has been the dimensionality of cognitive abilities. Is there essentially only one general cognitive ability for which people differ or are there many? These views are traditionally associated with such psychologists as Spearman (1904), on the one hand, and Guilford (1967), on the other. Currently there are many opinions, a number of which were represented at this workshop.

We believe the issue can be productively analyzed in terms of the level of analysis. If we are talking about what differentiates us from the apes, then some concept like Spearman's *g* will probably suffice. It would not much matter which aspect of cognition we used to measure intelligence, verbal or nonverbal; we would explain most, if not all, differences between species by some such measure. If any number of measures, different on the face of it but leading to the same conclusion, could be employed, in a sense they would all measure the same thing. The same might be also true of comparisons among very different individuals within a species. For certain purposes, a single index—say IQ—might still predict another variable—such as years of completed education—sufficiently well that we can, for that purpose, regard it as a functional unity. However, if people of the same IQ arrive at their score by a very different pattern of specific abilities, and this pattern is stable, either over time or in terms of genetic and environmental effects, and the pattern of subcomponents leads to very different skills, then there is a sense in which *g* is an illusion. What is the evidence from twin studies regarding the dimensionality of specific cognitive abilities?

Early studies were mixed in their conclusions. Vandenberg (1965), for example, analyzed twin data from two studies on verbal, spatial, number, reasoning, word fluency, and memory tests. He carried out a multivariate F-test on the MZ and DZ within-pair mean cross product matrices and found four significant roots indicating at least four independent genetic dimensions. Loehlin and Vandenberg (1968), analyzing the same data by a simple subtractive method, found more persuasive evidence for a large genetic general factor. The same data were again analyzed by Eaves and Gale (1974) employing a primitive principal component approach that nonetheless attempted to focus on hypothesis testing in a more appropriate manner. Their approach indicated heritable variation for both general and specific cognitive factors.

The first fully satisfactory reanalysis of these data—there was little other data in existence at the time—was by Martin and Eaves (1977) using a structural modeling approach that set the standard for the field. Their analysis confirmed the finding of a large, general genetic component but also at least three specific genetic components: verbal, spatial, and word fluency abilities. A much larger data set did become available from the National Merit Scholarship Qualifying Test twin study of Loehlin and Nichols

(1976), which involved English usage, math, vocabulary, social studies, and natural science tests given to over 2000 pairs of twins. A strong, general genetic factor and four significant specific genetic factors were also found. Rice et al. (1989), using parent-offspring data on four specific cognitive ability measures in the Colorado Adoption Project (CAP), found evidence of a general rather than for specific genetic factors.

The conclusion we draw from these studies is that both general and specific genetic variation exist and that how many significant specific components are detected is mainly a question of statistical power. This outcome would not be surprising in view of the likely underlying polygenic nature of the genetic control. Many genes underlying a trait may or may not be of equivalent phenotypic effect; most likely they would not, and the studies seem to confirm this.

One aspect of these earlier studies was somewhat neglected at the time, namely the differential picture according to whether one looked at the genetic or environmental components. The emphasis was on whether there were both general and specific genetic factors; however, the environmental components appeared in marked contrast not only to the genetic component but also to each other. In general, the shared environmental component appeared more unitary, supporting the notion of g, whereas the specific environmental component was highly specific, supporting the notion of highly specific cognitive abilities. Taken at face value this picture is telling us that the genes are responsible for the systematic complexity of the phenotype and that the environment acts in a simpler manner: the shared environment, presumably associated to some extent with rearing influences, appears to act in a generally beneficent or deleterious manner and the specific environment much more idiosyncratically.

One other early multivariate twin study of education, SES, and income carried out on over 2000 pairs of male twins revealed a very similar pattern regarding the differential importance of general and specific factors, depending on whether one looked at the genetic or environmental component of variation (Taubman 1976; Fulker 1978). The subjects were mailed questionnaires and asked about years of completed education, their initial occupations, and their current incomes at about 50 years of age. A structural model was fitted to the data using essentially the method of Martin and Eaves, and Joreskog, which partitioned genetic and environmental covariance components into general and time-specific factors. At the genetic level, there was strong continuity of genetic influence from the early period of educational experience, persisting through the period of the first job and into adulthood, strongly affecting all three variables. At each period, however, there were also marked specific genetic influences of about the same magnitude as that responsible for continuity. Presumably, a general genetic factor related to general ability acted in a similar manner to determine success or failure at each point in life. In marked contrast, at the shared environmental level (although influences became progressively less pronounced, as one might expect), they were accounted for by a single environmental factor without even a hint of any time-specific influences. We think of this factor as the waning effect of family environmental influence, becoming less with time but always attributable to the same favorable or unfavorable start in life. The specific environmental factors were entirely different again from those attributable to either genetic or shared environmental variation. They were entirely specific in character. Those important at any one time had no lasting effect, failing to influence the phenotype at all at later points in time. This study provides one of the clearest demonstrations of the contingent nature of the answer to the question: how many kinds of abilities are there? It depends on the domain of inquiry.

In the classical twin study the effects of shared environment may be, to some extent, confounded by the effects of assortative mating. Eaves et al. (1984) raised the interesting possibility that the apparent unitary nature of shared environmental variation, or c^2, in these studies might reflect the multivariate effects of assortative mating. If individuals chose mates based on an appreciation of ability but did not differentiate among these abilities specifically (i.e., so long as a mate had some ability but it didn't matter what), then the effect over generations would force the genes for specific traits into linkage disequilibrium and association and create a general genetic factor. A similar mechanism could underlie the puzzling association between IQ and height and the putative association between IQ and physical attractiveness, both of which are phenotypes subject to strong assortative mating (Eysenck 1979). Incidentally, our own research using parent-offspring data from the CAP does not tend to support this mechanism, although the subjects are still too young to be completely sure.

Whatever the mechanisms, the differential outcome of the structure of these specific ability measures in these studies fully substantiates our earlier contention that the issue of *g* or many abilities is not a question with a single answer. The answer depends on the domain under discussion. To ask if there is a single *g* or many specific abilities is to ask a misleading question. It all depends on the relationships we wish to focus upon. From a societal point of view there may only be *g*; from the specific environmental point of view, there may be very many specific abilities, perhaps as many as we can measure; genetically, the situation may well be more structured and complex. When we only observe phenotypic relationships, cognitive abilities appear to be both specific and general in nature. Twin and other behavioral genetic research strategies offer the opportunity to study quite different aspects of the organism.

There are now a number of sophisticated multivariate twin analyses in the literature. A series of twin methodology workshops, organized by a number of us, have been responsible for making the requisite methodology more readily available. Our collected wisdom has recently been brought together in a NATO-sponsored book, which also serves as a bibliography (Neale and Cardon 1992). However, few are concerned with the problems of cognitive abilities. The developmental studies that have been carried out with collaborators and students at the Institute for Behavioral Genetics, using both twins and adopted and nonadopted siblings and their parents, are an exception. It is still somewhat early—we have yet to fully develop our models, and computing power is a problem in spite of the availability of relatively fast machines— yet we would like to take this opportunity to present our preliminary thoughts and findings.

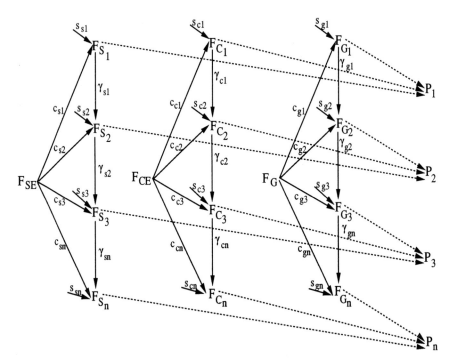

Figure 3.3 Longitudinal model of transmission effects and factor loadings.

One analysis that illustrates our approach to combining twin and sibling data is that involving continuity and change in general cognitive ability from 1 to 7 years of age (Cardon et al. 1992a). Twin data were available at ages 1, 2, and 3, with sibling data available at ages 1, 2, 3, 4, and 7. The tests employed were the Bayley Mental Development Index at ages 1 and 2, the Stanford-Binet at ages 3 and 4, and the WISC-R at age 7. At age 1, there were 106 full sibling pairs, 88 adopted sibling pairs, 73 MZ pairs, and 64 DZ pairs. Numbers dropped to approximately one-half at the last available age point. For this reason, our model-fitting was through a pedigree function based on the multivariate normal distribution

$$L_i = -\frac{1}{2} \ln |\Sigma_i| - \frac{1}{2} (x - \mu)' \Sigma_i^{-1} (x - \mu). \tag{3.1}$$

The genetic and environmental model is essentially that previously described, involving h^2, c^2, and e^2. Dominance variation and assortative mating, which cannot be assessed in this design, were ignored; however, even if they were present, they would not seriously bias estimates based on combined sibling and twin data. The longitudinal model involves imposing a simplex, or simple first-order time series, on

Table 3.1 Tests of unique environment, shared environment, and genetic developmental patterns.

Model Description	Log Likelihood	NPAR[a]	vs.	χ^2	df	p
Unique Environment						
1. Full model	−899.85	42				
2. γ_{si}, s_{si}; $c_{si} = 0$	−900.78	37	1	1.86	5	>0.70
3. c_{si}, s_{si}; $\gamma_{si} = 0$	−899.87	38	1	0.03	4	>0.70
4. s_{si}; $\gamma_{si} = c_{si} = 0$	−902.80	33	1	5.89	9	>0.70
Shared Environment						
5. γ_{ci}, s_{ci}; $c_{ci} = 0$	−902.80	28	4	0.01	5	>0.70
6. c_{ci}, s_{ci}; $\gamma_{ci} = 0$	−904.96	29	4	4.33	4	>0.30
7. c_{ci}; $\gamma_{ci} = s_{ci} = 0$	−904.36	24	6	1.20	5	>0.70
8. $c_{ci} = c_{cj}$; $\gamma_{ci} = s_{ci} = 0$	−906.10	20	7	3.47	4	>0.5
9. $c_{ci} = \gamma_{ci} = s_{ci} = 0$	−908.65	19	8	5.11	1	<0.025
10. γ_{ci}; $c_{ci} = 0$, $s_{ci} = 0$, $i \neq 1$	−905.87	24	5	6.14	4	>0.10
11. $\gamma_{ci} = \gamma_{cj}$; $c_{ci} = 0$, $s_{ci} = 0$, $i \neq 1$	−906.10	23	5	0.47	5	>0.70
Genetic						
12. γ_{gi}, s_{gi}; $c_{gi} = 0$	−908.28	15	8	4.37	5	>0.50
13. c_{gi}, s_{gi}; $\gamma_{gi} = 0$	−916.28	16	8	20.42	4	<0.001
14. γ_{gi}; $c_{gi} = 0$, $s_{gi} = 0$, $i \neq 1$	−989.38	11	12	162.20	4	<0.0001
15. $\gamma_{gi} = \gamma_{gj}$; $c_{gi} = 0$	−915.54	12	12	14.51	3	<0.01

[a]NPAR: number of parameters estimated.

these three components. This model represents the effects of continuity through direct transmission parameters from age to age and represents change, or new influences, as a specific factor arising at each age point independent of transmission. The full model is shown in Figure 3.3. This model of longitudinal transmission was first developed by Eaves et al. (1986) and has been further developed by Hewitt et al. (1988) and Phillips and Fulker (1989), who incorporated the effects of assortative mating.

The likelihood-ratio test was used to assess the most reasonable and parsimonious structure of the developmental model for each of the three components. To illustrate this point, these components are shown in Table 3.1. This method of model comparison is reasonably thorough and searching, and while not above reproach statistically, since it involves so many multiple comparisons, it does seem the best way to proceed compared to alternative procedures.

The outcome of this analysis was strikingly straightforward and very similar to that found in Taubman's study. The results of the reduced model are shown in Figure 3.4. The complexity of the developmental process was found to reside mainly in the genotype, with both continuity and change contributing to the rank order of phenotypes at ages 1, 2, and 3. At age 4, no new genetic variation was seen to arise; variation

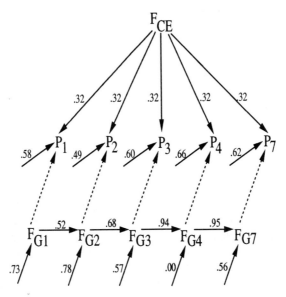

Figure 3.4 Final reduced developmental model.

established at ages 1, 2, and 3 was sufficient to account for individual differences at age 4.

At age 7, however, new variation again arises, perhaps indicating the effects of one year of schooling. The specific environmental variation only shows change and not continuity. We surmise that this component is largely reflecting error variance, although not entirely. In contrast, the shared environment shows a small but persistent common factor equally pervasive at each age, perhaps indicating the constancy of the effect of home background during this period of childhood. We have recently updated this analysis, including a telephone test of specific cognitive abilities at age 9, which yields a strong general factor. The picture is virtually the same. The only new information is that there is little fresh genetic variation at age 9, just as there is little at age 4 in the previous analysis. As in the case of the Taubman study, complexity or simplicity in cognitive abilities depends on whether one takes a genetic or environmental perspective. There is no one picture at the level of the phenotype.

Although interesting in themselves, longitudinal analyses tell us nothing about the relationships between specific cognitive abilities and general intelligence. However, with measures of specific cognitive abilities available at ages 3, 4, 7, and 9, we are able to begin to explore not only this relationship, but also how it changes over time.

Early analyses of specific cognitive abilities were rendered problematic by methodological shortcomings. Not all used suitable estimation procedures, and the issue of error of measurement and test-specific, as distinct from ability-specific, variance was not adequately considered. Our recent data analyses on young twins and adopted and nonadopted siblings attempted to remedy these deficiencies by employing a hierarchical factor model (Cardon et al. 1992b; Cardon 1992). In a sense, the hierarchical model of cognitive abilities tries to reconcile the two extreme positions of *g* or many specific abilities. Specific component abilities are seen as basic with higher-order factors integrating them into broader, more general cognitive abilities (Burt 1955). We are currently trying to combine this hierarchical approach with the developmental approach described above.

The approach to specific cognitive abilities was that adopted in the Hawaii Family Study (DeFries et al. 1979), in which a battery of 16 tests was constructed that primarily identified four abilities: verbal (V), spatial (S), perceptual speed (P), and memory (M). Details of the tests developed for the children may be found in Plomin et al. (1988). Briefly, the idea was to attempt to devise tests that were isomorphic with those in the Hawaii adult battery so far as this was possible. In general there were eight tests at each age point, with two tests representing each of the four ability factors.

Twin data were available at ages 3 and 4, there being approximately 50 pairs of MZ and DZ twins at each age; just over 100 pairs each of adopted and nonadopted siblings were available at age 3, with the numbers dropping to 45 and 54, respectively at age 9. Numerous adopted and nonadopted singletons and their parents together with some older siblings were also available, although results on these data will not be reported. Once again, a pedigree maximum-likelihood analysis was employed to make optimal use of all available data points in the face of the inevitable inequality of numbers at different ages inherent in a longitudinal study of this type.

A thorough analysis of cognitive structure within a single time point was under-taken at age 7 (Cardon et al. 1992b). The basic hierarchical model is shown in Figure 3.5. Each of the four first-order factors (V, S, P, and M) is shown with a pair of indicator tests. Each of the factors is due to two sources of variation: the higher-order general factor, which we call IQ, is responsible for the correlation among the first-order factors, and for each ability there is a unique component uncorrelated with IQ. This structure was imposed on the three components of covariation appropriate to the twin-sibling design: genetic, shared environment, and specific environment. For each component, a nested set of comparisons of the model was undertaken, starting with the full model and progressively testing for more specific features of the model. The outcome of these analyses was more informative than those of the earlier studies we have discussed, providing some justification for the hierarchical approach. The detailed model comparisons are given in Tables 3.2–3.4. Now that we have controlled for unreliability, the specific environment shows evidence of a general factor made up of spatial and memory first-order factors. Verbal and perceptual speed seem to play no part in defining this general factor. We know of no previous analyses that have convincingly shown a general factor operating at the level of the specific environment. Moreover, after allowing for this general factor, there are no independent first-order specific environment factors. Perhaps somewhat surprisingly, we found no significant variation due to shared environment, although this may be a question of power. At the level of the genetic component, both first- and second-order factors are clearly

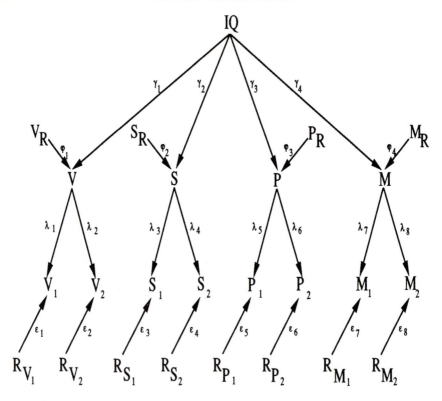

Figure 3.5 Hierarchical path model of specific abilities in the Colorado Adoption Project (CAP).

heritable, with the single exception of perceptual speed which seems to be identified only with *g*. The chi-squared tests of significance are striking. A very strong general factor is evident with loadings on all four first-order factors, and considerable first-order factor variance independent of *g* is clearly shown for verbal, spatial, and memory. Perhaps not surprisingly in view of the polygenic model, there are genes for specific cognitive abilities independent of those for IQ. Similar analyses at ages 3, 4, and 9 suggest essentially the same pattern.

The interesting question now, given the data we currently have, is how does this general and specific cognitive structure develop from ages 3 to 9? We can begin to investigate this question by combining the hierarchical and longitudinal models we have just discussed. The full model at the level of the phenotype is shown in Figure 3.6 (Cardon 1992). With data on twins and siblings, this diagram can be pulled apart into three identical components, as previously described. Before attempting to describe the outcome of the analysis, a few cautionary remarks should be made. First, there are about 300 free parameters in the model, which is many given the volume of

Table 3.2 Model comparisons for shared environmental effects on specific cognitive abilities at age 7.

Model Description		Log Likelihood	NPAR[a]	vs.	χ^2	df	p
1.	Full model	−1737.57	60	1			
2.	No common factor loadings	−1747.36	56	1	19.56	4	<0.001
3.	No verbal common factor loading	−1737.58	59	1	0.00	1	0.99
4.	No spatial common factor loading	−1741.45	59	1	7.74	1	<0.01
5.	No perceptual speed common factor loading	−1737.58	59	1	0.00	1	0.99
6.	No memory common factor loading	−1741.45	59	1	7.74	1	<0.01
7.	No verbal and perceptual speed common factor loadings	−1737.57	58	1	0.00	2	0.99
8.	No primary factors	−1738.69	56	1	2.22	4	0.60
9.	No verbal and perceptual speed common factor loadings and no primary factors[b]	−1738.69	52	1	2.22	8	0.95

[a]NPAR: number of parameters estimated.

[b]This model also requires omission of the first-order measurement loadings on verbal and perceptual speed abilities.

Table 3.3 Model comparisons for unique environmental effects on specific cognitive abilities at age 7.

Model Description		Log Likelihood	NPAR[a]	vs.	χ^2	df	p
1.	Model 9, Table 3.2	−1738.69	52				
2.	No common factor loadings	−1739.92	48	1	2.46	4	0.60
3.	No primary factors	−1739.57	48	1	1.76	4	0.80
4.	No residual factors	−1740.70	44	1	4.02	8	0.80
5.	No shared environmental effects	−1744.74	32	1	6.05	20	0.99
				2	9.64	16	0.80
				3	10.34	16	0.80
				4	8.08	12	0.70

[a]NPAR: number of parameters estimated.

data that we currently have. What parameter estimates we obtain may not be the most stable we could hope for. This state of affairs, of course, will improve as our longitudinal studies progress. Second, employing the pedigree log-likelihood approach with such incomplete data and so many parameters in the model places a very heavy demand on even the fastest available computers. Merely to establish the broad

Table 3.4 Model comparisons for genetic effects on specific cognitive abilities at age 7.

Model Description	Log Likelihood	NPAR[a]	vs.	χ^2	df	p
1. Model 5, Table 3.3	−1744.74	32				
2. No common factor loadings	−1797.02	28	1	104.56	4	<0.001
3. No verbal common factor loading	−1764.09	31	1	38.70	1	<0.001
4. No spatial common factor loading	−1773.71	31	1	57.94	1	<0.001
5. No perceptual speed common factor loading	−1788.12	31	1	86.76	1	<0.001
6. No memory common factor loading	−1754.86	31	1	20.24	1	<0.001
7. No primary factors	−1807.21	28	1	124.94	4	<0.001
8. No primary verbal factor	−1762.59	31	1	35.70	1	<0.001
9. No primary spatial factor	−1746.85	31	1	4.22	1	<0.05
10. No primary perceptual speed factor	−1745.40	31	1	1.32	1	0.20
11. No primary memory factor	−1780.68	31	1	71.88	1	<0.001
12. No residual factor	−1748.75	24	1	8.02	8	0.40
13. No primary perceptual speed or residual factors	−1748.75	23	1	8.02	9	0.50
			10	6.70	8	0.50
			12	0.00	1	0.99

[a]NPAR: number of parameters estimated.

outlines of an appropriate model by means of chi-squared tests of significance took approximately 1,300 hours of CPU time on an IBM RS/6000, one of the world's fastest workstations at the time of analysis. This limitation inhibits a thorough testing of competing models.

At the phenotypic level, the outcome in terms of factor loadings is shown in Figure 3.7. The most striking feature of this path diagram is the pattern of transmission of the four specific cognitive abilities, which is very strong for verbal, moderate for spatial, somewhat weaker for perceptual speed, and weaker still for memory. This transmission conveys both *g*, or IQ, and specific variance from the previous time point. Thus, for example, verbal ability at year 3 is roughly attributable in equal amounts to IQ (0.67) and to an independent specific factor (0.74). Both contributions strongly transmit to year 4 through the transition parameter 0.84. Thus, the bulk of variation in year 4 verbal is attributable in equal amounts to year 3 IQ and year 3 specific verbal ability. The additional effects of IQ and verbal ability at year 4 are relatively modest. The picture is somewhat different for spatial ability, where year 3 IQ and specific variation both transmit, but a substantial amount of additional variation for spatial ability at year 4 comes from both year 4 IQ and new specific variation. For perceptual speed and memory, new variation, largely of a specific nature, emerges at year 4 with modest contribution from IQ. Thus, the diagram shows that it is the combined effects of IQ and truly specific variation from previous time points that influence later specific cognitive ability, augmented by later IQ and specific variation. The implication is that specific cognitive abilities are a complex combination of general and specific abilities at earlier ages and new specific abilities at the later time point. The diagram enables us to trace relative influences of either a general or specific nature from any of the earlier time points and assess their importance relative to those new influences that emerge.

One feature of the diagram that needs clarification concerns the pattern of loadings from IQ to specific cognitive abilities at each of the age points. These loadings represent, to a large degree, the new contribution of IQ and do not represent the loadings that would be obtained if cross-sectional analyses were carried out. The cross-sectional loadings can be calculated as the cumulative correlation between IQ and any given specific ability and are shown in the lower part of the diagram. These loadings, as expected, are much more typical of those for a group of specific cognitive abilities defining IQ. Longitudinal analysis, however, makes clear that what looks like simple stability of IQ over different age points is really the result of many complex interactions.

When analysis was carried out at the genetic and environmental level, shared environmental influences seemed negligible and the developmental process seemed to be largely driven by genetic variation. In the unique environment, there was no transmission for specific cognitive abilities but a small amount of transmission for IQ. In the genetic component, IQ at year 3 largely drove specific cognitive abilities both at year 3 and at year 4. At year 4, IQ played a minor role in determining specific cognitive abilities at that age. A similar picture emerged between years 7 and 9, with year 7 being the determining factor. Overall, what was fairly apparent was a much more uniform, reasonably high transmission from specific to specific at all ages. It appears that specific cognitive abilities are approximately equally influenced by general and specific factors at years 3 and 7, and through strong transmission determining the picture at years 4 and 9. Overall, specific cognitive abilities appear to play a greater role in the developmental process than does IQ.

In which direction should we be looking in our future research? What seems to be the case is that genetic variation plays a major part in the development of cognitive ability. In addition, specific cognitive abilities seem to have a substantial existence in terms of genetic control independent of IQ. Therefore, one obvious avenue to pursue is the attempt to identify, at a more fundamental level, the nature of the genetic control of specific cognitive abilities. Two major strategies are available for pursuing this aim: (a) the use of saturated genetic markers to identify regions of a chromosome where specific genes may exist, and (b) the testing of specific loci to see what part, if any, they may play in determining variation for a specific trait. These are strategies of linkage and that of the candidate gene, respectively. Both represent a formidable challenge and may never be wholly successful if a very large number of genes are involved. The advantage of the linkage approach is that we can soon hope for a very large number of markers that will cover most of the genome. The disadvantage is that in order to detect genetic variation, the gene must contribute to a sizeable proportion

48

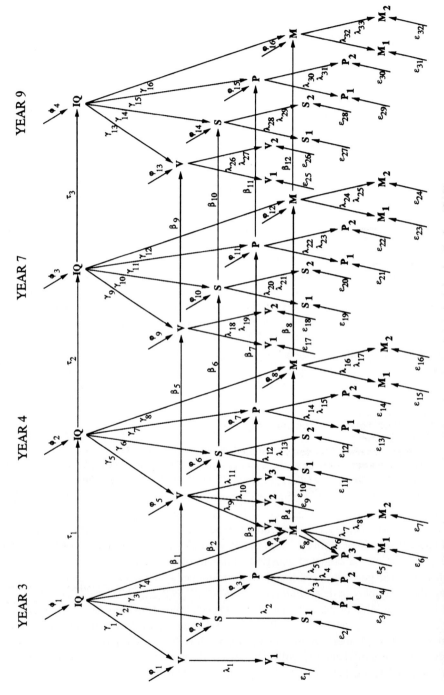

Figure 3.6 Longitudinal hierarchical model of specific abilities in the Colorado Adoption Project (CAP).

49

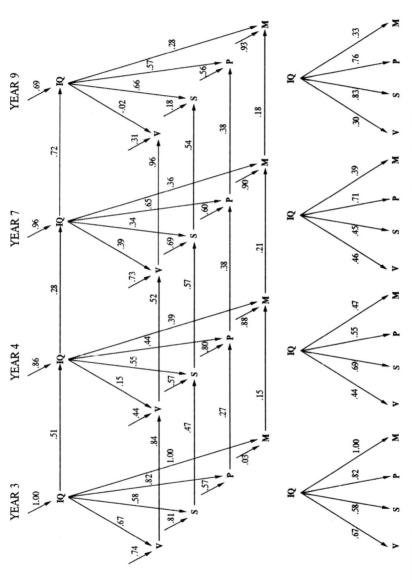

Figure 3.7 Path diagram showing standardized parameter estimates for longitudinal model of specific ability phenotypes.

of observed variance. Genes that contribute less than 10% of the variance are going to require enormous sample sizes, probably beyond existing resources—both financial and motivational. The advantage of the candidate gene approach is that sources of variation around the 1% level could be detected. The disadvantage is that the number of potential candidates may well run into the 100,000s.

The candidate gene approach shows promise for elucidating the genetic control of cognitive abilities, and the pedigree maximum-likelihood methodology we have described in this chapter lends itself admirably to the problem of detecting candidate genes in the presence of continuous variation. Martin et al. (1987) provide an elegant demonstration of the effectiveness of the approach applied to a component of blood chemistry (α-1-antitrypsin). The multitude of genes identified with brain structure may be a place to start searching for candidates. To our knowledge, none have been identified to date that relate to cognitive ability.

The linkage approach, not just to cognition but to behavior in general, has also little to offer by way of findings at the current time. It has, of course, been very successful in identifying the basis for a number of physical diseases. One possible finding of interest is the location of a major locus for dyslexia on chromosome 15 suggested by Smith et al. (1983) using a major marker. More recently, Fulker et al. (1991), employing a more extensive number of RFLP markers and a simplified sib-pair approach based on multiple regression, found evidence for three major genes on chromosome 15 influencing reading ability. The analysis used a continuous measure of reading and a sample selected for low scores. Some of the subjects overlapped with the earlier study of Smith et al. We were able to show that selected samples resulted in a considerable increase in power, suggesting this approach may have some utility in searching for major loci for specific cognitive abilities.

ACKNOWLEDGEMENTS

The research described in this paper was supported in part by grants HD–18426, HD–19802, and training grant HD–07289 from the National Institute of Child Health and Development (NICHD), grant MH–43899 from the National Institute of Mental Health (NIMH), grant RR–0713–25 awarded to the Univ. of Colorado by the Biomedical Research Support Grant Program, Division of Research Resources, National Institutes of Health (NIH), and by a grant from the John D. and Catherine T. MacArthur Foundation.

REFERENCES

Burt, C. 1955. The evidence for the concept of intelligence. *Br. J. Ed. Psychol.* **25**:158–178.
Cardon, L.R. 1992. Multivariate path analysis of specific cognitive abilities in the Colorado Adoption Project. Ph.D. diss., Univ. of Colorado, Boulder.

Cardon, L.R., D.W. Fulker, J.C. DeFries, and R. Plomin. 1992a. Continuity and change in general cognitive ability from 1 to 7 years. *Devel. Psychol.* **28**:64–73.

Cardon, L.R., D.W. Fulker, J.C. DeFries, and R. Plomin. 1992b. Multivariate genetic analysis of specific cognitive abilities in the Colorado Adoption Project at age 7. *Intelligence*, in press.

DeFries, J.C., R.C. Johnson, A.R. Kuse, G.E. McClearn, J. Polovina, S.G.Vandenberg, and J.R. Wilson. 1979. Familial resemblance for specific cognitive abilities. *Behav. Genet.* **9**:23–43.

Eaves, L.J., and J.S. Gale. 1974. A method for analyzing the genetic basis of covariation. *Behav. Genet.* **4**:253–267.

Eaves, L.J., A.C. Heath, and N.G. Martin. 1984. A note on the generalized effect of assortative mating. *Behav. Genet.* **14**:371–376.

Eaves, L.J., J. Long, and A.C. Heath. 1986. A theory of developmental change in quantitative phenotypes applied to cognitive development. *Behav. Genet.* **16**:143–162.

Eysenck , H.J. 1979. The Structure and Measurement of Intelligence. New York: Springer.

Fulker, D.W. 1973. A biometrical genetic approach to intelligence and schizophrenia. *Soc. Bio.* **20**:266–275.

Fulker, D.W. 1978. Multivariate extensions of a biometrical model of twin data . In: Progress in Clinical and Biological Research, vol. 24A, Twin Research: Psychology and Methodology, ed. W.E. Nance, pp. 217–236. New York: Alan R. Liss.

Fulker, D.W. 1979. Nature and nurture: Heredity. In: The Structure and Measurement of Intelligence, ed. H.J. Eysenck, pp. 102–132. Berlin: Springer.

Fulker, D.W., L.A. Baker, and R.D. Bock. 1983. Estimating components of covariance using LISREL. *Data Analyst* **1**:5–8.

Fulker, D.W., L.R. Cardon, J.C. DeFries, W.J. Kimberling, B.F. Pennington, and S.D. Smith. 1991. Multiple regression analysis of sib-pair data on reading to detect quantitative trait loci. *Read. Writ. J.* **3**:299–313.

Gottesman, I.I., and J. Shields. 1966. Contributions of twin studies to perspectives in schizophrenia. In: Progress in Experimental Personality Research, ed. B.A. Maher, pp. 1–84. New York: Academic.

Guilford, J.P. 1967. The Nature of Human Intelligence. New York: McGraw-Hill.

Hewitt, J.K., L.J.Eaves, M.C. Neale, and J.M. Meyer. 1988. Resolving causes of developmental continuity or "tracking." I. Longitudinal twin studies during growth. *Behav. Genet.* **18**:133–151.

Jinks, J.L., and D.W. Fulker. 1970. Comparison of the biometrical genetical, MAVA, and classical approaches to the analysis of human behavior. *Psychol. Bull.* **73**:311–349.

Joreskog, K.G. 1978. Structural analysis of covariance and correlation matrices. *Psychometrika* **36**:409–426.

Loehlin, J.C., and R.C. Nichols. 1976. Heredity, Environment, and Personality. Austin: Univ. of Texas Press .

Loehlin, J.C., and S.G. Vandenberg. 1968. Genetic and environmental components in the covariation of cognitive abilities: An additive model. In: Progress in Human Behavior Genetics, ed. S.G. Vandenberg, pp. 261–285. Baltimore: Johns Hopkins Univ. Press.

Martin, N.G., D.I. Boomsma, and M.C. Neale. 1989. Foreword to special issue on structural modeling with LISREL. *Behav. Genet.* **19**:5–7.

Martin, N.G., P. Clark, A.F. Ofulue, L.J. Eaves, L.A. Corey, and W.E. Nance. 1987. Does the PI polymorphism alone control α-1-antitrypsin expression? *Am. J. Hum. Genet.* **40**:267–277.

Martin, N.G., and L.J. Eaves. 1977. The genetical analysis of covariance structure. *Heredity* **38**:79–95.

Neale , M.C., and L.R. Cardon. 1992. Methodology for Genetic Studies of Twins and Families. Dordrecht: Kluwer.

Phillips, K., and D.W. Fulker. 1989. Quantitative genetic analysis of longitudinal trends in adoption designs with application to IQ in the Colorado Adoption Project. *Behav. Genet.* **19**:621–658.

Plomin, R., J.C. DeFries, and D.W. Fulker.1988. Nature and Nurture during Infancy and Early Childhood. New York: Cambridge Univ. Press.

Rice, T., G. Carey, D.W. Fulker, and J.C. DeFries. 1989. Multivariate path analysis of specific cognitive abilities in the Colorado Adoption Project: Conditional path model of assortative mating. *Behav. Genet.* **19**:195–207.

Smith, S.D., W.J. Kimberling, B.F. Pennington, and H.A. Lubs. 1983. Specific reading disability: Identification of an inherited form through linkage analysis. *Science* **219**:1345–1347.

Spearman, C. 1904. General intelligence objectively determined and measured. *Am. J. Psychol.* **15**:201–293.

Taubman, P. 1976. The determinants of earnings: Genetics, family and other environments—A study of white male twins. *Am. Econ. Rev.* **66**:858–870.

Vandenberg, S.G. 1965. Multivariate analysis of twin differences. In: Methods and Goals in Human Behavior Genetics, ed. S.G. Vandenberg, pp. 29–43. New York: Academic .

Vogler, G.P., and D.W. Fulker. 1983. Human behavior genetics. In: Handbook of Multivariate Experimental Psychology, 2nd ed., ed. J.R. Nesselroade and R.B. Cattell, pp. 475–503. New York: Plenum.

4

How Do Social, Biological, and Genetic Factors Contribute to Individual Differences in Cognitive Abilities?

C.G.N. MASCIE-TAYLOR
Department of Biological Anthropology, University of Cambridge,
Downing Street, Cambridge CB2 3DZ, U.K.

ABSTRACT

This chapter demonstrates that variation in cognitive abilities is associated with a large number of variables, including differences in home environment, obstetric complications, maternal smoking behavior prior to and during pregnancy, occupational group, gender, handedness, genetic polymorphisms, family and school moves, regional variation, brain size, height and acquired myopia, nutritional status and nutritional supplementation, and exposure to lead. Many of these variables are intercorrelated, and it is therefore difficult to disentangle the underlying causative mechanisms.

INTRODUCTION

Individual differences in cognitive abilities have been shown to be related to social and family variables, personal biological attributes as well as to the quality of the home and the physical environment. In this chapter I review the main variables associated with cognitive abilities and will focus primarily on studies conducted in developed countries.

HOME ENVIRONMENT

Home environment refers to the combined effects of the physical and psychosocial aspects of the environment. Thus family size, birth order, extent of crowding, type of

Twins as a Tool of Behavioral Genetics
Edited by T.J. Bouchard, Jr. and P. Propping © 1993 John Wiley & Sons Ltd.

accommodation, availability of household amenities, parental involvement, and financial hardship all form part of the complex of variables contributing to variation in home environment.

For Britain, the effect of many of these variables has been reviewed by Rutter and Madge (1976). They showed that poor housing, together with low income, is associated with a wide range of personal troubles as well as social disadvantage; children living in overcrowded homes tended to have lower educational attainment.

Although poor housing, overcrowding, lack of amenities, and poor financial status are commonly found together, there has been a tendency to examine the impact of each of these variables in isolation without taking into account the contribution of other correlated variables. McManus and Mascie-Taylor (1983) and Mascie-Taylor (1984) attempted to resolve this problem using data from a longitudinal survey of British children studied from birth (the National Child Development [NCD] study). Both publications concentrated on assessing cognitive abilities at age 11, when a reading test, a mathematics test, and tests of verbal and nonverbal IQ were given. The last two yield a total IQ score and nearly 14,000 children completed all four tests.

Initial findings suggested significant differences in mean IQ scores of children living in varying degrees of crowding and with access to different numbers of household amenities, such as sole use of kitchen and bathroom. The difference in IQ score between those living in uncrowded conditions (< 1 person/room) and crowded conditions (> 2 persons/room) was over 12 IQ points. For amenities, children in homes with sole use of amenities scored 9 IQ points higher on average than children who had sole use of no amenity.

After multiple regression analyses (which took into account differences due to birth weight, gestation period, blood group, sex of child, parental age, birth order, family size, type of accommodation, occupation, height, handedness, and the number of family and school moves) significant crowding and amenity effects on IQ score remained, although the difference between children living in uncrowded and crowded conditions fell to just over 2 IQ points. For amenities, the difference was just under 2 IQ points.

Rutter and Madge (1976) suggested that overcrowding may influence school performance through lack of play space, the unavailability of a quiet room to study, and perhaps through disturbance of sleep by other family members. The NCD study also collected information on the number of people sharing a bedroom (56.3% of the sample) and the number of people sharing a bed (17.3% of the sample). Both variables showed a significant association with IQ: the mean difference in IQ score between one child in a bedroom and three or more persons sharing one was over 6 IQ points, while children who shared a bed scored over 7 IQ points on average lower than those who did not. After multiple regression analysis the effect of bed sharing remained highly significant; the difference between the two groups fell to 1.8 IQ points, while the independent effect of number of persons sharing a bedroom was negligible.

The type of housing people live in and poverty are also major contributors to disadvantage. The NCD study collected information on whether the accommodation a child lived in was owner-occupied (54.2% of the sample) or rented (45.8%) and the type of dwelling: whole house (89.8%) or rooms/static mobile homes (10.2%). Exact income was not obtained; however, several measures of financial status were noted, including whether the child received free school meals (10.4%), if the family suffered financial hardship (11.4%), if the father was unemployed (7.7%), and whether the mother was working outside the home (61.4%).

The analyses indicated that after multiple regression analysis, the type of dwelling showed hardly any association with IQ, while children living in owner-occupied accommodation scored on average 4 IQ points higher than children living in rented accommodation. Each financial index showed small significant effects after multiple regression analyses.

In summary, the results obtained from this large cohort study indicate that factors associated with disadvantage show a statistically independent relationship to IQ score. Consequently a child who receives free school meals lives in rented, overcrowded accommodation which lacks basic amenities, whose father is unemployed and where there is financial hardship, will on average have an IQ score 16 points below a child who is not disadvantaged in any of these ways. The study supports other research conducted in the 1980s, which suggested that physical aspects of the environment contribute to cognitive development.

The other two variables commonly included as part of home environment are birth order and family size. The disadvantages of coming from a large British family appear to start at birth and continue throughout childhood. For school-aged children there is considerable evidence that individuals from larger families tend to have a lower verbal IQ and reading attainment than children from smaller families. The association is less marked for mathematics attainment, and there is only a slight relationship for nonverbal IQ.

A number of hypotheses have been put forward to account for these findings. A purely genetic explanation is unlikely. The finding of a reduced association in the professional and managerial occupations and among Catholic families does imply that there is something specific about the sort of people who have large families. In the NCD study, I found a strong family size and total IQ relationship with differences in mean IQ of nearly 7 IQ points between families with up to two children compared with families of four or more (Mascie-Taylor 1984). These results were prior to multiple regression analysis; however, afterwards no significant differences were apparent.

The relationship between birth order and ability have been discussed extensively. In the NCD study, the differences between first-born and fifth-born or later exceeded 6 IQ points after multiple regression analysis. For reading attainment (also standardized to a mean of 100 and standard deviation of 15), the difference was nearly 3 points.

A number of reasons have been put forward to account for the birth order/IQ association. These include the stimulus of competition from younger siblings, the greater responsibility given to the firstborn, and hence their greater need for achievement. Parents might also interact more with their firstborn child and may stimulate them more.

The confluence model (Zajonc 1983) postulates that an individual's intellectual development is a function of the intellectual environment he/she is exposed to in the home. Families with many children spaced close together are presumed to have intellectual environments inferior to those of families with fewer children spaced apart. Although the model is good at explaining the variance of aggregated data (over 90%), it is poor at explaining individual scores (about 1%). The "bright career" of the confluence model appears to have declined. This has led some researchers to suggest that family structure variables (i.e., birth order, family size, and mean spacing between children) are not related to individual intellectual development (Galbraith 1983; Retherford and Sewell 1991).

Wachs (1984, 1992) argued that different aspects of the environment influence different aspects of development (environmental specificity hypothesis) and that environmental effects are mediated by the individual characteristics of the child (organismic specificity hypothesis). These hypotheses imply that environmental effects do not act globally and that it is not possible to differentiate simply between "good" and "bad" environments. Recently, several, but not all, studies have reported results consistent with these hypotheses.

OCCUPATIONAL GROUP

The relationship between occupational group and cognitive abilities is difficult to assess simply because of the known interdependence between occupational group and many of the proximal home environmental variables. In Britain, nearly all studies have shown a marked association between social class (socioeconomic group) and IQ that far exceeds that observed between IQ or attainment and any variety of reproductive abnormality. In the NCD study, social class variation was associated with a 17-month difference in reading attainment at 7 years of age (after holding other factors constant), whereas low birth weight was associated with only a 4-month difference. By age 11, the difference in mean IQs between children in social class I (professional) and class V (unskilled workers) averaged 17 IQ points.

After multiple regression analysis there remained a highly significant IQ/social class effect, with a spread of 10 IQ points between classes I and V. The existence of such large socioeconomic differences may be due to parental style, genetic differences, or schooling effects.

The British findings contrast with some American results, which show smaller differences between blue and white collar workers. Thus distal variables (e.g., socioeconomic status) are poorer at predicting cognitive development than proximal home environment variables in many, but not all, American studies (White 1982).

FAMILY AND SCHOOL MOVES: FAMILY DISRUPTION

A London literacy survey showed that children who had attended at least three schools by the age of eight or nine had lower reading scores than those who had attended only one or two schools. Interestingly, however, children of naval families who change residence and school frequently showed no educational deficit. In the NCD study, multiple regression analysis revealed that family moves have no significant effect on a child's IQ, and the effect of school moves was only significant beyond three moves. There was a deficit of 3.5 IQ points between a child who had moved four or more times and a child who had moved only once.

There has been considerable interest in the effects of parental death, divorce, and single parenthood on children's cognitive abilities. Paternal death had no apparent effect on educational attainment. The effects of parental divorce have recently been reviewed (Amato and Keith 1991), and a meta-analysis of 92 studies of children living in divorced single-parent families compared with children living in intact families indicated a median well-being deficit of 0.14 standard deviations. They ascribed this deficit to parental absence, economic disadvantage, and family conflict.

Illegitimate children read less well on average than legitimate children. Illegitimacy is commonly associated with social disadvantages and family disruption. It is generally held that these factors are mainly responsible for the poorer attainment rather than illegitimacy per se.

REGIONAL VARIATIONS

Studies in the United States have shown that reading difficulties are considerably more common in areas of low socioeconomic status, especially in the inner city areas. Such findings may reflect the low occupational status of the parents living in such areas. This reason is unlikely to account for regional differences in Britain. Scottish children have superior reading attainments (but lower nonverbal IQ scores) than English children despite fewer middle class families in Scotland. It has been speculated that better school teaching of reading skills in Scotland and parental interest in helping their children learn to read may explain the British findings.

OBSTETRIC COMPLICATIONS AND SMOKING IN PREGNANCY

It is commonly suggested that obstetric complications lead to decreased cognitive abilities in children. For example, it has been shown that the number of complications occurring either antenatally, intrapartum, or immediately postpartum correlate with IQ score, with a reduction of just over 1 IQ point between zero complications and 6 complications after holding a number of other variables constant.

A much more marked effect can be shown with birth weight, although many studies did not differentiate between low birth weight at full term and a shortened gestation period. I found a curvilinear relationship between birth weight and IQ: the difference in IQ between children whose birth weight was 2 standard deviations below the mean and those 2 standard deviations above the mean was, on average, 6 IQ points

(Mascie-Taylor 1984). These results were obtained after multiple regression analysis, which took into account 27 variables, including length of gestation.

Smoking during pregnancy has been shown to be associated with lowered birth weight and increased risk of perinatal death. In the NCD survey, children born to mothers who smoked during pregnancy were shorter and read less well at 7 years of age than did children whose mothers did not smoke. McManus and Mascie-Taylor (1983) and Mascie-Taylor (1984) showed that after multiple regression there was a gradient in IQ scores of children: higher scores were found in children whose mothers did not smoke at all, next came children whose mothers stopped smoking when they knew they were pregnant, with the lowest IQ scores from mothers who continued to smoke. A mother's smoking history was obtained at birth; no further information was available as to subsequent smoking history.

NUTRITIONAL STATUS

The most consistent finding is that children who suffer from protein-energy malnutrition, particularly during the first year of life, show significant cognitive deficiencies compared to well-nourished children. For instance, a Jamaican study found that severely mal-nourished children were 8 IQ points lower than control children. Such children are more likely to be found in developing countries where poor nutritional status is compounded by quality of the home environment and other detrimental features.

Relatively few studies have determined the effects of mild to moderate malnutri-tion. Thus the interplay between nutrition, environment, and cognitive development remains unclear. One recent study, carried out on a rural sample of Philippine preschool children, investigated the relationship between home environment, nutri-tional status, and maternal intelligence as determinants of children's intellectual development (Church and Katigbak 1991). No significant relationship between intellectual performance and nutritional status was found, even though up to 80% of the children were classified as being mild to moderately malnourished.

There have also been attempts to determine the effects of nutritional supplementa-tion on IQ. One of the earliest studies gave vitamin and mineral supplements during pregnancy to a group of New York City women of low socioeconomic status. At four years of age, the children of these women had IQs averaging 8 IQ points higher than a control group of children whose mothers were given a placebo during pregnancy.

An association between helminth infection and educational achievement has also been reported. Recently, using a double-blind clinical trial on Jamaican school children, Nokes et al. (1993) found that moderate to high burdens of *Trichuris trichuria* (whipworm) were associated with reduced cognitive function. They showed that after multiple regression analysis, removal of worms led to a significant improvement in auditory short-term memory, scanning, and retrieval of long-term memory. After nine weeks of treatment, no significant differences were noted between the treated and uninfected control groups.

In Britain, the debate over the importance of nutritional supplementation was rekindled after a study conducted in a school in Wales. Benton and Roberts (1988) found a significantly greater enhancement of performance on a nonverbal intelligence test for the group receiving vitamin and nutritional supplementation compared with a placebo group. No differences were found between groups on a verbal test. There were some problems with the interpretation of these results, and replication is clearly required.

Nelson et al. (1990) failed to reproduce the vitamin effect. This study involved a much shorter supplementation period and differed from the Benton and Roberts study in several other respects, which diminishes its usefulness as a replication. A full replication, with some additional features designed to deal with criticisms of the original study, was carried out in Scotland by Crombie et al. (1990). One of the modifications was to use a series of nonverbal and verbal tests in addition to those utilized by Benton and Roberts. Crombie et al. also failed to show the effect reported in the Welsh study, although there were small nonsignificant effects in the predicted direction on the original nonverbal test and on one of the other nonverbal tests used.

In developed countries, the direct effect of subnutrition on intellectual development appears to be slight. Nevertheless, skeletal development and height correlate positively with cognitive ability, and subnutrition is an important cause of reduced growth and development. Thus subnutrition might be an indirect contributor to lowered cognitive ability.

The debate over the importance of nutritional supplementation is not over. Eysenck (1991) recently suggested that the observed increase of 10–12 IQ points of adoptees, which is usually ascribed to better education, more mental stimulation, and other social factors, is much more likely to be the result of improved nutrition received after adoption.

LEAD EXPOSURE

Childhood lead poisoning was first described over a century ago. Recent studies in the United States, United Kingdom, Europe, and New Zealand have examined the relationship between low levels of lead exposure (low level refers to an exposure to lead that is below the clinical signs of lead poisoning) and childhood neurobehavioral effects. A Massachusetts study showed that a high lead group (> 20 ppm) scored significantly lower on IQ, speech and language processing, and attention, than a low lead reference group (< 10 ppm). The mean difference in full-scale IQ between the low and high lead groups was just over 4 IQ points. A pilot study on British school children reported similar findings, with significantly lower IQ, reading, and spelling scores in children with elevated blood lead levels after controlling for age, social class, and sex of the child. Fergusson et al. (1988) conducted a longitudinal study of dentine lead levels in 724 New Zealand children. They found that high tooth lead levels were strongly related to lower reading scores, poorer spelling, lower mathematics scores, and poorer handwriting. The association between lead and IQ was negative but not significant after multiple regression analysis. They concluded that there was most probably a small causative relationship between lead and IQ; however, the evidence was not conclusive.

Longitudinal studies of the effect of lead on infant development have been carried out. Lead crosses the placenta, and umbilical cord blood levels are correlated with maternal concentrations. Several studies have shown a relationship between cord blood lead levels and lower scores on the Bayley scales of infant development. A few studies have looked to see whether the effects of lead exposure are lasting. The largest study (an 11-year follow-up of 132 American subjects; Needleman and Bellingen 1991) found that high dentine lead levels at the beginning of the project were associated with a sevenfold increased risk of failure to graduate from high school and a sixfold increased risk of reading disability. In addition, students with high lead levels had poorer vocabulary and grammatical reasoning scores, longer reaction times, poorer hand-eye coordination, and slower finger tapping. Needleman and Bellinger (1991) concluded that "the effects of lead are enduring and are likely to be predictors of life success." One problem with lead studies of this kind is that lower socioeconomic status and poorer families are much more likely to live in homes with peeling paint and in neighborhoods with more lead from petrol.

BRAIN SIZE

A number of studies (e.g., Susanne 1979) have found a modest relationship between external head size and intelligence. Large variation in skull thickness, however, makes it difficult to extrapolate from external skull dimensions to predict cranial capacity. Recently, Willerman et al. (1991) overcame the limitations of previous work by studying brain morphometric correlates of intelligence using magnetic resonance imaging. Their sample of 40 healthy college students was equally divided by sex and into high versus average IQ groups. Analysis of covariance contrasted high versus average IQ and male versus female, controlling for body size. Their results showed that the high IQ group had a greater brain size, as did the men (although the latter finding was not significant after the removal of an outlier). A stepwise multiple regression revealed that IQ correlated with brain size in males before and after correcting for body size (before $r = 0.51$, $p < 0.05$; after $r = 0.65$, $p < 0.01$) and for women the correlations were 0.33 and 0.35 (both not significant). The IQ/brain size correlation pooled from the sexes was 0.51 ($p < 0.01$). Willerman suggests that since brain size is correlated with cortical surface area, the larger size may lead to more efficient information processing.

HEIGHT AND ACQUIRED MYOPIA

A large number of studies have reported a modest positive correlation of about 0.20 between stature and children's IQ (Jensen and Sinha 1993). Height also correlates with academic grades, occupation, social class status, and social mobility, and these variables also correlate with intelligence. I found a highly significant positive but curvilinear relationship between height and total IQ score after multiple regression analysis (Mascie-Taylor 1984). Short children (those more than 2 standard deviations below the mean height) scored, on average, 6 IQ points lower than tall children (those more than 2 standard deviations above the mean).

Various explanations of the height/IQ association have been put forward, including a between-family genetic hypothesis resulting from assortative mating, prenatal environmental effects, and poor nutrition after birth. Some studies have shown sex differences in the relationship between height and intelligence. In a longitudinal survey, Humphreys et al. (1985) found that individual differences in girls' height at 8 and 9 correlated 0.40 with later intelligence at ages 11 and 12. There was little evidence for similar anticipation in boys.

Children with acquired myopia have been shown to be taller on average than those with normal vision. In the NCD study, children with myopia had higher scores on the 11-year reading comprehension, arithmetic, and general ability tests. The gains in mean score were equivalent to 1.9 years on the reading test and 1.2 years on the arithmetic test and 1.4 years on the general ability test. Since myopic children are more likely to have fathers in nonmanual occupations and to come from small families—factors that are also associated with educational performance—allowance was made in the statistical analyses for social class background, family size, height, and a number of other variables. After adjustment for these factors, the myopic children still showed striking advantages, and the difference in IQ score between normal and myopic children was over 4 IQ points.

SEX DIFFERENCES

Sex differences in cognitive abilities have been recognized for many decades. Maccoby and Jacklin (1975) concluded that there was a gender difference favoring girls in verbal ability and boys in quantitative and spatial abilities. A meta-analysis by Linn and Petersen (1985) found large sex differences only on measures of mental rotation, with small differences on measures of spatial perception. Feingold (1988) compared the norms for the four standardizations of the Differential Aptitude Test (DAT), conducted between 1947 and 1980, and the four standardizations of the Preliminary Scholastic Aptitude Test (PSAT), conducted between 1960 and 1983. He found evidence for a narrowing of the gender difference: by 1980, boys were not different from girls on PSAT-verbal and had halved the difference in clerical speed and accuracy. The sex difference favoring girls was still apparent for English language skills but the gap had narrowed. For the DAT aptitudes, girls had closed the gap for verbal reasoning, abstract reasoning, and numerical ability and had halved the difference on mechanical reasoning and space relations. Although girls had halved the gap for PSAT-math, the gender difference at the highest levels of performance on high school mathematics remained constant from 1960 to 1983.

For at least 30 years it has been suggested that the sex differences in spatial ability might reflect a sex-linked (X-linked) gene. Some early studies reported parent-child correlations consistent with the X-linked hypothesis; however, these results were not replicated in other studies, and the general consensus by the late 1970s was that the X-linked hypothesis was dead. Nonetheless, Thomas (e.g., 1988) has revived the hypothesis and has suggested a model of an X-linked gene with two alleles: the recessive form facilitating performance but not the dominant form. Males and females who possess the facilitating allele might have greater right-hemisphere specialization, which in turn is responsible for better performance on spatial tasks.

HANDEDNESS AND LATERALITY

A large number of studies have examined the relationship between handedness and cognitive abilities. Some found left-handedness to be associated with impaired nonverbal ability. Others found left-handedness to be associated with improved spatial ability, while still others found no effect of handedness. One explanation for these contrasting findings is handedness interaction, due to a particular sex. Harshman et al. (1983) analyzed data on cognitive abilities from three large samples of normal subjects. For subjects with above median reasoning ability, they found that the spatial scores of left-handed males was reduced while those of left-handed females was raised. The opposite pattern was observed for subjects with below median reasoning ability. They went on to suggest that differences in cognitive abilities are partly neurological in origin.

GENETIC POLYMORPHISMS

There have been attempts to study the relationship between segregating genetic markers and IQ as a means of analyzing continuous variables in human populations. A study carried out in a group of Oxfordshire (England) villages found a significant association between IQ score and ABO blood groups, as did a similar study of families living in an urban area of Cambridge, England (Gibson et al. 1973; Mascie-Taylor 1977). In the Oxford survey, individuals of blood group O were about 3 IQ points higher, on average, than those of A phenotype. In the Cambridge study it was possible to differentiate between individuals of AA from AO genotypes, and the difference between O phenotype and AA genotype exceeded 8 IQ points. Associations between haptoglobin phenotypes and spatial ability, and the ability to taste phenylthiocarba-mide (PTC) and total IQ score have also been reported (Mascie-Taylor et al. 1985). The cause of such associations could be due to linkage, pleiotropic effects, or to population admixture.

FAMILIAL CORRELATIONS: NONSHARED ENVIRONMENT

The results of 111 studies of familial resemblance in measured intelligence were reviewed by Bouchard and McGue (1981). The simple broad heritability value obtained from twin studies is now thought to be about 45–50% (Chipeur et al. 1990) compared with much higher values (70–80%) estimated from earlier analyses.

Recently, results of one more IQ study of twins reared apart has been reported (the Minnesota Study by Bouchard et al. 1990). Bouchard and his colleagues studied more than 100 sets of twins or triplets reared apart from the United States and United Kingdom over the past 11 years, although the reported results only concern 56 monozygotic twins reared apart. All twins completed about 50 hours of medical and psychological assessment, and separate examiners administered the IQ tests. The twins had spent their formative years apart. The total contact time of the twins was quite variable, with a mean of 112.5 weeks. For the WAIS test, results suggest that about 70% of the observed variation in IQ can be attributed to genetic factors.

Plomin and Daniels (1987) discussed the evidence for nonshared (or within family, individual or unique) environmental influences on cognitive abilities, personality, and psychopathology. They suggested that shared environmental influences are important in childhood, but their influences wane to negligible levels during adolescence. Support for their hypothesis comes from adoption data. Four recent studies of older adoptive siblings yield IQ correlations of zero on average. A 10-year follow-up of the Texas Adoption Project suggests a shared environment of 0.25 for the children at 8 years of age and an estimate of -0.01, 10 years later.

CONCLUSIONS

The partitioning of variation into genetic and environmental components has been the goal of behavioral geneticists and workers from associated disciplines for a long time. In this chapter I have pointed out that a large number of biological, social, familial, and physical environmental variables associate with IQ. Their relative importance is difficult to quantify since so many of them intercorrelate. The observational study design commonly used in many studies on human subjects does not allow underlying causative mechanisms to be determined.

There remains a pressing need to determine the influence of the shared and nonshared environment on cognitive ability variation. How such influences change with increasing age of the child also requires further study. The importance of nutritional supplementation and lead exposure remains controversial. Large-scale longitudinal studies, which measure and quantify the environment and which use experimental designs, are needed to answer many of the outstanding questions.

REFERENCES

Amato, P.R., and B. Keith. 1991. Parental divorce and the well being of children: A meta-analysis. *Psychol. Bull.* **110**:26–46.

Benton, D., and G. Roberts. 1988. Effect of vitamin and mineral supplementation on intelligence of a sample of children. *Lancet* **i**: 140–143.

Bouchard, T.J., Jr., D.T. Lykken, M. McGue, N.L. Segal, and A. Tellegen. 1990. Sources of human psychological differences: The Minnesota study of twins reared apart. *Science* **250**:223–228.

Bouchard, T.J., Jr., and M. McGue. 1981. Familial studies of intelligence. *Science* **212**:1055–1059.

Chipuer, H.M., M. Rovine, and R. Plomin. 1990. LISREL modelling: genetic and environmental influences on IQ revisited. *Intelligence* **14**:11–29.

Church, A.T., and M.S. Katigbak. 1991. Home environment, nutritional status, and maternal intelligence as determinants of intellectual development in rural Philippine preschool children. *Intelligence* **15**:49–78.

Crombie, I., J. Todman, G. McNeill, C. du V. Florey, I. Menzies, and R.A. Kennedy. 1990. Effect of vitamin and mineral supplementation on verbal and nonverbal reasoning of children. *Lancet* **335**:744–747.

Eysenck, H.J. 1991. Improvement of IQ and behavior as a function of dietary supplementation. *Pers. Indiv. Diff.* **12**:329–365.

Feingold, A. 1988. Cognitive gender differences are disappearing. *Am. Psychol.* **43**:95–103.

Fergusson, D.M., J.E. Fergusson, L.J. Horwood, and N.G. Kinzett. 1988. A longitudinal study of dentine lead levels, intelligence, school performance and behavior. Part II: Dentine lead and cognitive ability. *J. Child Psychol.* **29**:793–809.

Galbraith, R.C. 1983. Individual difference in intelligence: A reappraisal of the confluence model. *Intelligence* **7**:185–194.

Gibson, J.B., G.A. Harrison, V.A. Clarke, and R.W. Hiorns. 1973. I.Q. and ABO blood groups. *Nature* **246**:498.

Harshman, R.A., E. Hampson, and S.A. Berenbaum. 1983. Individual differences in cognitive abilities and brain organization. Part I: Sex and handedness differences in ability. *Can. J. Psychol.* **37**:144–192.

Humphreys, L.G., T.C. Davey, and R.K. Park. 1985. Longitudinal correlation analysis of standing height with intelligence. *Child Devel.* **56**:1465–1478.

Jensen, A.R., and S.N. Sinha. 1992. Physical correlates of human intelligence. In: Biological Approaches to the Study of Human Intelligence, ed. P.A. Vernon. Norwood, NJ: Ablex.

Linn, M.C., and A.C. Petersen. 1985. Emergence and characteristics of sex differences in spatial ability: A meta-analysis. *Child Devel.* **56**:1479–1498.

Maccoby, E.E., and C.N. Jacklin. 1975. The Psychology of Sex Differences. London: Oxford Univ. Press.

Mascie-Taylor, C.G.N. 1977. Mirgration and gene flow in *Drosophila* and man. Ph.D. diss., Univ. Cambridge.

Mascie-Taylor, C.G.N. 1984. Biosocial correlates of IQ. In: The Biology of Human Intelligence, ed. C.J. Turner and H.B. Miles. London: The Eugenics Society.

Mascie-Taylor, C.G.N., J.B. Gibson, R.W. Hiorns, and G.A. Harrison. 1985. Associations between some polymorphic markers and variation in IQ and its components in Otmoor villagers. *Behav. Genet.* **15**:371–383.

McManus, I.C., and C.G.N. Mascie-Taylor. 1983. Biosocial correlates of cognitive abilities. *J. Biosoc. Sci.* **15**:289–306.

Needleman, H.L., and D. Bellinger. 1991. The health effects of low level exposure to lead. *Ann. Rev. Pub. H.* **12**:111–140.

Nelson, M., D.J. Naismith, V. Burley, S. Gatenby, and N. Geddes. 1990. Nutrient intakes, vitamin and mineral supplementation and intelligence in British school children. *Br. J. Nutr.* **64**:13–22.

Nokes, C., S.M. Grantham-McGregor, A.W. Sawyer, E.S. Cooper, and D.A.P. Bundy. 1993. Parasitic helminth infection and cognitive function in school children. *Proc. Roy. Soc. Lond. B*, in press.

Plomin, R., and D. Daniels. 1987. Why are children in the same family so different from one another? *Behav. Brain Sci.* **10**:1–60.

Retherford, R.D., and W.H. Sewell. 1991. Birth order and intelligence: Further tests of the confluence model. *Am. Sociol. Rev.* **56**:141–158.

Rutter, M., and N. Madge. 1976. Cycles of Disadvantage. London: Heinemann.

Susanne, C. 1979. On the relationship between psychometric and anthropometric traits. *Am. J. Phys. Anthro.* **51**:421–424.

Thomas, H. 1988. Simple tests implied by a genetic X-linked model. *Br. J. Math. Stat. Psychol.* **41**:179–191.

Wachs, T.D. 1984. Proximal experience and early cognitive-intellectual development: The social environment. In: Home Environment and Early Cognitive Development: Longitudinal Research, ed. A.W. Gottfried. Orlando, FL: Academic.

Wachs, T.D. 1992. The Nature of Nurture. Newbury Park, CA: Sage.

White, R.K. 1982. The relation between socioeconomic status and academic achievement. *Psychol. Bull.* **91**:461–481.

Willerman, L., R. Schultz, J.N. Rutledge, and E.D. Bigler. 1991. *In vivo* brain size and intelligence. *Intelligence* **15**:223–228.

Zajonc, R.B. 1983. Validating the confluence model. *Psychol. Bull.* **93**:457–480.

5

Current Status and Future Prospects in Twin Studies of the Development of Cognitive Abilities: Infancy to Old Age

D.I. BOOMSMA

Department of Psychonomics, Vrije Universiteit, De Boelelaan 1111,
1081 HV Amsterdam, The Netherlands

ABSTRACT

In this chapter, structural models for the genetic analysis of longitudinal data are introduced and several generalizations discussed that pertain to the estimation of genetic and environmental individual scores and mean trends. Cross-sectional and longitudinal twin and adoption studies of cognitive development are reviewed. The most important changes in the genetic architecture of IQ that can be observed over time are an increase in heritability from infancy to childhood and a decrease in common environmental influences during adolescence. From age 6 onward, heritability for general intelligence is around 50%, and the high longitudinal stability for IQ seems largely mediated by genetic factors.

INTRODUCTION

There may be a priori reasons to expect age-dependent changes in the contributions of genetic and environmental effects to individual IQ differences. Scarr and Weinberg (1978), for example, expect developmental differences in the size of environmental influences. Younger children may resemble their parents more on environmental grounds before they enter schools and other social institutions, and the influence of genetic factors may increase as they grow older. Of course, some want to deny the role of heredity at any age, either for emotional reasons (e.g., John Stuart Mill who wrote "Of all the vulgar modes of escaping from the consideration of the effect of social and moral influences upon the human mind, the most vulgar is that of attributing the

Twins as a Tool of Behavioral Genetics
Edited by T.J. Bouchard, Jr. and P. Propping © 1993 John Wiley & Sons Ltd.

diversities of conduct and character to inherent natural differences" [in Gould 1980])
or for lack of convincing evidence (e.g., Roubertoux and Capron 1990). According to
Nash (1990), the absence of genetic influences on intelligence is highly unlikely
because of the need to account for the evolution of cognitive processing capabilities.
Without some genetic variation, it is impossible to understand how the evolution of
functional brain structures involved in cognitive performance could have occurred.

During development, changes in a quantitative trait may be due to distinct subsets
of genes turning on and off, whereas continuity may be due to stable environmental
causes. Contrary to popular points of view, genetically determined characters are not
always stable, nor are longitudinally stable characters always influenced by heredity
(Molenaar et al. 1991). In this chapter I discuss developmental models that are
concerned with the disentanglement of genetic and nongenetic causes of stability and
change. Two important generalizations of the multivariate extension of these models
concern (a) the estimation of genetic and environmental time-dependent profiles for
individual subjects and (b) the inclusion in the model of genetic and environmental
mean trends.

The literature review of genetic studies of cognitive development addresses the
following questions:

1. Are heritabilities for cognitive abilities age specific?
2. How are genetic and environmental processes involved in stability and change
   in individual differences in intelligence and specific cognitive abilities?
3. Are changes in the environmental contributions to individual differences in
   intelligence informative as to how environment shapes cognitive development?

Twin Analyses in General

The classical twin study of monozygotic (MZ) and dizygotic (DZ) twins does not
permit simultaneous estimation of additive (A) and nonadditive or dominant (D)
genetic effects, common (C) or between-family environmental and individual (E) or
within-family environmental influences (see Hewitt 1989 and references to the work
of Eaves and colleagues therein). If nonadditive genetic effects are present, a model
in which they are not specified will overestimate A and underestimate C. If there is
assortative mating, common environmental effects will be overestimated. However,
a two-parameter AE model will be rejected if $C > 2D$ or if $C < 0.5D$. Equally, an
environmental EC model will fail in the presence of A or D. Hewitt (1989) summarizes
the value of a twin study as follows: " ... it leads to testable hypotheses about
appropriate variance decomposition for a particular measurement or multivariate set
of measures, it permits a test of sex differences in the expression of genetic and
environmental influences, allows us to test causal hypotheses for the relationships
between variables in both cross-sectional and longitudinal designs, and when augmented by other family members provides the nucleus for exploring issues as wide

ranging as the mechanisms of marital assortment and marital interaction through to the estimation of and control for rater bias."

Use of Twins in Intelligence Research

Are twins representative of singletons, although it has been claimed that they have a lower average IQ? According to Storfer (1990), lower average IQ scores of (identical) twins can be explained by their lower birth weights. The heavier twin of a pair is likely to have an IQ score equal to that of a nontwin of comparable birth weight and gestational age. Moreover, the twin disadvantage may disappear with age. Kallmann et al. (1951) present results from white twin pairs aged 60 or more who do as well as a standardization group aged 50–54 on Wechsler subtests.

Intelligence Tests

Genetic studies of cognition have almost exclusively used traditional IQ tests, and specific abilities have usually been defined as subtests. Results from tests developed from other perspectives, e.g., Piaget or information-processing, are scarce. The most frequently employed general intelligence tests are the Wechsler, the Stanford-Binet, Raven's Progressive Matrices, and the Bayley scales. The Wechsler consists of three tests: the WPPSI for children aged 4 to 6.5, the WISC for ages 7 to 16, and the adult WAIS. All three have 11 subscales, divided into performance and verbal tests, that are similar but not identical at different age levels. The Stanford-Binet is most suitable for ages 4 to 17 and gives an intelligence score that is heavily weighted with verbal abilities. For infants 1 to 30 months, the Bayley Mental Development Index and the Bayley motor scales are frequently used. A group intelligence test based on figure-analogy that is suitable for both children (from 5 years on) and adults is the Raven's Progressive Matrices test, developed by J.C. Raven and the geneticist L.S. Penrose.

Scores for IQ tests are usually constructed to be normally distributed with the same mean and standard deviation in each age group. Information on growth in means and variances is thus lost. Alternative ways of scoring standard tests have sometimes been considered. McArdle (1988) analyzed longitudinal WISC scores from children aged 6, 7, 9, and 11. An analysis of percentage correct scores clearly shows an increase in both means and variances with age.

ARE HERITABILITIES FOR COGNITIVE ABILITIES AGE SPECIFIC?

Bouchard and McGue (1981) have summarized IQ correlations obtained in family and adoption studies. The pattern of correlations strongly suggests polygenic inheritance without consistent sex differences. Comparing parent-offspring and twin correlations may offer a first suggestion about age-dependent genetic and environmental effects. As explained in Plomin et al. (1988), significant parent-offspring resemblance implies significant heritabilities, both in childhood and adulthood, and a substantial genetic correlation across time. Average weighted MZ (N = 4672 pairs) and DZ (N = 5546) correlations are 0.86 and 0.60. Doubling the difference between MZ and DZ correlations gives a heritability estimate close to 50%. The parent-offspring correlation for parent and offspring reared in the parental home equals 0.42 (N = 8433) and the correlation of adoptive parent and offspring is 0.19 (N = 1397). Doubling the difference between these correlations gives an estimate of heritability of 46%, so that at first glance little evidence seems to exist for intergenerational differences in heritability. In the few studies in which parents and offspring received the same IQ test at the same age, correlations are not higher than the average parent-offspring correlation reported by Bouchard and McGue. Vroon and Meester (1986) observe a correlation of 0.34 for Raven's Progressive Matrices in 2847 father-son pairs tested by the Dutch army. McCall (1970) reports IQ correlations on parents and children (N = 35) who were both tested from age 3.5 to 11. Parent-offspring correlations are much lower, with a median value of 0.29, than sibling correlations from the same study (median r = 0.55, N = 100). The twin results are suggestive of common environmental influences. In twin studies, however, these cannot be distinguished from the effect of assortative mating, which is quite substantial for IQ. Bouchard and McGue (1981) report a spouse correlation of 0.33 (N = 3817). However, the adoptive parent-offspring correlation of 0.19 also indicates the presence of common environmental influences on IQ.

A developmental meta-analysis of twin similarities in personality and intelligence was published in 1990 by McCartney et al. (1990). The average MZ and DZ correlations for total IQ from 42 studies are 0.81 and 0.57, closely resembling the values reported by Bouchard and McGue (1981). Average correlations for specific cognitive abilities (verbal, quantitative, and performance) show the same pattern as for total IQ, with the exception of perception, where MZ and DZ correlations from 11 studies average 0.55 and 0.45. To study age as a moderator of twin resemblances, mean age from each study was correlated with the intraclass correlations from that same study. For total IQ and specific abilities, except verbal IQ, DZ twins become more dissimilar over time than MZ twins. For total IQ (results from 16 independent studies), the correlations of twin resemblance with age are 0.15 for MZ and – 0.25 for DZ twins. Decrease in twin similarities is largest for perception IQ: – 0.64 for MZ and – 0.79 for DZ twins (results based on five studies). Correlating estimates for heritability, common and unique environment with mean age shows correlations of 0.36, – 0.37, and – 0.15 for total IQ (no separate results are given for specific abilities). Excluding studies where mean age is less than 5 years yields even higher correlations (0.52, – 0.50, and – 0.28). These last analyses, however, must be viewed with caution, because components of variance are less reliable than the intraclass correlations on which they are based, and the analyses are carried out on few data points (exactly how many is unclear). There is no accepted significance test for these correlations, and according to McCartney et al. (1990) they should be interpreted as effect-size estimates. An analysis of age contrasts that allows for significance testing, however, shows only one significant age contrast for MZ twins (for performance IQ) and one for DZ twins (for quantitative IQ), although most contrasts for specific abilities are negative. This means that there is decreasing concordance between twins as they get older. For total IQ, only the DZ age contrast is negative. These results suggest that heritability for IQ increases over time, although the effects do not seem very large.

McCartney et al. (1990) recognize that differences between age groups can result from either age or cohort effects. Heath et al. (1985) addressed the question of cohort effects on education data in Norwegian twins born between 1915 and 1960 and their parents. Sundet et al. (1988) analyzed cohort effects on general ability in male twins from the same country, born between 1931 and 1960. For males educational attainment is subject to secular change, showing an increase in heritability. Results for general ability and for females offer no evidence for cohort effects.

DeFries et al. (1976) review studies of specific cognitive abilities. Evidence from six medium-sized U.S. studies and a Swedish twin study suggests that heritability decreases in spatial, vocabulary, word fluency to arithmetic speed, and reasoning abilities, while in parent-offspring studies verbal IQ seems more heritable than performance IQ. Plomin (1986, 1988) summarizes genetic studies of IQ and specific cognitive abilities from infancy to senescence. In infancy, heritabilities are low (15%), while the influence of common environment is large. This result, however, was mainly based on studies with the Bayley Mental Development Index, which some argue is not a good indicator of general intelligence (Bornstein and Sigman 1986; Storfer 1990). On the other hand, animal studies have also found that genetic variation in behavior develops postnatally (Scott 1990). From age 6 onward—for which age Wilson (1983) finds MZ and DZ correlations of 0.86 and 0.59—heritability estimates for IQ are 50% in adolescence and in adulthood. Heritabilities for specific abilities may increase during adolescence. Fischbein (1979) applied verbal and inductive tests to male twins at ages 12 and 18 and mathematical tests to twins of both sexes at ages 10 and 13. Although the same twins were measured twice, no bivariate analyses are given. Correlations for verbal ability increase slightly from 0.70 to 0.78 for MZ and decrease from 0.60 to 0.50 for DZ. For inductive tests, correlations increase for both MZ (0.59 to 0.78) and DZ twins (0.46 to 0.56). For mathematical abilities differences between MZ and DZ correlations increase (from 0.08 to 0.21 for boys and from – 0.04 to 0.14 for girls). For boys, heritabilities thus seem to increase from around 20% to 40%. It has been suggested that heritabilities increase in old age; however, only three studies of aging twins have been conducted. Kallmann et al. (1951) studied 120 twin pairs aged 60 or more on WAIS subtests, Stanford-Binet Vocabulary, and a tapping test. Their between/within pair F ratios may easily be converted into intraclass correlations. Most measures show heritabilities around 50%, with the possible exception of memory tasks, where heritabilities seem lower. Plomin (1986) reports a similar result for memory tests in childhood and adolescence. Swan et al. (1990) studied 267 aging male twins (mean age 63 years). Two cognitive screening tests were administered: the Iowa Screening Battery for Mental Decline (rMZ = 0.47, rDZ = 0.36) and the Mini-Mental State examination (rMZ = 0.51, rDZ = 0.24), which correlates reasonably highly with verbal IQ. Subjects also received the Digit Symbol subtest from the WAIS (rMZ = 0.72, rDZ = 0.50). Heritability estimates that take into account differences in variances between zygosities were 22%, 38%, and 76%, respectively. Pedersen et al. (1992) report resemblances in Swedish twins with an average age of 65 years. The sample consisted of MZ pairs reared together (N = 67) and apart (N = 46, separated before 11 years of age), and DZ twins together (N = 89) and apart (N = 100). For a principal component score based on 13 subtests, MZa and MZt correlations were 0.78 and 0.80, DZa and DZt correlations were 0.32 and 0.22. These correlations suggested a broad heritability of 80%, which includes nonadditive genetic variance. Correlations for subtests are lower, and average heritabilities for verbal, spatial, perceptual speed, and memory were 58%, 46%, 58%, and 38%. Taken together, these studies do not suggest large changes in heritabilities as people grow older.

Correlates of Intelligence

Galton (1883) was the first one to propose that reaction time (RT) is correlated with general intelligence and may be used as a measure of it. Vernon (1991) reports correlations of – 0.44 between IQ and RT from 2 studies that also measured nerve conduction velocity (NCV), which is a measure of the speed with which electrical impulses are transmitted by the peripheral nervous system. Correlations of RT and NCV were – 0.28 and – 0.18, while IQ and NCV correlated at r = 0.42 and 0.48. No twin studies of NCV are available; however, Reed (1984) found NCV heritable in mice. He suggests that this forms a sufficient basis for asserting that there is genetic determination of variation in human intelligence. Twin studies suggest heritabilities for RT that are of the same magnitude as those for IQ. McGue and Bouchard (1989) observed heritabilities of 54% and 58% for basic and spatial speed factors in a sample of MZ (N = 49) and DZ (N = 25) twins reared apart. For a general speed factor based on eight complex RT tests, Vernon (1989) found a heritability of 49% in 50 MZ and 52 DZ twins. Vernon also found that RT tests requiring more complex mental operations show higher heritabilities. A bivariate analysis of these data with IQ in 50 MZ and 32–SS DZ pairs (15 to 57 years) was reported by Baker et al. (1991). Phenotypic correlations of verbal and performance IQ with general speed were both – 0.59 and were entirely mediated by genetic factors. Genetic correlations were estimated to have absolute values of 0.92 and 1.0. Rose et al. (1981) estimated heritability as 76% for a perceptual speed measure in 74 MZ and 127 DZ college-aged twins and genetic half-siblings (MZ twin offspring). Boomsma and Somsen (1991) measured RT in 12 MZ and 12 DZ adolescent twins. For choice RT higher heritabilities (20%) were seen for shorter than for longer (7%) interstimulus intervals. Heritabilities of almost 50% were seen for RT measured in double task trials. Ho et al. (1988) analyzed WISC-IQ and speed measures in 30 MZ and 30 DZ pairs (8–18 years). Speed measures were rapid automatic naming and symbol-processing factors. Heritabilities for these factors are 0.52 and 0.49. Multivariate results indicate that the correlation between IQ and speed measures (both r's 0.42) is mainly due to genetic correlations of 0.46 and 0.67. Willerman et al. (1979) correlated problem-solving speed in parents with WISC subtests and IQ scores of adopted (average age 8.3) and natural (average age 9.9) children. Mother-child correlations were low; however, father–natural-child correlations were always higher than father–adopted-child correlations: for block design these correlations are 0.24 and 0.17; for object assembly 0.27 and 0.16; and for PIQ 0.31 and 0.18. Other biological correlates of intelligence that are at least partly genetically determined include evoked potentials, glucose uptake in the brain, pupillary dilatation during mental activity, myopia, allergies and other immune disorders, left-handedness, and uric acid levels (Storfer 1990).

CONTINUITY AND CHANGE IN INDIVIDUAL IQ DIFFERENCES

Special issues of the journals *Child Development* (1983, vol. 54, nr. 2) and *Behavior Genetics* (1986, vol. 16, nr. 1) give overviews of the most important current longitudinal studies with twins and adoptees and of developments in the field of longitudinal structural equation modeling.

Developmental Genetic Models

In longitudinal studies, the same set of variables is measured repeatedly over time on the same subjects. The correlation matrix of such data often displays a simplex pattern, that is, a simple order of complexity, where correlations are maximal near the main diagonal (i.e., among adjoining occasions) and decrease as the time between measurements increases. Such a structure was already noticed around the turn of this century by Pearson and co-workers with respect to physical space relations and was called the "Rule of Neighborhood" (Guttman and Guttman 1965). A simplex pattern can be generated by a first-order autoregressive process, where the partial correlation $r_{ik.j} = 0$, whenever $i < j < k$.

In addition to autoregressive models, so-called growth curve models may be used to analyze repeated measures. Growth curve models often include both the mean trend and the covariance structure. From the perspective of the covariance structure (discarding the mean trend), the growth curve model can be viewed as a confirmatory common factor model in which individual scores are determined by a constant base of "true" or common factor scores (Kenny and Campbell 1989). The growth curve model has the following form:

$$y(t) = \eta(0) + \lambda(t) \times \eta(1) + \varepsilon(t) , \qquad (5.1)$$

where $y(t)$ is the observed score at occasion t. The scores on the latent common factors $\eta(0)$ and $\eta(1)$ represent the true scores that are constant over time.

In autoregressive models, by contrast, random change within the true score is introduced at each time. The true score continually changes—either increasing or decreasing—making adjacent time periods more similar than more remote ones. It can be written as (discarding the subject subscript to ease presentation):

$$y(t) = \eta(t) + \varepsilon(t) \text{, and } \eta(t) = \beta(t) \times \eta(t-1) + \zeta(t) \text{,} \qquad (5.2)$$

where $y(t)$ is the observed score measured from its mean at time t, $\beta(t)$ is the autoregressive coefficient, which is the correlation between $\eta(t)$ and $\eta(t-1)$ if the variables are standardized. $\eta(t)$ is the true score at occasion t that is subject to change over time because it depends both on the previous occasion and on a random innovation term $\zeta(t)$. The stability of individual differences over time, which can be expressed as the correlation between the variables $\eta(t)$ and $\eta(t+1)$, equals

$$cor\,[\eta(t), \eta(t+1)] = [\beta(t+1)] \times var\,[\eta(t)]/SD\,[\eta(t)] \times SD\,[\eta(t+1)] \text{.} \quad (5.3)$$

Thus stability depends on $\beta(t+1)$, and on the variances of $\eta(t)$ and $\zeta(t+1)$, because $var\,[\eta(t+1)] = \beta^2(t+1) \times var\,[\eta(t)] + var\,[\zeta(t+1)]$.

It is important to realize that, if variables are not standardized, β may be greater than one and this may lead to misleading interpretations of stability. In this model, parameters are invariant across persons, but may change over time. In a time-series model, where a single person or a few people are measured at many occasions, parameters may differ between persons, but are invariant across time. Only in this last model must the absolute value of β be less than or equal to one.

The autoregressive simplex model can be generalized to the genetic analysis of longitudinal data (Boomsma and Molenaar 1987). Let Equation 5.2 define the latent genetic and environmental time series, and let the basic genetic model for the observations be:

$$P(t) = \lambda(t)\,\eta(t) + \varepsilon(t) = \lambda_g(t)\,G(t) + \lambda_e(t)\,E(t) + \varepsilon(t) \text{,} \qquad (5.4)$$

where $t = 1,...,T$ are the number of time points that need not be equidistant; P is the observed phenotype that can be univariate or multivariate; $G(t)$ and $E(t)$ are series of genetic and environmental factor scores that are uncorrelated; the λs are loadings of observed variables on latent factors; and ε represents influences unique to each variable and individual. Estimates of λ, β, and the variances of ζ and ε can be used to construct individual genetic and environmental profiles across time by means of Kalman filtering (Boomsma et al. 1991). Such individual profiles enable the attribution of individual phenotypic change to changes in the underlying genetic or environmental processes. Simulations have shown that these individual estimates can be reliably obtained. Estimation of $G(t)$ and $E(t)$ permits identification of sources of underlying deviant development in individual subjects.

The role of genetic and environmental influences on average growth does not usually feature in behavior genetic studies. McArdle (1986) was one of the first to model phenotypic means in a longitudinal analysis of twin data. His model can be viewed as a restricted common factor model in which variation in level (*L*) and shape (*S*) factors is decomposed into second order, but zero mean, latent genetic and environmental factors. Means are modeled as:

$$E[P(t)] = E[L(t)] + w(t) E[S(t)] , \qquad (5.5)$$

where $w(t)$ is the factor loading of $P(t)$ on $S(t)$ at occasion t. Application of this model to data from the Louisville twin study (Bayley scale at 6, 12, 18, and 24 months) shows a strong, largely linear change over age, strong common environmental and small genetic effects.

The model proposed by McArdle does not result in the decomposition of longitudinal means into a genetic and environmental part. Dolan et al. (1991) suggested an alternative that involves estimating the contribution of genetic and environmental factors to changes in means over time. The mean structure of a univariate time series is decomposed by the following model:

$$E[P(t)] = v + E[G(t)] + E[E(t)] , \qquad (5.6)$$

where v is a constant intercept term. The latent means are in part attributable to the preceding occasion and in part independent thereof:

$$E[G(t)] = \beta_g(t) E[G(t-1)] + \Delta_g \, var[\zeta_g(t)] \text{ and}$$
$$E[E(t)] = \beta_e(t) E[E(t-1)] + \Delta_e \, var[\zeta_e(t)] . \qquad (5.7)$$

The same factors (*G* and *E*) contribute to both means and individual differences as is expressed in the dual function of the autoregressive coefficients (β) and the residual variance components ($var[\zeta]$). The Δs are time-invariant coefficients of proportionality that relate the mean to the standard deviation of the innovations. The model can be tested by fitting it to the covariance structure with and without including the means. If the goodness of fit does not decline and the parameter estimates are stable, this is taken as support for the validity of the model.

Applications of Developmental Models to Longitudinal Data

In general, analyses of longitudinal data do not allow observed time series to be decomposed into more than one underlying series. Twin data are unique in that they do allow such a decomposition. An interesting feature of such data is also the possibility of simultaneously fitting factor and simplex structures to the data, e.g., a factor structure for common environmental influences and a simplex for the genetic process.

Eaves et al. (1986) present a general developmental model in which both factor and simplex processes are incorporated together with the possibility of phenotypic transmission. More general versions of this model have been presented subsequently (e.g., in Hahn et al. 1990). Application of the original model to cognitive data from the Louisville twin study (3 months through 15 years) shows initially small but persisting and accumulating effects of a single set of genes and an appreciable influence of common environmental effects that are also persistent as well as age-specific input. Unique environmental influences are occasion-specific.

Loehlin et al. (1989) tried to apply the Eaves et al. model to IQ data from two occasions approximately 10 years apart. On the first occasion, adoptive children were between 3 and 14 years old. Correlations for repeated testing were 0.66 for 258 adoptive and 0.70 for 93 biological children. Parents were measured once. A model with only phenotypic transmission gave a good fit and reasonable parameter estimates (Eaves et al. 1986; note that such a model is equivalent to specifying genetic and environmental transmission parameters to be the same). On the second occasion, no evidence for common environment was found. Heritability at time 1 was 26%, at time 2, 37%; however, Loehlin et al. (1989) do not want to put too much emphasis on numerical estimates. In fitting the model the genetic correlation of parents and offspring equals 0.5 at both times 1 and 2. This seems correct only if the genetic correlation between both occasions equals unity.

Wilson (1983) analyzes the pattern of spurts and lags in mental development by analysis of variance of repeated measures in twins. Two correlations are obtained from this analysis: one for the sum of the repeated measures (which is, however, seriously affected by autocorrelation) and one for the pattern of changes over time (i.e., the interaction of pairs × occasions). Results from the Louisville twin study show that heritability for developmental profiles increases with age. The MZ–DZ difference at 3 to 12 months is only 0.07, but increases to 0.32 at years 8 to 15. This result does not reveal, however, how genes operate throughout development, as did the model-fitting approach of Eaves et al. (1986).

Plomin et al. (1988) combine IQ data from the Louisville study of twins measured at ages 1, 2, 3, and 4 and data on scholastic abilities in young adult twins from a study by Loehlin and Nichols with IQ data from adoptive and control children aged 1, 2, 3, and 4 and their parents in a longitudinal analysis. Estimates for twin-shared environment are high, yet transmission from parental phenotype to a child's environment is not significant. The heritability estimate in adults is 50%; in children it increases from 10%, 17%, and 18% to 26% at 1, 2, 3, and 4 years. Estimates for the genetic stability parameter from childhood to adulthood also increase from around 0.60 to 1 from ages 1 to 4. These results lend support to the developmental amplification model proposed by DeFries (in Plomin and DeFries 1985). In this model the effects of genes that are relevant to mental development during infancy and childhood are amplified during adulthood. Cardon et al. (1992) analyze Bayley, Stanford-Binet and Wechsler data from adopted and nonadopted siblings measured from 1 to 7 years and twins measured from 1 to 3 years. They find higher heritabilities than Plomin et al. (1988) and also evidence for common environmental influences in siblings as well as twins. Heritabilities are estimated at 55%, 68%, 59%, 53%, and 52% for ages 1, 2, 3, 4, and 7 years; the influence of common environment is 10% at each age. The genetic part of the model shows increasing transmission parameters and substantial genetic innovations at all ages, except age 4. Common environment functions as a single, constant background factor.

Carey (1988) warns against simple interpretations of genetic correlations in terms of sets of genes common or specific to variables or times. Two occasions may have all their genes in common and show low genetic correlations, while systems with only a few genes in common can have high genetic correlations. Genetic correlations depend on the rank order of genic effects and the type of polygenic system. Carey suggests that we must distinguish between biological pleiotropism, in which the same genes physically underlie different traits and statistical pleiotropism, and in which allelic effects on one trait predict allelic effects on other characters.

New Infant Measures of General Intelligence

Bornstein and Sigman (1986) challenge the belief that there is little (phenotypic) association between cognitive performance in infancy and adulthood. Part of this belief stems from the fact that one of the most frequently used measures of infant IQ, the Bayley Mental Development Index, is a poor predictor of later IQ scores. The Bayley motor scales and the Gesell infant development scales also show no association with later IQ (Storfer 1990). Promising new measures of infant cognitive function are decrement of attention or habituation and recovery of attention or novelty preference. Bornstein and Sigman report 15% and 22% (r^2) common variance for cognition in childhood with habituation and novelty preference, respectively, in a large number of studies. Individual differences in habituation and novelty preference date from the earliest months of life; however, few studies have analyzed these differences. Bornstein and Sigman refer to an unpublished study in which maternal IQ correlated with attention at term in preterm infants. DiLalla et al. (1990) obtained measures of novelty preference in twins 7, 8, and 9 months. Midtwin scores were regressed on midparent WAIS-IQ. Significant regressions were observed for novelty preference at 9 but not at 7 months (β of 0.22 and 0.33 for immediate and retest at 9 months; these regressions are analogous to heritability estimates). A Bayley composite measure showed regressions of 0.13 and 0.06 at 7 and 9 months.

CHANGES IN ENVIRONMENTAL CONTRIBUTIONS AND SHAPING OF COGNITIVE DEVELOPMENT

Gottfried and Gottfried (1986) review ten longitudinal studies of home environment and cognition. They consider it an empirical fact that family environment correlates with young children's cognitive development. An advantaged home environment is associated with higher SES, with being first-born, but not with sex of the child. Later, as compared to earlier, home environment measures are more highly correlated with cognitive development. After SES has been controlled for, correlations between home factors and cognitive development persist. However, SES also correlates with cognitive development independent of home factors. Storfer (1990) discusses environmental factors that may shape IQ in the upward direction. These include having an older father, being a single child, and child-rearing practices of Jewish and Japanese families. Plomin and DeFries (1985) compare relationships between environmental measures and Bayley scores in adoptive and control families. If heredity affects this relationship, correlations will be larger in control than in adoptive families. At 24 months, relationships are stronger than at 12 months and are mediated environmentally to a significant extent, although there also is genetic mediation. Environmental measures at 12 and 24 months predict Bayley scores at 24 months almost equally well. Plomin and DeFries find that heredity is not involved in this longitudinal relationship.

Gottfried and Gottfried (1986) also report high correlations between early home environment and academic achievement. This seems difficult to reconcile with the absence of common environmental influences in adoption studies of adolescents and adults. At ages 4 to 7, Scarr and Weinberg (1977) find correlations of 0.39 for adopted siblings ($N = 53$), 0.30 for adopted-natural pairs ($N = 134$) and 0.42 for biological siblings ($N = 107$). No resemblance in IQ was found, however, in adoptees aged 16–22 (Scarr and Weinberg 1978). A correlation of – 0.03 was observed in 84 adopted sibling pairs, while the correlation for biological siblings was 0.35 ($N = 168$). Teasdale and Owen (1984) report data on adult (18–26 years) adoptees from a Danish adoption register. They find genetic but no common environmental influences on intelligence, whereas for educational attainment both factors are of importance. Loehlin et al. (1989) also conclude that in early adulthood there is no influence of common environment on IQ.

CONCLUSIONS AND FUTURE PROSPECTS

Heritability for cognition differs as a function of age. It increases from around 15% in infancy to around 50% at age 6 but does not seem to change very much during adolescence and adulthood. Not much can be said about genetic influences on IQ in old age because few studies are available. A genetic analysis of individual IQ differences in elderly subjects raises interesting problems because of the possibility of dealing with selected samples in which the selection process is directly associated with the dependent variable. In contrast to the increasing influence of heredity on IQ, the large influence of the shared family environment that is seen early in life rapidly decreases in adolescence. Wilson (1983), in his longitudinal study of twins, finds that common environment accounts for 70% of the variance at age 3 and for 20% at age 15.

The developmental course of genetic and nongenetic influences on cognition has been studied in infancy and childhood and results indicate that genetic factors are stable across time and that their influences are possibly amplified as children grow older. Beyond childhood we do not know what causes stability in intelligence, although parent-offspring data are suggestive of high genetic correlations across age. Most tests of specific cognitive abilities indicate significant genetic influence, with some evidence that heritability for memory is lower than for other abilities. A few studies suggest that intercorrelations between tests arise because of genetic covariance (e.g., Labuda et al. 1987). Martin et al. (1984), however, show that phenotypic correlations between subtests of the National Merit Scholarship Qualifying Test also arise because of a single underlying between-families environmental factor, and they suggest that a single dimension of mate selection or cultural inheritance accounts for a significant part of the phenotypic covariance.

It is almost completely unknown to what extent the relations between IQ and its biological correlates are genetically or environmentally mediated, and there is a clear need for multivariate analyses of twin data in this area (Vernon 1991).

Application of the techniques of molecular genetics are now being considered to look for multiple loci that affect quantitative traits such as intelligence (Plomin and Neiderhiser 1991). Twin data may be of use in this respect, as they offer the possibility to estimate individual genetic and environmental scores (Boomsma et al. 1990). Genetic scores can be used to investigate their relationship with RFLPs and other genetic markers. It is conceivable that the power of these types of analyses will greatly increase when genetic instead of phenotypic scores can be used.

In addition, knowledge about the reasons why certain subjects exhibit high phenotypic scores may be of practical interest. Risk assessment may be improved by the knowledge that a high phenotypic score is caused by a high genetic or a high environmental deviation.

ACKNOWLEDGEMENT

I would like to thank P.A. Vernon, C.V. Dolan, P.C.M. Molenaar, and N.G. Martin for their helpful comments.

REFERENCES

Baker, L.A., P.A. Vernon, and H.Z. Ho. 1991. The genetic correlation between intelligence and speed of information processing. *Behav. Genet.* **21**:351–368.

Boomsma, D.I., and P.C.M. Molenaar. 1987. The genetic analysis of repeated measures. I. Simplex models. *Behav. Genet.* **17**:111–123.

Boomsma, D.I., P.C.M. Molenaar, and C.V. Dolan. 1991. Individual latent growth curves: Estimation of genetic and environmental profiles in longitudinal twin studies. *Behav. Genet.* **21**:241–253.

Boomsma, D.I., P.C.M. Molenaar, and J.F. Orlebeke. 1990. Estimation of individual genetic and environmental factor scores. *Genet. Epidem.* **7**:83–91.

Boomsma, D.I., and R.J.M. Somsen. 1991. Reaction times measured in a choice reaction time and a double task condition: A small twin study. *Pers. Indiv. Diff.* **12**:519–522.

Bornstein, M.H., and M.D. Sigman. 1986. Continuity in mental development from infancy. *Child Devel.* **57**:251–274.

Bouchard, T.J. Jr., and M. McGue. 1981. Familial studies of intelligence: A review. *Science* **212**:1055–1059.

Cardon L., D.W. Fulker, J.C. DeFries, and R. Plomin. 1992. Continuity and change in general cognitive ability from 1 to 7 years of age. *Devel. Psychol.* **28**:64–73.

Carey, G. 1988. Inference about genetic correlations. *Behav. Genet.* **18**:329–338.

DeFries, J.C., S.G. Vandenberg, and G.E. McClearn. 1976. Genetics of specific cognitive abilities. *Ann. Rev. Genet.* **10**:179–207.

DiLalla, L.F., L.A. Thompson, R. Plomin, K. Phillips, J.F. Fagan, M.M. Haith, L.H. Cyphers, and D.W. Fulker. 1990. Infant predictors of preschool and adult IQ: A study of infant twins and their parents. *Devel. Psychol.* **26**:759–769.

Dolan, C.V., P.C.M. Molenaar, and D.I. Boomsma. 1991. Simultaneous genetic analysis of longitudinal means and covariance structure in the simplex model using twin data. *Behav. Genet.* **21**:49–65.

Eaves, L.J., J. Long, and A.C. Heath. 1986. A theory of developmental change in quantitative phenotypes applied to cognitive development. *Behav. Genet.* **16**:143–162.

Fischbein, S. 1979. Intra-pair similarity in IQ of monozygotic and dizygotic male twins at 12 and 18 years of age. *Ann. Hum. Bio.* **6**:495–504.

Galton, F. 1883. Inquiries into Human Faculty and Its Development. London: MacMillan.

Gottfried, A.W., and A.E. Gottfried. 1986. Home environment and children's development from infancy through school entry years: Results from contemporary longitudinal investigations in North America. *Child. Envir.* **3**:3–9.

Gould, S.J. 1980. Ever since Darwin: Reflections in Natural History, 2nd ed. Harmondsworth: Penguin.

Guttman, R., and L. Guttman. 1965. A new approach to the analysis of growth patterns: The simplex structure of intercorrelations of measurements. *Growth* **29**:219–232.

Hahn, M.E., J.K. Hewitt, N.D. Henderson, and R. Benno, eds. 1990. Developmental Behavior Genetics: Neural, Biometrical and Evolutionary Approaches. Oxford: Oxford Univ. Press.

Heath, A.C., K. Berg, L.J. Eaves, M.H. Solaas, L.A. Corey, J. Sundet, P. Magnus, and W.E. Nance. 1985. Education policy and the heritability of educational attainment. *Nature* **314**:734–736.

Hewitt, J. 1989. Of biases and more in the study of twins reared together. *Behav. Genet.* **19**:605–608.

Ho, H.Z., L.A. Baker, and S.N. Decker. 1988. Covariation between intelligence and speed of cognitive processing: Genetic and environmental influences. *Behav. Genet.* **18**:247–261.

Kallmann, F.J., L. Feingold, and E. Bondy. 1951. Comparative adaptional, social and psychometric data on the life histories of senescent twin pairs. *Am. J. Hum. Genet.* **3**:65–73.

Kenny, D.A., and D.T. Campbell. 1989. On the measurement of stability in over-time data. *J. Pers.* **57**:445–481.

Labuda, M., J.C. DeFries, and D.W. Fulker. 1987. Genetic and environmental covariance structures among WISC-R subtests: A twin study. *Intelligence* **11**:233–244.

Loehlin, J.C., J.M. Horn, and L. Willerman. 1989. Modeling IQ change: Evidence from the Texas adoption project. *Child Devel.* **60**:993–1004.

Martin, N.G., R. Jardine, and L.J. Eaves. 1984. Is there only one set of genes for different abilities? A reanalysis of the National Merit Scholarship Qualifying Test (NMSQT) data. *Behav. Genet.* **14**:355–370.

McArdle, J.J. 1986. Latent variable growth within behavior genetic models. *Behav. Genet.* **16**:163–201.

McArdle, J.J. 1988. Dynamic but structural equation modeling of repeated measures data. In: Handbook of Multivariate Experimental Psychology, ed. J.R. Nesselroade and R.B. Cattell, pp. 561–614. New York: Plenum.

McCall, R.B. 1970. Intelligence quotient pattern over age: Comparison among siblings and parent-child pairs. *Science* **170**: 644–648.

McCartney, K., M.J. Harris, and F. Bernieri. 1990. Growing up and growing apart: A developmental meta-analysis of twin studies. *Psychol. Bull.* **107**:226–237.

McGue, M., and T.J. Bouchard. 1989. Genetic and environmental determinants of information processing and special mental abilities: A twin analysis. In: Advances in the Psychology of Human Intelligence, vol. 5., ed. R. Sternberg, pp. 7–45. Hillsdale, NJ: Lawrence Erlbaum.

Molenaar, P.C.M., D.I. Boomsma, and C.V. Dolan. 1991. Genetic and environmental factors in a developmental perspective. In: Problems and Methods in Longitudinal Research: Stability and Change, ed. D. Magnussen, L.R. Bergman, G. Rudinger, and B. Thorestad, pp. 250–273. Cambridge: Cambridge Univ. Press.

Nash, R. 1990. Intelligence and Realism: A Materialist Critique of IQ. London: Macmillan.

Pedersen, N.L., R. Plomin, J.R. Nesselroade, and G.E. McClearn. 1992. A quantitative genetic analysis of cognitive abilities during the second half of the life span. *Psychol. Sci.* **3**:346–353.

Plomin, R. 1986. Development, Genetics and Psychology. Hillsdale, NJ: Lawrence Erlbaum.

Plomin, R. 1988. The nature and nurture of cognitive abilities. In: Advances in the Psychology of Human Intelligence, vol. 4., ed. R.J. Sternberg, pp. 1–33. Hillsdale, NJ: Lawrence Erlbaum.

Plomin, R., and J.C. DeFries. 1985. Origins of Individual Differences in Infancy: The Colorado Adoption Project. New York: Academic.

Plomin, R., J.C. DeFries, and D. Fulker. 1988. Nature and Nurture during Infancy and Early Childhood. Cambridge: Cambridge Univ. Press.

Plomin, R., and J. Neiderhiser. 1991. Quantitative genetics, molecular genetics, and intelligence. *Intelligence* **15**:369–387.

Reed, T.E. 1984. Mechanism for heritability of intelligence. *Nature* **311**:417.

Rose, R.J., J.Z. Miller, and D.W. Fulker. 1981. Twin-family studies of perceptual speed ability: II. Parameter estimation. *Behav. Genet.* **11**:565–575.

Roubertoux, P.L., and C. Capron. 1990. Are intelligence differences hereditarily transmitted? *Cahiers Psychol. Cogn.* **10**:555–594.

Scarr, S., and R.A. Weinberg. 1977. Intellectual similarities within families of both adopted and biological children. *Intelligence* **1**:170–191.

Scarr S., and R.A. Weinberg. 1978. The influence of "family background" on intellectual attainment. *Am. Sociol. Rev.* **43**:674–692.

Scott, J.P. 1990. Foreword. In: Developmental Behavior Genetics: Neural, Biometrical and Evolutionary Approaches, ed. M.E. Hahn, J.K. Hewitt, N.D. Henderson, and R. Benno. Oxford: Oxford Univ. Press.

Storfer, M.D. 1990. Intelligence and Giftedness: The Contributions of Heredity and Early Environment. San Francisco: Jossey-Bass Publ.

Sundet, J.M., K. Tambs, P. Magnus, and K. Berg. 1988. On the question of secular trends in the heritability of intelligence test scores: A study of Norwegian twins. *Intelligence* **12**:47–59.

Swan, G.E., D. Carmelli, T. Reed, G.A. Harshfield, R.R. Fabsitz, and P.J. Eslinger. 1990. Heritability of cognitive performance in aging twins. *Arch. Neur.* **47**:259–262.

Teasdale, T.W., and D.R. Owen. 1984. Heredity and family environment in intelligence and educational level: A sibling study. *Nature* **309**:620–622.

Vernon, P.A. 1989. The heritability of measures of speed of information-processing. *Pers. Indiv. Diff.* **10**:573–576.

Vernon, P.A. 1991. Studying intelligence the hard way. *Intelligence* **15**:389–395.

Vroon, P.A., and A.C. Meester. 1986. Distributions of intelligence and educational level in fathers and sons. *Br. J. Psychol.* **77**:137–142 .

Willerman, L., J.C. Loehlin, and J.M. Horn. 1979. Parental problem solving speed a correlate of intelligence in parents and their adopted and natural children. *J. Ed. Psychol.* **71**:627–634.

Wilson, R.S. 1983. The Louisville twin study: Developmental synchronies in behavior. *Child Devel.* **54**:298–316.

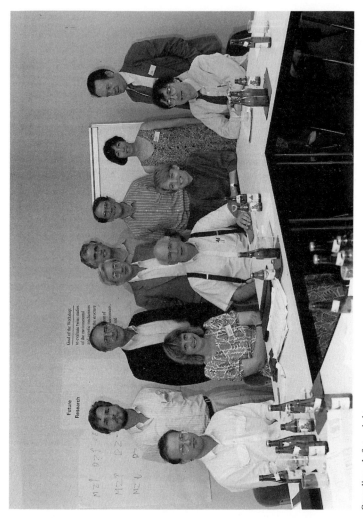

Standing, left to right:
J. Asendorpf, M. Kinsbourne, D. Fulker, N.G. Martin, C.G.N. Mascie-Taylor, L.A. Baker, C. R. Brand

Seated, left to right:
U. Lindenberger, D. Bishop, T.J. Bouchard, Jr., D.I. Boomsma, H. Gardner

6

Group Report: Intelligence and Its Inheritance—A Diversity of Views

L.A. BAKER, Rapporteur

J. ASENDORPF, D. BISHOP, D.I. BOOMSMA, T.J. BOUCHARD, Jr.,
C.R. BRAND, D. FULKER, H. GARDNER, M. KINSBOURNE,
U. LINDENBERGER, N.G. MARTIN,
C.G.N. MASCIE-TAYLOR

INTRODUCTION

What does the study of twins tell us about the structure and origins of intelligent behavior in humans? Agreement on an explicit definition of "intelligence" would seem a reasonable starting point. As history would predict, however, little consensus was reached by a group of scientists whose opinions were surely as diverse as their individual hereditary constitutions and environmental histories. A clash of paradigms was evident in these attempts to define what we set out to explain, and there was little fundamental resolution after days of fervent deliberations.

We agreed that behavioral genetic research, especially the biometrical analysis of twin data, can be extremely useful in helping to understand complex psychological variables, and even in defining them. This research has helped validate existing measures of intellectual ability and has provided support for both general and specific dimensions of these abilities. Moreover, methods of structural equation modeling in twin designs allow evaluation of measurement properties. Researchers have been able to propose and test statistical models that uncover processes underlying cognitive development. As illuminated in the background papers by Boomsma and by Fulker and Cardon (both this volume), recent statistical advancements in the analysis of longitudinal twin and adoption data have proved particularly informative about the nature and development of intelligence.

Twins as a Tool of Behavioral Genetics
Edited by T.J. Bouchard, Jr. and P. Propping © 1993 John Wiley & Sons Ltd.

BRIEF PRELUDE

The usefulness of biometrical analysis of covariance structures using twin data is highlighted first in this report. How do such methods address the validation of existing measures or the development of alternative measures of intelligence? The majority of twin and adoption studies of cognition have relied on traditional measures of intelligence (IQ tests and psychometric batteries of abilities). Recent studies, however, are beginning to consider more process-oriented models of cognition, such as those that refer to the basis of information-processing abilities.

The positions of various proponents of how to characterize abilities are then described. These range from the longstanding view that a general ability factor, as defined through psychometric tests, is still of prime importance in explaining individual differences in various aspects of success in humans, to the more radical perspective that current measures are of limited validity and that intelligence measures of greatest social significance are yet to be developed.

An evolutionary framework for considering both the social and biological significance of intellectual abilities is then discussed from two points of view: biometrical genetics and evolutionary psychology. The two approaches differ primarily in their level of analysis (phenotypic vs. latent genetic and environmental), although efforts to combine these approaches may prove useful in understanding the evolution of intellectual abilities.

Other issues reviewed include (a) the conceptualization and empirical study of the environment, specifying the nature of genetic effects by incorporating molecular genetics into biometrical twin analyses, (b) the secular rise in average IQ test performance and its relevance to behavioral genetic research, and (c) the role of speed of processing in cognition.

BIOMETRICAL ANALYSES OF COVARIANCE STRUCTURES IN COGNITION

Heritability of Intelligence

Early research on the origins of individual differences in human intelligence focused on single (univariate) measures of ability, and the primary question of interest was on the relative importance of genetic and environmental sources of variance (i.e., heritability and environmentality of IQ test variation).

Even in the study of specific cognitive abilities, reflected by subscales of IQ tests or other psychometric tasks, questions about heritability predominated. It was thought that through such univariate analyses of various abilities, it would be determined whether general and/or specific abilities existed. Some cognitive variables would show little genetic variance, some almost exclusively genetic influence, some gene-environment (GE) correlations, and so forth. If each ability had its own "psychogenetic story," this might reveal the structure of human abilities (e.g., see Horn and Cattell 1966).

The issue of whether or not cognitive abilities are heritable and the magnitude of the effects of environment and genes are today of much less interest than they were in the past. Most participants agreed that the majority of existing (reliable) measures of cognitive performance show some genetic variance, with heritability for a general ability factor being around 50–60%, although some estimates, based on adult samples, may be slightly higher (e.g., Bouchard et al. 1990; Pedersen et al. 1992; Tambs et al. 1984). Yet most agree that there are more interesting questions to be addressed than the magnitude of heritability of any particular aspect of intelligence measured at a given age.

Genetic and Environmental Architecture of Intelligence: Support for Both *g* and Specific Abilities

Fulker and Cardon, and Boomsma (both this volume) provide a sampling of the more interesting questions about cognition being addressed today. Beyond a simple partitioning of variance into genetic and environmental sources, more sophisticated analyses allow the study of *factorial structures* of genetic and environmental influences across a group of ability tests. Moreover, continuities and changes in genetic and environmental influences over time may also be determined from longitudinal and cross-sectional age comparisons of twin data.

These questions stem from multivariate biometrical analyses (see Martin and Eaves 1977; Fulker and Eysenck 1979; Heath et al. 1989), which focus on the sources of *covariation* among abilities rather than on the sources of variation of any one ability. This approach allows the examination of relationships among traits (e.g., between IQ and educational attainment, or among several specific cognitive ability tests) and the partitioning of covariance into genetic and environmental sources. The new question becomes one of what drives the correlation between two variables. For example, one may determine the importance of correlated genetic effects across different abilities, which may stem either from pleiotropy or linkage disequilibrium.

Twin studies have provided support for both general and specific abilities by examining factor structures of genetic effects and environmental effects for several tests of ability, as reviewed by Fulker and Cardon (this volume; see also Vandenberg 1965; Eaves and Gale 1974; Martin and Eaves 1977; Loehlin and Nichols 1976). Genetic influences for many different cognitive abilities are highly correlated (suggesting a general factor of genetic influences), which lends support to the notion of a general ability factor. The additional finding of independent, specific genetic variation for separate abilities, however, simultaneously validates the existence of specific factors of intelligence beyond this general factor.

Longitudinal data on twins and adoptees provide additional insights into developmental processes and give further support for the existence of both general and specific abilities, as illustrated in Fulker and Cardon's analyses. The considerable stability throughout childhood of genetic effects specific to each cognitive ability (and independent of genetic effects on a general ability factor) suggest that developmental processes are driven largely by heritable factors.

Although the value of biometrical analyses of twin data was undisputed, a number of methodological issues arose. These are discussed below.

Studying Development: Longitudinal and Cross-sectional Designs

The usefulness of longitudinal twin and adoption designs is made apparent by the background papers from Fulker and Cardon and from Boomsma, as well as other papers in this book. They allow the examination of intraindividual change and its genetic and environmental bases. A more limited set of questions about development may still be addressed through cross-sectional age comparisons, such as the extent to which relative effects of genes and environment change over time. There are, however, some instances where the most informative designs include both longitudinal and cross-sectional comparisons.

In psychometric research on adult intellectual development and cognitive aging, for example, researchers now agree that the combined use of evidence from cross-sectional and longitudinal data leads to a more veridical estimate of age-related changes in cognitive abilities than the use of one of the two types of data sets alone (cf. Kruse et al. 1993). On the one hand, later cohorts tend to score at a higher level than earlier cohorts on some cognitive ability measures. In these instances, cross-sectional comparisons result in an overestimation of negative age trends. On the other hand, the effects of repeated testing and of nonrandom sample attrition limit the generalizability of longitudinal studies. In the field of cognitive aging, dropout is not random but reflects conditions that may be closely related to the dependent variable (i.e., ill health or death). For this reason, longitudinal data tend to underestimate negative age trends in cognitive abilities.

Recent analyses of adult age changes in cognitive abilities (Hertzog and Schaie 1986, 1988; Hertzog 1989; Schaie 1983) combined cross-sectional and longitudinal evidence and used latent structural modeling as well as hierarchical regressions to arrive at a better description and understanding of age-related change. The reanalysis of the Seattle Longitudinal Study (Schaie 1983) by Hertzog and Schaie (1986, 1988) was particularly instructive because the cohort-sequential design of this study allows for the estimation of age trends based on both longitudinal and cross-sectional comparisons. The following picture has emerged from this reanalysis and related work:

1. Interindividual differences in intellectual abilities are highly stable over the adult years (i.e., with seven-year interval stability coefficients greater than 0.90).
2. Between the ages of 57 and 63, most individuals begin to show significant decrements in intellectual performance.

3. Negative age trends in measures of perceptual speed seem to mediate most of the negative age trends in other cognitive abilities (Hertzog 1989; Schaie 1989).

These findings with normal (i.e., genetically noninformative) samples are well in agreement with the high heritability estimates found for intellectual functioning among older adults (Jarvik et al. 1972; Kallmann et al. 1951; Pedersen et al. 1992; Swan et al. 1990). The high stability of interindividual differences in adult cognition may reflect, to a large degree, the continuous influence of genetic factors. The analysis of longitudinal data sets of aging twins will help to clarify the extent to which interindividual differences in the onset and course of age-associated cognitive decline are also under genetic control. Recent findings reporting relatively high heritability estimates for measures of perceptual speed (Pedersen et al. 1992; Swan et al. 1990) are important in this regard. If perceptual speed is at least as heritable as other cognitive abilities, and if decline reflected in measures of speed drives the decline in other cognitive abilities, as suggested by recent evidence (Hertzog 1989; Schaie 1989), then genetic influence is likely to explain a major portion of the interindividual variability in age-related decline. Thus, longitudinal studies of aging twins are of prime importance to determine sources of interindividual differences in the onset and course of age-related cognitive decline during late adulthood.

GE CORRELATION AND G × E INTERACTION

Those not totally familiar with the mechanics of biometrical analyses often criticize these models as being invalid for not routinely including terms for GE correlation and G × E interaction. Several papers by behavioral geneticists (Plomin et al. 1977; Scarr and McCartney 1983; Loehlin and DeFries 1987), which point to the importance of these effects have been criticized by other behavioral geneticists (i.e., those employing rigorous model-fitting techniques) because their approach is often misrepresented.

Nonetheless, some developmental psychologists continue to find that these papers provide an attractive way to think about how both genetic and environmental factors may work together in shaping individual intelligence in that they try to side-step variance estimations by focusing on (so far psychological) mechanisms that contribute to GE correlation: selection, evocation, and manipulation of specific environments ("active" social interaction); a partner's responses to one's personality ("reactive" selection); evocation and manipulation of one's rearing environment by one's parents. This process-oriented point of view is highly attractive both to personality psychologists and to developmental psychologists because both fields view people as active producers of their own environment and development. The empirical implementation of this approach is solely concerned with the study of person-environment (PE) correlations and their changes over time. The fact that P depends in part on G is theoretically recognized but not empirically studied. There may be two advantages of this approach. First, it can specify psychological mechanisms that may explain GE covariance to some extent. Second, the notion that genetic and environmental differences can covary at the population level due to a continuous genotype-environment transaction at the individual level overcomes the pitting of genetic versus environmental influences and thereby helps developmental psychologists to accept that genetic differences must be considered in any study of individual differences.

Others, however, regard this as mostly wordy sophistry, largely intractable to experimental investigation. They point to a considerable body of rigorous theoretical and empirical work in this area (e.g., Baker 1989; Carey 1986; Eaves 1976; Eaves, Last, Martin et al. 1977; Eaves, Last, Young et al. 1977; Jinks and Fulker 1970), the sum total of which is that some, but not all, forms of G × E interaction and GE correlation *can* be detected in principle, but that few good replicated examples exist in the literature, even from studies that have had the power to detect these effects if they were there. This literature is largely ignored by developmentalists.

It was agreed, finally, that GE correlations and G × E interactions may be studied more explicitly by incorporating molecular genetic technology into classical twin designs, as discussed below more fully in this report.

DEFINITION AND CHARACTERIZATION OF HUMAN INTELLIGENCE

The standard claim that there *exists* a general ability factor (g), at least as traditionally defined through psychometric tests, was generally accepted by all discussants within this group. Disagreement was apparent, however, as to whether such a general factor is broad enough to capture the richness of intelligent behavior in humans. Some even argued vehemently that g is so narrow a concept that it should not be referred to as intelligence, per se. Others viewed the question of whether general and specific abilities exist as moot, and considered g itself to be an outmoded concept that has outlived its practical utility. Finally, some pragmatists emphasized that most researchers study what is already well-defined and measurable. Thus, g has considerable practical utility in behavioral genetic research and will continue to be used until alternative measuring instruments are available.

The major positions taken during the discussion are summarized below, along with reactions and criticisms from advocates of other positions. It must be realized, however, that some of these positions were already well-stated in the background papers (especially by Brand and by Fulker and Cardon) and are not emphasized in this chapter in proportion to the amount of time spent discussing them. Other positions not stated in those papers are given greater emphasis here.

The Existence and Importance of g

Brand (this volume) provides an overview of this position, which is representative of the "London School" of thought. In short, this position states that if a general factor can explain so much variation and covariation among test scores and other measures of success in life[1], then *g* must be important and worth studying in its own right.

Brand's argument for considering general intelligence (*g*) was that virtually all reliable measures of cognitive abilities show substantial positive intercorrelation. This pattern of relationships among tests was originally referred to by Spearman as a "positive manifold." Despite many twentieth-century efforts to identify independent dimensions of intellectual variation (e.g., by Thurstone, Guilford, Hudson, Gardner), *g*-factors continue to account for some 50% of the variance in matrices of correlations between abilities. Beyond *g*, other independent dimensions each typically account for less than 10% of the variance of abilities. From this perspective, *g* may be to cognitive psychology what carbon is to organic chemistry.

Many psychologists have tried to invent mental assessments that reflect features going beyond the "narrow, scholastic" assessments provided by *g*-loaded tests. J.P. Guilford proposed and tried to develop no less than 150 different ability measures, including assessments of "social intelligence" and divergent thinking. Today, R.J. Sternberg proposes hundreds of distinct mental processes (and their interaction effects) that might be tapped, and H. Gardner (see below) more modestly proposes seven mental abilities, akin to Thurstone's (ca. 1935). Yet none of these theorists has so far produced a battery of reliable and valid tests that are empirically independent of each other and of *g*. Some assert that most recent proposals (e.g., by Gardner) to develop elaborate observational methods for studying sources of variation would be expensive in terms of time, effort, and research monies, and would not yield any construct different than those already tapped by traditional IQ mental ability tests.

Although most agreed on the importance of *g* from both pragmatic and theoretical perspectives, some argued strongly for the importance of ability-specific variance in addition. Using confirmatory factor analyses with a hierarchical solution, one may explicitly *test* (i.e., via nested comparisons) whether the first-order (specific ability) factors capture any reliable variance that is not represented at the second-order (*g*) level. (This is precisely the logic underlying the Fulker and Cardon developmental model [see chapter 4, this volume]). Based on this approach, Horn and Cattell (1966), for example, have unambiguously demonstrated that there is reliable ability-specific variance.

Alternative Definitions of Intelligence and Cognition

It was strongly argued that the view of cognition and intelligence put forth by the London School is so widely accepted in its broad outlines that it is difficult for many people to envision an alternative position. Critics of this view assert that the concept of intelligence and the instrument of the intelligence test were the results of a particular

---

[1] O'Toole and Stankov (1992) show that IQ is a significant predictor of mid-life mortality, giving this statement literal meaning.

history and set of circumstances obtained a century ago. Researchers in England and France, in particular, were searching for ways of predicting failure in a specific institution: the turn-of-the-century school. These schools were said to stress the kinds of literacies that were important for administrators or bureaucrats in large organizations and particularly in far-flung empires.

How might intelligence—or cognate concepts—be defined or assessed in a very different kind of culture, for example, one without schooling at all or one in which the skills of the hunter, sailor, shaman, orator, mystic, or trader might be at a premium? How might such skills be assessed if paper-and-pencil or other short-answer instruments had not been invented or were for some reason proscribed?

Theory of Multiple Intelligences (MI)

In the last decade, in response to new work in the cognitive sciences and as a result of growing dissatisfaction with the limitations of traditional instruments, psychologists have put forth quite different views of intelligence (Baron 1985; Ceci 1990; Gardner 1983; Olson and Bruner 1974; Pea 1992). These views incorporate greater attention to the particular operations involved in problem-solving and product-making, the differences in skills across different domains, the role of cultural expectations and reward systems, and the social context of much intellectual activity. While standard psychometric activities continue, and sometimes incorporate new technologies, there is a tension between the London School and these new, broader, and more contextualized views of intelligence (Sternberg 1985, 1988; Sternberg and Detterman 1986).

One influential theory has been put forth by Gardner (1983, 1992). Rejecting psychometric techniques altogether, Gardner bases his theories principally on two lines of evidence: (a) the diverse roles or "end-states" that are valued in cultures around the world; (b) knowledge about the evolution and current organization of the cerebral cortex. Gardner defines *an* intelligence as the capacity to solve problems or fashion products that are valued in at least one culture. Drawing on eight distinct lines of evidence, Gardner proposes that human beings as a species have evolved to be able to carry out at least seven forms of information-processing or "intelligences"; in addition to the linguistic and logical faculties highlighted in schools and in most standardized measures of intelligence, Gardner proposes spatial intelligence, musical intelligence, bodily kinesthetic intelligence, interpersonal intelligence, and intrapersonal intelligence.

Gardner and his colleagues have attempted to develop techniques for assessing these intelligences in "intelligence-fair" ways. Among the most promising ways are the creation of new environments—resembling children's museums, in some respects—in which individuals reveal their intellectual proclivities and strengths by their patterns of interaction with various materials, puzzles, etc. It is possible to create reliable systems for scoring the intelligences of participating young children. Other methods for assessing intelligences include teaching new languages or games, with appropriate scaffolding, in order to determine potential in different intellectual domains, or the opportunity for subjects to engage in various simulations, which might be delivered by computer technology.

Misconceptions Regarding MI Theory

Contrary to the implications of Brand's paper (this volume), MI theory makes no claim about the hereditary or nonhereditary nature of the several intelligences. Nor are the intelligences related to social class, except insofar as lack of exposure to, say, music, will prevent the development of musical intelligence. Neither is MI theory committed to the independence or nonindependence of the several intelligences; their relation (or lack of relation) is an empirical matter to be determined by intelligence-fair instruments.

At the workshop, much discussion focused on the relationship between g and MI theory. Gardner did not dispute the existence of substantial estimates of heritability for g. With other participants, he believes that the nature of psychometric g has yet to be understood and may be, to some extent, an artifact of the kinds of instrumentation used and of the background of testing as we know it today. Critics pointed out that once MI theory is fully instrumented, it may well turn out that scores on the various intelligences will yield a "positive manifold." Gardner argued that were "tests" of intelligence to be constituted quite differently, or to be used in very different environments (e.g., non-schooled environments), g might either loom much less important, or a new g, only loosely related to the current scholastically oriented g, might emerge through a suitable kind of factorial analysis. Critics of MI theory pointed out that many new kinds of tests have been tried and have fallen by the wayside, and that tests with good reliability and validity yield a substantial g.

Gardner argued that one implication of the "g perspective" is that individuals can be aligned in terms of a single general ability and that education should be carried out differently for various individuals, based on the intellectual potential assessed by the IQ test. He argued for an aptitude-by-trait interaction view but acknowledged lack of evidence for this view. He also argued that the belief that an educational approach will be effective can itself be very powerful (Dweck and Licht 1980). The Japanese Suzuki Talent Education is able to evoke powerful—even prodigious—musical performances from youngsters, independent of assessed intellectual level or potential. Critics of his view maintained that claims of potentially powerful interventions seldom stand the test of time and rigorous evaluation.

Multiple Views on the Meaning and/or Causes of g

There was considerable discussion of what mechanisms g might represent. A common view among psychologists is that g represents the general problem-solving capacity of the brain. Its existence is supported substantially through statistical analyses at the phenotypic level (see Brand, this volume). Quantitative geneticists view g as a phenotypic trait that is to a considerable degree under the influence of many genes.

This *g,* however, may be, in whole or in part, an evolutionary "artifact" of cross-character assortative mating for many specific cognitive abilities (see Eaves et al. 1984) that have separate adaptive functions.

An alternative view, from a neurological perspective advocated by Kinsbourne, is that *g* simply reflects the overall effectiveness of a well-formed brain not subjected to any disruption pre- or postnatally and which has had sufficient opportunity to acquire skills necessary to score well on an IQ test. This view expressly denies that *g* can be attributed to any particular location or module of the brain and is highly consistent with the argument that evolution selects for specialized rather than general purpose mechanism (see below). The existence of mentally retarded individuals with special abilities (idiot savants) shows that specific mental capacities or skills can exist even in individuals who do not have well-formed brains and have quite low IQs.

THE EVOLUTIONARY FRAMEWORK OF COGNITION

There was a great diversity of opinion regarding how evolutionary issues in human cognition should be addressed. Biometrical analyses address the issue via the genetic and environmental architecture of the trait(s). Information is gained about gene action (e.g., genetic dominance, epistasis), the effects of different mating systems (e.g., assortative mating, inbreeding). These analyses suggest that both social and biological significance attach to traditional measures of intelligence. Evolutionary psychology takes an entirely different approach and analyzes the adaptive significance of various aspects of mental abilities. This approach rejects, on broad evolutionary grounds, the possibility that evolution would have generated a general purpose problem-solving mechanism.

Biometrical Analyses and Inbreeding Studies:
Evidence for Directional Dominance in Intelligence

The evolutionary significance of cognitive abilities can be illuminated through the partitioning of variances and covariances in biometrical analyses. In addition, some predictions can be tested through inbreeding studies. Fitness characters have predictable patterns of genetic architecture, which can be revealed through quantitative genetic designs. According to Fisher's fundamental theorem, if intelligence has been subjected to constant natural selection throughout evolutionary history, additive genetic variance for cognition should have vanished by now and only nonadditive genetic variance should remain. In fact, this extreme case is not found in nature; considerable additive genetic variance is always detected for a wide range of important fitness characters in both animals and plants. One reason why Fisher's asymptote is not attained may be that the environment never stays constant long enough for all additive variance to be fixed. Thus, the interesting criticism of *g* that some might be tempted to assert, i.e., that *g* cannot be important because there is too much additive variance, does not hold up when examined in the context of a broad array of empirical evidence.

Studies of inbreeding do suggest directional dominance (or epistatic) effects, whereby children of cousin marriages (who are partially inbred) show a depression in mean IQ of about 2–3 points compared to non-inbred children (Bashi 1977; Jensen 1983). In addition, children from incestuous matings not only show a variety of physical defects but also evidence for cognitive deficits (Bouchard 1993). Inbreeding reveals recessive alleles in homozygous combinations (and directly reveals effects of directional dominance). The phenomenon of inbreeding depression is common in many species and is a clear indication that we are dealing with a fitness characteristic. It suggests that recessive alleles are associated with low rather than high IQ, and Fisher's theory of the evolution of dominance, supported by evidence from plant and animal experiments, suggests that there has indeed been selection for high IQ through much of human evolution. Thus, whatever the views of educational psychologists about the importance of IQ, it is clear that it has been an important trait in evolution (Mather 1974).

Do we find evidence for directional dominance in areas other than inbreeding studies? Bouchard and McGue's (1981) data for twins do not suggest dominance, but only additive genetic variance and common environmental effects. However, twin studies alone have very low power to detect dominance (Martin et al. 1978), so this is not surprising. Even worse, dominance and shared environmental variance are completely negatively confounded in the classical twin study and, if both are acting, it is impossible to estimate the relative importance of either. However, with the inclusion of other relationships, such as parents and offspring, half siblings, or biological and adopted siblings, the presence of dominance could be inferred. Unfortunately, the power to detect dominance as a variance component is much less in these designs than as a mean effect in inbreeding studies, provided that it has a directional component.

An extra complication is that any dominance effects may be masked by developmental trends in genetic influences on IQ (which make the parent-offspring correlation lower than the sibling correlation, and the sib correlation lower than the DZ twin correlation) and by the effects of assortative mating (which mimic shared environment and are thus negatively confounded with dominance unless there are spousal data). Assortative mating increases additive genetic variation between families and may accentuate social stratification—if IQ is socially important. The analysis of dominance effects depends to a large extent (except for inbreeding studies) on resemblance among relatives who are not of the same age (e.g., parent-offspring). This highlights the need to consider developmental issues in modeling research. The background papers concerned with the modeling of development clearly show the conditional nature of genetic variation at different age points. Genetic and environmental variations are not fixed; the patterns may change over time and, more importantly, the organization of cognitive abilities may change with time due to developmental processes. The only way to address these important issues is to use a multivariate longitudinal design.

Evolutionary Psychology

Taking the functional approach of evolutionary psychology, Buss argued for greater significance of specific cognitive abilities, as opposed to a "*g*-factor," in adaptation of humans (see also Buss, this volume). Consistent with this view, it was pointed out by the population geneticists that the highly correlated genetic influences across many diverse cognitive abilities may be simply an artifact of cross-character assortative mating for these specific abilities (see Eaves et al. 1984).

Because of the numerous complex and specialized adaptive problems that humans have faced over evolutionary history, evolutionary psychologists expect human cognitive abilities to be numerous, specialized, complex, and domain-specific (Tooby and Cosmides 1990, 1992). By analogy, the body has a heart to pump blood, a liver to filter impurities, and sweat glands for thermal regulation—numerous, specialized mechanisms, each tailored to perform different functions and solve different adaptive problems, yet all working together within the organism.

A heuristic set of evolutionary psychological criteria for validating the existence of different intellectual or cognitive mechanisms was proposed by Buss:

1. Are there different mechanisms activated by different input (i.e., different adaptive problems confronted)?
2. Do different information-processing rules apply?
3. Do the outputs solve different adaptive problems?
4. Are different parts of the brain more active when solving different problems (e.g., as shown through PET scans)?
5. Do the sexes differ in observed ability in ways that correspond to the different adaptive problems each sex has presumably faced?

These criteria are merely heuristic and provisional, but do suggest a set of nonarbitrary standards for identifying specialized cognitive mechanisms.

Buss also advanced the possibility that there may be specialized cognitive mechanisms, each of which is uniquely configured to process information about different adaptive problems. These adaptive problems could include gathering and scavenging (spatial location memory) (Silverman and Eals 1992); hunting (spatial rotation) (Silverman and Eals 1992); cheater detection in social exchange (Cosmides 1989); reciprocal alliance formation (Buss 1986); hierarchy negotiation (Buss 1986); coalition formation and maintenance (Tooby and Cosmides 1990); infidelity detection (Buss et al. 1992); identification of beliefs and desires in others (concepts of mind) (Wellman 1989). Evolutionary psychology provides a heuristic for identifying what many of these adaptive problems are and hence a guide to uncover the possibly numerous cognitive mechanisms and abilities that humans possess.

Should these cognitive mechanisms be considered as "modules"? The neuroscientists argue that the term "modularity" is ill-conceived when applied to complex behaviors. Focal brain lesions do not selectively eliminate capabilities such as lie detecting or hierarchy negotiation. Such activities result from the coordinated use of more primitive underlying mental operations. Only the latter are selectively eliminated by focal lesions. Modularity reduces to the differential coordination of various parts of the brain to accomplish different cognitive operations. Phrased differently, complex activities are generated by neural "software" that coordinates particular subsets of primitive operations. Whether such software itself is differentially available to the two genders or is a genetically determined individual variable is an issue that deserves further study.

THE ROLE OF TWIN STUDIES IN ADDRESSING THESE ISSUES

What do these positions on the definition and nature of intelligence have to offer each other? Can their differences be resolved through the use of behavioral genetic studies? Here, the extent to which biometrical analyses of twin and adoption data can address these issues is summarized.

The lack of well-defined measuring instruments available to assess Gardner's constructs was a major criticism of his position. Thus, *g,* especially if it is defined as a higher-order factor of intercorrelated cognitive abilities, maintains great usefulness in genetic designs: it is well measured and shows importance in predicting various kinds of success in life (Barrett and Depinet 1991; Behrman et al. 1980). It remains to be seen whether Gardner's alternative measures, when they appear, will be useful in our study of individual differences.

Those who advocate alternative measures of intelligence might find genetically informative designs to be of some use in the actual development of such new constructs and their operational measures. They might conceptualize their variables somewhat differently if their observations were made on MZ and DZ twins rather than on nongenetically informative individuals. For example, if the hypothesis states that musical intelligence has a core of abilities, the best way to find that out is through the study of MZ twins reared apart. Certain pairs of abilities may have a genetic correlation, as opposed to being independent factors. This is just to say that definitions of intelligence might be *different* (not necessarily *better*) if developed with information about genes, environment, and evolutionary significance.

Some behavioral genetic studies of more contemporary measures of cognitive abilities have already begun. For example, a wide range of information-processing tasks are being investigated in twins, in an effort to study more closely the component processes involved in complex mental functioning (e.g., Baker et al. 1991; McGue and Bouchard 1989; Vernon 1989). These studies may possibly help to identify just what are the specific "core operations" within a general area like verbal or spatial performance. Extensions of these studies may also prove useful in understanding basic operations related to musical talent or mathematical ability. Presumably these "core operations" are most likely to be linked to specific gene complexes. Educational interventions are most likely to be effective if the structure of these complex faculties is better understood and if one has the options of "supplementing" operations that are impaired.

We might also consider how biometrical analyses of genetic and environmental architecture might enter into the evolutionary psychology perspective. It is incumbent upon evolutionary psychologists to show that there is specific variation for these abilities not explained by *g,* and that there is genetic variation for these specifics with nonadditive variation acting in the predicted direction.

OTHER ISSUES: PAST AND FUTURE

Studying the Environment

Twin studies are usually thought of as providing information about genetic influences. They can, however, with equal validity be used to examine environmental influences on the phenotype. First, a twin study can demonstrate that an environmental influence exists, and the variance can be partitioned into shared and nonshared environmental influences.

Mascie-Taylor (this volume) documents the numerous correlations between what are commonly, but erroneously, called "environmental" factors with respect to intelligence. These correlations are difficult to interpret, however, because of the inevitable confusion between genetics and environment in nuclear families and the covariation between the environmental factors. The same pregnant woman may, for example, abuse alcohol, continue to smoke, use illicit drugs, suffer from poor nutrition, and have inadequate hygiene and medical care during pregnancy. An adverse outcome for her child is not immediately interpretable in terms of environmental factors because these characteristics may be more frequent in women with a genetic predisposition to low intelligence and maladaptive behaviors such as an undue sensation-seeking tendency. Thus, the fetal outcome may consequently reflect, in part, those antecedent genetic conditions as opposed solely to the intervening environmental variables.

The importance of both prenatal and postnatal factors in the study of environment is discussed elsewhere in this volume (see Macdonald et al.). In our own discussions, there was particular concern that prenatal factors may not entirely be considered as shared environmental effects. Although prenatal factors that are typically classifiable under shared environment may include maternal nutrition, fetal oxygenation, and the effects of circulating toxins, there may be a differential placental vasculature, particularly in MZ twins (e.g., the MZ twin transfusion syndrome), that produces large differences between co-twins (e.g., in birthweight) (see also Bryan, this volume).

Genetic variance may appear as environmental if an environmental effect arises from a genetically controlled risk factor. The following (hypothetical) sequence of events is illustrative: An immune-dysfunctional mother becomes sensitized to a fetal antigen. The resulting immune reaction damages fetal brain development (Adinolfi

1976). (Neuromigrative and synaptic pruning impairments in dyslexics may be a case in point.) So the product—dyslexia, though arising from an adverse intrauterine event—may result from a maternal effect based on a genetically transmitted variable, immune dysfunction.

Bishop was concerned with the polarization between genetics and environment implicit in Brand's paper (this volume). As Rutter (1991) pointed out, it is a common misconception that if a trait shows strong heritability, then environmental factors are unimportant. High heritability indicates that genetic factors account for a high proportion of variance within the population, but it does *not* imply that the trait cannot be altered by environmental manipulations. Estimates of heritability will depend on the range of environments in the population: as environments become more uniform within a population, genetically based individual differences will account for a greater proportion of the variance. To take an extreme example, one can demonstrate high heritability for spelling ability in Western societies (e.g., Stevenson et al. 1987). However, one might expect to find lower heritability in a society such as rural Papua New Guinea, where access to schooling is much more haphazard. Environmental factors that are common to all individuals in a society (e.g., schooling in a Western society) may have an impact on the mean level of a trait, without opposing findings from studies demonstrating secular changes in the average level of a strongly heritable trait (Rutter uses the example of height, although IQ is also relevant here). One would expect this to occur if the whole population was exposed to a change in environment. Clearly, it is naive to assume that affording everyone comparable environmental opportunity and support will eliminate individual differences in intelligence. However, we should beware of concluding that a highly heritable trait is nonmodifiable. Rutter (1991) points to the example of phenylketonuria as a genetically determined form of mental retardation that can be treated by imposing dietary restrictions outside the normal range of environmental variation. Far from being irrelevant, modification of the environment may be especially important for children with a genetic predisposition to low ability.

The same kinds of issues raised with respect to the conceptualization and measurement of cognition were also raised by Gardner with respect to the environment. He urged caution with respect to the sampling and range of environments studied by behavioral geneticists, and emphasized a need to take into account what is known about radically different cultures (Geertz 1973). In addition to a focus on the home and school, it may be useful to observe and measure personality and social features in unfamiliar environments (like a new city), in rich environments (like a "hands-on" museum), and across environments (how consistent is behavior at home and in school?). Behaviors between siblings, or between siblings and parents, may also vary substantially across situations (Mischel 1968), and need to be regarded in more than an unequivocal way.

These criticisms suggest the possibility of interactions between *persons* and *situations*. Critics of these ideas, however, argued that it remains incumbent on those who insist on the importance of *person* × *situation* interactions to show that the interaction variance is large relative to the main effect, and that it is adaptively important. Failing this, whatever the range of tasks and environments proposed in speculations of Gardner and others, g remains the best single predictor of success, provided the latter is defined within a moderately sensible Darwinian framework.

Gardner also raised the issue of reliance on self-reports and other short-answer measures of environmental factors, which often show low reliability and validity (see Nisbett and Ross 1980; Kahneman et al. 1982, concerning criticisms of introspective reports). Systematic and reliable obervations of sibling-sibling and parent-child relations, however, have showed considerable consensus among family members and have been used successfully in exploring the nature of environment effects in genetically informative designs (see Baker and Daniels 1990; Plomin and Daniels 1987).

One conclusion that received consensus was that environmental effects can be studied most appropriately in biometrical designs. By including measured indices of the environment in model-fitting analyses of twin and adoption data, one can evaluate the full extent of latent sources of their variation and covariation with cognitive outcomes.

Speed of Processing

A particularly unlikely correlate of IQ is inspection time (discovered by Nettlebeck in Australian retarded individuals; see Nettlebeck 1987). The shortest exposure for which an individual can make a judgment of difference in lines correlates moderately with scores on IQ tests (see Vernon 1987). Simple measures of information-processing speed deserve further attention in twin research.

Arguably, individual differences in basic information processing speed underlie surface differences in psychometric g. One theoretical model is provided by Anderson (1992), where g is akin to a tape recorder, and specific abilities (e.g., verbal, spatial, memory) are akin to tapes: a good tape recorder allows differences between the quality of tapes to be detected, and relatively high g is similarly envisaged to allow the differentiation of specific abilities to be observed. Anderson's model accords with Detterman and Daniel's (1989) observation that, in psychometric data, g factors appear stronger (i.e., s-factors emerge less clearly and strongly) in testees ranging across lower values of g. Thus, perhaps we can reconcile the general omnipresence and explanatory power of g with the apparent differentiation of specific abilities in people with higher levels of g. Might Spearman's g thus be partly reconciled with Gardner's MIs?

Some argued against such optimism, however, in explaining IQ differences as a function of inspection time. Working memory (Baddeley 1986, 1992) was mentioned as another possible candidate for explaining individual differences in IQ (see Kyllonen and Christal 1990).

Kinsbourne argued that such diverse procedures as reasoning, discriminative reaction time, estimates of working memory, stability of the configuration of the evoked potential and even peripheral nerve conduction time all exhibit high correlations with psychometric *g*, making it unlikely that *g* represents a separate mental domain, like speed of processing. Rather, it could represent the general integrity or well-formedness of the nervous system, just as the presence or absence of minor congenital abnormalities may index the well-formedness of the body. A better-developed brain better houses specialized processors (such as those listed by Gardner).

Deviations from well-formedness can be local in the brain and impact differentially on differential mental skills. Thus, an uneven skill profile could arise from selective deficit, as readily as from selective ability development. Neither origin can simply be assumed. They may, however, converge. Differences in neural synaptic organization may characterize genetically disparate brains as well as distinguish damaged from undamaged brains.

Secular Rise in Average IQ

The worldwide rise in IQ test raw scores has been documented by Flynn (1987). Based primarily on military sources of young men tested in the 1950s to 1970s in 14 advanced Western countries, Flynn claimed there was a massive IQ rise of 5 to 25 points in a single generation. In his review, Flynn therefore criticized IQ tests as not highly scholastic, but being limited to abstract rather than real-world problem-solving ability. Flynn claimed, thus, that intelligence must be a socially unimportant characteristic. Perhaps teachers do not take notice of individual abilities, and educational attainment is based on hard work rather than individual intelligence.

Alternative explanations for the secular change in IQ performance were discussed:

1. Intelligence and educational levels might have risen due to twentieth-century improvements in schooling. There are, however, no undisputed educational improvements, and pupil-teacher ratios are uncorrelated with pupils' success; Flynn himself rejects this explanation outright and calls attention to the loss in SAT scores over the same time period.

2. The rise may only have occurred on some types of IQ tests, especially those involving multiple-choice format, time-limited administration, no guessing correction, and built-in penalties for persistence in trying to solve hard items (see Brand 1990). Flynn believes that the very slight rise in Scottish IQ subtest scores since 1962 hides, because of methodological errors, a large increase in real IQ (which Flynn holds, however, to reflect "mere problem-solving ability").

3. Improved nutrition is another explanation provided by Schoenthaler (1991; see also Blinkhorn 1990). However, this involves a marked shift from previous hereditarian positions allowing nutritional effects on IQ only at very low levels of either nutrition, IQ, or both.

4. Better maternal and infant health in recent years leads to a more well-formed brain.

Can twin and adoption research shed light on the causes of these secular changes in IQ test performance? The conventional wisdom in behavioral genetics is that group means and individual differences represent two different realms of variation so that inferences about the etiology of group differences have no implications for inferences about individual differences. Rowe argued that this is false. There are two realms of variance—between and within groups—but only one realm of developmental process. A similar point has recently been made by Turkheimer (1991). Thus, developmental process should affect both group means and individual differences, and findings for group differences are relevant for individual differences and vice versa. This is true except in certain specialized conditions, such as when there is restricted sampling or the introduction of new influences. The outcomes for group and individual variation may be different, but exceptions to a rule do not falsify it. For example, lead exposure influence on IQ can be detected in a study of unrelated siblings reared together if some pairs are lead exposed and some are not. It will also be detected in a comparison of group means when one group has more lead exposure than another. Hard scientific questions, such as race differences in IQ, cannot be avoided by claiming that group means and individual differences are intrinsically different. They are not, because there is only one seamless process of development.

Several attempts have been undertaken to arrive at models allowing the simultaneous analysis of mean variation and individual differences (Dolan et al. 1992; Turkheimer 1991). These models represent steps towards the empirical study of the possible relationship between group means and individual differences.

The Role of Molecular Genetics in Quantitative Genetic Designs

What will be the future of the twin method in light of rapid developments in molecular genetics and its application to studying continuous traits like cognitive abilities? Some biologists believe that the rapid advances in molecular genetics will soon render obsolete biometrical studies of twins, families, and adoptions. There was a general consensus that nothing could be further from the truth, for as soon as a single gene effect has been found, one will want to estimate the proportion of variance, or risk, that it accounts for. While this can be done on a sample of unrelated individuals, traditional twin or family design will allow a much richer range of hypotheses to be tested, as well as help to avoid some of the pitfalls of artifactual associations produced, for example, by population stratification. We may ask how much of the remaining variance is genetic (Martin et al. 1987). Is there evidence of epistatic interaction with the polygenic background and, if so, what kind (Martin et al. 1987)? Comparing the intrapair variance of MZ twins of different measured genotypes allows one to test whether environmental sensitivity is under genetic control (G × E interaction). For example, MZ twins of blood group M have lower intrapair variance in serum lipid levels than those of blood group N (Martin et al. 1983). This suggests that environmental factors (e.g., diet) may play a more important role in N than M individuals. In searching for "IQ genes" in the framework of a twin study, it would be just as important to look for genes with effects on environmental sensitivity using this technique, as effects on the mean. Of course, since the former relies on a variance ratio test (Martin et al. 1983), it has much less power than tests for mean differences.

With respect to the effects of a measured genotype on a phenotype such as IQ, even while this can be done on unrelated individuals, the use of twins is very attractive because of the possibility to partition the phenotype into a genetic and a nongenetic part at the level of the individual. Attention was drawn to the work of Boomsma, Molenaar, and Orlebeke (1990) and Boomsma, Molenaar, and Dolan (1991), who extended the genetic analysis of covariance structure to the estimation of individual environmental and genetic factor scores. In principle, and under favorable circumstances in the context of twin studies, it should be possible to assign each individual twin both a genetic and an environmental factor score. In this case, it might prove more powerful to look for measured gene effects on estimated genetic factor scores than on phenotypic measurements.

Analysis of Extremes

Alternative approaches to genetic analyses of normal variation in cognition were also highlighted. Examination of extremes of talent using the Fulker and DeFries (1985) regression method, tetrachoric correlations (e.g., Baker 1986), or ordinal regression techniques could prove particularly useful in understanding sources of abnormal variation. These might also prove particularly useful in exploratory analyses of alternative measures of intelligence, as one does not even need well-validated constructs to work with genetically informative designs. Those of Gardner's persuasion could use ratings or ordered variables or a "cafeteria of opportunity" measurement system in twin design, e.g., as a preliminary, exploratory analysis.

CONCLUSIONS AND SUGGESTIONS FOR FUTURE RESEARCH

Our discussions reflected communication among, and challenges offered to, several different research traditions and schools of thought. It is hoped that developmental psychologists, neuropsychologists, evolutionary psychologists, and cognitive psychologists will soon recognize that behavioral geneticists are keenly aware of many theoretical issues they allegedly ignore (e.g., G × E covariation, effects of the environment, developmental processes). More importantly, behavioral genetic methods are actually the most appropriate methods to study such issues.

Several suggestions for future research are apparent in this report. These are briefly reiterated below.

1. Study the decline in aging and its relationship to speed of processing.
2. Use biometrical models to study "environmental" variables, such as presented by Mascie-Taylor (this volume).

3. Use twin and adoption methods to develop and evaluate new measures of intelligence.
4. Examine abnormal variation by analyzing extremes of talent, e.g., using the Fulker and DeFries (1985) regression method for twin data.
5. Use linkage and candidate gene approaches to study the nature of genetic influences on specific abilities. This will also allow more detailed study of GE correlation and interaction (e.g., see Martin et al. 1987a,b).
6. Study differences between MZ twins as a means of specifying environmental effects. This has been nicely illustrated by Kinsbourne's examples, as a means to explicate neurological mechanisms (which could be genetic or environmental in origin) related to cognitive development.
7. Continue use of longitudinal twin and adoption studies (illustrated by Fulker and Cardon and by Boomsma, this volume) to examine whether stability in genetic factor scores at the level of *g* is better predicted by *g* itself or by cognitive abilities underlying *g*; that is, is stability higher for first-order abilities than for *g*?

REFERENCES

Adinolfi, M. 1976. Neurologic handicap and permeability of the blood CSF barrier during foetal life to maternal antibodies and hormones. *Devel. Med.* **18**:243–246.

Anderson, M. 1992. Intelligence and Development: A Cognitive Theory. Oxford: Oxford and Blackwell.

Baddeley, A.D. 1986. Working Memory. Oxford: Clarendon.

Baddeley, A.D. 1992. Working memory. *Science* **255**:556–559.

Baker, L.A. 1986. Estimating genetic correlations among discontinuous phenotypes: An analysis of criminal convictions and psychiatric-hospital diagnoses in Danish adoptees. *Behav. Genet.* **16**:127–142.

Baker, L.A. 1989. Genotype-environment covariance for multiple phenotypes: A multivariate test using adopted and nonadopted children. *Mult. Behav. Res.* **24**:415–430.

Baker, L.A., and D. Daniels. 1990. Nonshared environment and personality in adult twins. *J. Pers. Soc. Psychol.* **58**:103–110.

Baker, L.A., P.A. Vernon, and H.-Z. Ho. 1991. The genetic correlation between intelligence and speed of information processing. *Behav. Genet.* **21**:351–367.

Baron, J. 1985. Rationality and Intelligence. New York: Cambridge Univ. Press.

Barrett, G.V., and R.L. Depinet. 1991. A reconsideration of testing for competence rather than intelligence. *Am. Psychol.* **46**:1012–1024.

Bashi, J. 1977. Effects of inbreeding on cognitive performance. *Nature* **226**:440–442.

Behrman, J.R., Z. Hrubec, P. Taubman, and T.J Wales. 1980. Socioeconomic success. Amsterdam: North Holland Publ. Co.

Blinkhorn, S. 1990. A dose of vitamins and a pinch of salt. *Nature* **350**:6313.

Boomsma, D.I., P.C.M. Molenaar, and C.V. Dolan. 1991. Individual latent growth curves: Estimation of genetic and environmental profiles in longitudinal twin studies. *Behav. Genet.* **21**:241–253.

Boomsma, D.I., P.C.M. Molenaar, and J.F. Orlebeke. 1990. Estimation of individual genetic and environmental factor scores. *Genet. Epidem.* **7**:83–91.

Bouchard, T.J., Jr. 1993. The genetic architecture of human intelligence. In: Biological Approaches to the Study of Human Intelligence, ed. P.E. Vernon.. New York: Plenum, in press.

Bouchard, T.J., Jr., D. Lykken, M. McGue, N. Segal, and A. Tellegen. 1990. Sources of human psychological differences: The Minnsesota Study of Twins Reared Apart. *Science* **250**:223–228.

Bouchard, T.J., Jr., and McGue, M. 1981. Familial studies of intelligence: A review. *Science* **212**:1055–1059.

Brand, C.R. 1990. A "gross" underestimate of a "massive" IQ rise? A rejoinder to Flynn. *Irish J. Psychol.* **11**:52–56.

Buss, D.M. 1986. Can social science be anchored in evolutionary biology? *Rev. Euro. Soc.* **24**:41–50.

Buss, D.M., R. Larsen, D. Westen, and J. Semmelroth. 1992. Sex differences in jealousy: Evolution, physiology, and psychology. *Psychol. Sci.* **3**:251-255.

Carey, G. 1986. Sibling imitation and contrast effects. *Behav. Gen.* **16**:319–341.

Ceci, S. 1990. On Intelligence...More or Less. Englewood Cliffs, NJ: Prentice Hall.

Cosmides, L. 1989. The logic of social exchange: Has natural selection shaped how humans reason? *Cognition* **31**:187–276.

Detterman, D.K., and M.H. Daniel. 1989. Correlations of mental tests with each other and with cognitive variables are highest for low IQ groups. *Intelligence* **13**:349–359.

Dolan, C.V., P.C.M. Molenaar, and D.I. Boomsma. 1992. Decomposition of multivariate phenotypic means in multigroup genetic covariance structure analysis. *Behav. Genet.* **22**: 319–335.

Dweck, C., and B.G. Licht. 1980. Learned helplessness and intellectual achievement. In: Human Helplessness: Theory and Applications, ed. J. Garber and M.E.P. Seligman. New York: Academic.

Eaves, L.J. 1976. A model for sibling effects in man. *Heredity* **36**:205–214.

Eaves, L.J., and Gale, J.S. 1974. A method for analysing the genetic basis of covariation. *Behav. Genet.* **4**:253–267.

Eaves, L.J., A.C. Heath, and N.G. Martin. 1984. A note on the generalized effects of assortative mating. *Behav. Genet.* **14**:371–376.

Eaves, L.J., K. Last, N.G. Martin, and J.L. Jinks. 1977. A progressive approach to non-additivity and genotype-environmental covariance in the analysis of human differences. *Br. J. Math. Stat. Psychol.* **30**:1–42.

Eaves, L.J., K.A. Last, P.A. Young, and N.G. Martin. 1977. Model-fitting approaches to the analysis of human behaviour. *Heredity* **41**:249–320.

Eysenck, H.J., ed. 1979. The Structure and Measurement of Intelligence. New York: Springer.

Flynn, J.R. 1987. Massive IQ gains in 14 nations: What IQ tests really measure. *Psychol. Bull.* **101**:171–191.

Fulker, D.W. 1979. Nature and nurture: Heredity. In: The Structure and Measurement of Intelligence, ed. H.J. Eysenck, pp. 102–132. New York: Springer.

Fulker, D.W., and J.C. DeFries. 1985. Multiple regression analysis of twin data. *Behav. Genet.* **15**:467–474.

Gardner, H. 1983. Frames of Mind: The Theory of Multiple Intelligences. New York: Basic Books.

Gardner, H. 1992. Multiple Intelligences: The Theory in Practice. New York: Basic Books.

Geertz, C. 1973. The Interpretation of Cultures. New York: Basic Books.

Heath, A.C., M.C. Neale, J.K. Hewitt, L.J. Eaves, and D.W. Fulker. 1989. Testing structural equation models for twin data using LISREL. *Behav. Genet.* **19**:9–36.

Hertzog, C. 1989. Influences of cognitive slowing on age differences in intelligence. *Devel. Psychol.* **25**:636–651.

Hertzog, C., and W. Schaie. 1986. Stability and change in adult intelligence: 1. Analysis of longitudinal covariance structures. *Psychol. Aging* **1**:159–171.

Hertzog, C., and W. Schaie. 1988. Stability and change in adult intelligence: 2. Simultaneous analysis of longitudinal means and covariance structures. *Psychol. Aging* **3**:122–130.

Horn, J.L. 1982. The theory of fluid and crystallized intelligence in relation to concepts of cognitive psychology and aging in adulthood. In: Aging and Cognitive Processes, ed. F.I.M. Craik and S. Trehub, pp. 237–278. New York: Plenum.

Horn, J.L., and R.B. Cattell. 1966. Refinement and test of the theory of fluid and crystallized intelligence. *J. Ed. Psychol.* **57**:253–270.

Jarvik, L.F., J.E. Blum, and A.O. Varma. 1972. Genetic components and intellectual functioning during senescence: A 20-year study of aging twins. *Behav. Genet.* **2**:159–171.

Jensen, A.R. 1983. Effects of inbreeding on mental ability factors. *Pers. Indiv. Diff.* **4**:71–87.

Jinks, J.L., and D.W. Fulker. 1970. Comparison of the biometrical genetical, MAVA, and classical approaches to the analysis of human behavior. *Psychol. Bull.* **73**:311–349.

Kahneman, D., P. Slovic, and A. Tversky, eds. 1982. Judgment under Uncertainty: Heuristics and Biases. New York: Cambridge Univ. Press.

Kallmann, F.J., L. Feingold, and E. Bondy. 1951. Comparative, adaptational, social, and psychometric data of the life histories of senescent twin pairs. *Am. J. Hum. Genet.* **3**:65–73.

Kruse, A., U. Lindenberger, and P.B. Baltes. 1992. Longitudinal research on human aging: The power of combining real-time, microgenetic, and simulation approaches. In: Methodological and Research Strategical Issues in Longitudinal Research, ed. D. Magnusson. Cambridge: Cambridge Univ. Press, in press.

Kyllonen, P.C., and R.E. Christal. 1990. Reasoning ability is (little more than) working memory capacity?! *Intelligence* **14**:389–433.

Loehlin, J.C., and J.C. DeFries. 1987. Genotype-environment correlation revisited. *Behav. Genet.* **17**:263–278.

Loehlin, J.C., and P. Nichols. 1976. Heredity, Environment, and Personality. Austin: Univ. of Texas Press.

Martin, N.G., P. Clark, A.F. Ofulue, L.J. Eaves, L.A. Corey, W.E. Nance. 1984. Does the PI polymorphism alone control alpha-1-antitrypsin expression? *Am. J. Hum. Genet.* **40**:267–277.

Martin, N.G., and L.J. Eaves. 1977. The genetical analysis of covariance structure. *Ann. Hum. Genet.* **39**:219–229.

Martin, N.G., L.J. Eaves, and A.C. Heath. 1987. Prospects for detecting genotype × environment interaction in twins with breast cancer. *Acta Genet. Med. Gem.* **36**:5–20.

Martin, N.G., L.J. Eaves, M.J. Kearsey, and P. Davies. 1978. The power of the classical twin study. *Heredity* **40**:97–116.

Martin, N.G., D.M. Rowell, and J.B. Whitfield. 1983. Do the MN and Jk systems influence environmental variability in serum lipid levels? *Clin. Genet.* **24**:1–14.

Mather, K. 1974. The Genetical Structure of Populations. London: Chapman and Hall.

McGue, M., and T.J. Bouchard, Jr. 1989. Genetic and environmental determinants of information processing and special mental abilities: A twin analysis. In: Advances in the Psychology of Human Intelligence, ed. R.J. Sternberg. Hillsdale, NJ: Erlbaum.

Mischel, W. 1968. Personality and Assessment. New York: Wiley.

Nettlebeck, T. 1987. Inspection time and intelligence. In: Speed of Information Processing and Intelligence, ed. P.A. Vernon, pp. 295–346. Norwood, NJ: Ablex.

Nisbett, R.J., and L. Ross. 1980. Human Inference: Strategies and Shortcomings of Social Judgment. Englewood Cliffs, NJ: PUBLISHER.

Olson, D., and J.S. Bruner. 1974. Learning through experience and learning through media. In: Media and Symbols: The Forms of Expression, Communication, and Education, ed. D. Olson. Chicago: Univ. of Chicago Press.

O'Toole, B.I., and L. Stankov. 1992. Ultimate validity of psychological tests. *Per. Indiv. Diff.* **13**:699–716.

Pea, R. 1992. Distributed cognitions and education. In: Distributed Cognitions, ed. G. Salomon. New York: Cambridge Univ. Press.

Pedersen, N., R. Plomin, J. Nesselroade, and G. McClearn. 1992. A quantitative genetic analysis of cognitive abilities during the second half of the life span. *Psychol. Sci.* **3**:346–353.

Plomin, R., and D. Daniels. 1987. Why are children in the same family so different from one another? *Behav. Brain Sci.* **10**:1–59.

Plomin, R., J.C. DeFries, and J.C. Loehlin. 1977. Genotype-environment interaction and correlation in the analysis of human behavior. *Psychol. Bull.* **84**:309–322.

Rutter, M. 1991. Nature, nurture, and psychopathology: A new look at an old topic. *Devel. Psychopathol.* **3**:125–136.

Scarr, S., and K. McCartney. 1983. How people make their own environments: A theory of genotype-environment effects. *Child Devel.* **54**:424–435.

Schaie, K.W. 1983. The Seattle Longitudinal Study: A 21-year exploration of sychometric intelligence in adulthood. In: Longitudinal Studies of Adult Psychological Development, ed. K.W. Schaie, pp. 64–135. New York: Guilford.

Schaie, K.W. 1989. Perceptual speed in adulthood: Cross-sectional and longitudinal studies. *Psychol. Aging* **4**:443–453.

Schoenthaler, S.J. 1991. Abstracts of early paper on the effects of vitamin and mineral supplementation on IQ and behavior. *Pers. Indiv. Diff.* **12**:335–342.

Silverman, I., and M. Eals. 1992. Sex differences in spatial abilities: Evolutionary theory and data. In: The Adapted Mind: Evolutionary Psychology and the Generation of Culture, ed. J. Barkow, L. Cosmides, and J. Tooby. New York: Oxford.

Sternberg, R.J. 1985. Beyond IQ. New York: Cambridge Univ. Press.

Sternberg, R.J. 1988. The Triarchic Mind. New York: Viking.

Sternberg, R.J., and D. Detterman, eds. 1986. What is Intelligence? Hillsdale, NJ: Erlbaum.

Stevenson, J., P. Graham, G. Fredman, and V. McLoughlin. 1987. A twin study of genetic influences on reading and spelling ability and disability. *J. Child Psychol.* **28**:229–247.

Swan, G.E., D. Carmelli, T. Reed, G.A. Harshfield, R.R. Fabsitz, and P.J. Eslinger. 1990. Heritability of cognitive performance in aging twins. *Arch. Neur.* **47**: 259–262.

Tambs, K., J.M. Sundet, and P. Magnus. 1984. Heritability analysis of the WAIS subtests: A study of twins. *Intelligence* **8**:283–293.

Tooby, J., and L. Cosmides. 1990. On the universality of human nature and the uniqueness of the individual: The role of genetics and adaptation. *J. Pers.* **58**:17–68.

Tooby, J., and L. Cosmides. 1992. The psychological foundations of culture. In: The Adapted Mind: Evolutionary Psychology and the Generation of Culture, ed. J. Barkow, L. Cosmides, and J. Tooby. New York: Oxford.

Turkheimer, E. 1991. Individual and group differences in adoption studies of IQ. *Psychol. Bull.* **110**:392–405.

Vandenberg, S.G. 1965. Multivariate analysis of twin differences. In: Methods and Goals in Human Behavior Genetics, ed. S.G. Vandenberg, pp. 29–43. New York: Academic.

Vernon, P.A. 1987. Speed of Information Processing and Intelligence. Norwood, NJ: Ablex.

Vernon, P.A. 1989. The heritability of measures of speed of information-processing. *Pers. Indiv. Diff.* **10**:573–576.

Wellman, H. 1989. Children's Concept of Mind. New York: Academic.

7

What Has Behavioral Genetics Told Us about the Nature of Personality?

J.C. LOEHLIN

Department of Psychology, University of Texas, Austin, TX 78712, U.S.A.

ABSTRACT

Behavioral genetic analysis can be a useful first step toward understanding personality differences among individuals. It can be carried out at four levels: (1) a general allocation of variance to genes and environment, (2) subdivision within these, (3) correlation and interaction between genotype and environment, and (4) underlying mechanisms. Considerable empirical research at levels 1 and 2 is summarized for the traits *extraversion* and *emotional stability*. Some examples of work at level 3 are described, and some strategies for attacking level 4 are presented.

INTRODUCTION

Individuals differ in a very large number of ways. How many? One approach is to count the number of characteristics that people have found important enough to recognize and name. Modern languages contain literally thousands of descriptive words for personality. One search of an unabridged dictionary (Allport and Odbert 1936) found 17,953 words in English that could be used to describe personality: 4504 were considered personality trait names as such, with the rest describing temporary states (e.g., frantic), evaluations of character (e.g., worthy), capacities or talents (e.g., gifted), etc.

There is much redundancy in any such list, of course, and personality psychologists have attempted to reduce it in various ways. Some have proceeded through fairly direct empirical approaches, such as factor or cluster analysis. Thus we have the sixteen personality factors of R.B. Cattell, the so-called "Big Five" factors, currently attracting much interest as well as Eysenck's three superfactors, and so on (for a review, see John 1990). Others have developed more or less elaborate theories

Twins as a Tool of Behavioral Genetics
Edited by T.J. Bouchard, Jr. and P. Propping © 1993 John Wiley & Sons Ltd.

J.C. Loehlin

involving hypothetical systems or mechanisms underlying and accounting for be-havioral tendencies—psychoanalytic theories would be an example. Such theories typically focus on the causation of a few key traits, such as anxiety or impulsivity, that tend to be particularly problematic for their possessors or those who interact with them.

Whatever behavioral tendencies we elect to observe, and however we define, classify, and measure them, let us take it as given that there are complex dynamic systems underlying the traits—systems that, in principle, are describable in physio-logical terms. Some theorists—though not all—find it convenient to describe these systems in psychological language, as needs, superegos, defense mechanisms, or whatever. Let us assume that it is variation in the development of such systems in individuals that accounts for the great variety of ways in which people differ in their behavior. In the long run, we would like to understand such systems fully, and thus the causation of individual differences in personality. Obviously, this is a very long-term goal.

One potential clue to the understanding of personality systems is to ask about their genetic and environmental origins. Thus, a reasonable preliminary step is the approach taken by behavioral genetics, namely, to seek to classify the observed variation on particular traits into that which is related to genetic differences among individuals and that which is related to differences in the environments to which they have been exposed. This is, I emphasize, a preliminary step, and miracles should not be expected. Nevertheless, the methods of behavioral genetics *do* allow us to make bottom-line estimates of the relative genetic and environmental contributions to the variation of particular traits in a given population. They may let us take further steps as well, such as dividing the genetic variation into a transmissible and nontransmissible portion (the geneticist's "additive" and "nonadditive" effects), or dividing the environmental variation into a part that is shared by family members and a part that is idiosyncratic to individuals. These methods may also let us begin to investigate correlations and interactions between genetic and environmental sources of influence.

How do behavioral geneticists accomplish these things? Essentially, they observe resemblances on a particular trait in various informative groups such as monozygotic (MZ) and dizygotic (DZ) twins reared together or apart, adoptive and biologically related members of families, the families of identical twins, and so on. From the way that these resemblances vary with different degrees of genetic and environmental resemblance between the individuals studied, deductions can be made concerning genetic and environmental contributions to the variation of the trait. For example, if unrelated adopted children reared in the same family show no resemblance to one another on some trait; we conclude that shared family environmental factors are not major determinants of that trait—if they were, we would expect children reared together from birth in that family to be alike. For another example, if MZ twins, who share all their genes, are no more alike than DZ twins, who share only some of theirs, we conclude that genetic variation is not a primary determinant of individual dif-ferences on this trait in this population.

Classically, behavioral geneticists have tended to work with relatively simple comparisons among relevant groups, such as MZ and DZ twins reared together, or adopted and biological siblings, or MZ twins reared together and apart. Today, they often fit models simultaneously to data from a number of such relationships. I will illustrate both approaches later for some particular traits.

FOUR LEVELS OF BEHAVIORAL GENETIC ANALYSIS

First, I will summarize the behavioral genetic approach to the study of personality in the form of a table (Table 7.1), in which we can conceive of behavioral genetic analyses as occurring at several levels of refinement. At the first level, we have a rough, overall allocation of the variation of a trait into a portion that is associated with genetic differences among individuals and a portion that is not (and thus by default is assigned to environmental influences). This is the level represented by the traditional estimate of "heritability," based on, say, a comparison of MZ and DZ twins.

At the second level, we subdivide these, typically into additive and nonadditive genetic components and familial and nonfamilial environments. Analysis at this second level is fairly standard in contemporary behavioral genetic treatments of personality (e.g., Eaves et al. 1989 or Loehlin 1992). Further subdivisions are possible in principle, such as dividing the nonadditive genetic portion into the effects of dominance and epistasis, or the family environment into the effects associated with social class and those unique to particular families. These further subdivisions, however, are only occasionally pursued in practice.

At the third level, we deal with correlations and interactions between genotype and environment, either as a whole or involving particular subdivisions. Generally speaking, this level is dealt with more in theory than in practice, so far as personality is concerned, although a few exceptions will be mentioned later.

Finally, the fourth level looks at the actual mechanisms involved in the interplay of genotypes and environments that creates individuals and hence underlies and accounts for the variation among them. Today's behavioral genetics does not say a

Table 7.1   Four levels of behavioral genetic analysis.

1st level	Net overall genetic and environmental contributions
2nd level	Subdivision: additive and nonadditive genetic; familial and nonfamilial environmental
3rd level	Correlations and interactions between genotype and environment
4th level	Underlying causal mechanisms

great deal about such issues, except in an occasional speculative aside. Nevertheless, we may distinguish at least three general directions of approach to these problems, approaches that are reflected in the three other background papers to this group (see Buss, Schepank, and Pedersen, all this volume).

The first is a broad evolutionary or sociobiological approach, exemplified by Buss (this volume). Here he asks about the conditions that may have shaped a gene-based "human nature" over evolutionary time, and the further conditions that have permitted or encouraged the variations on that fundamental genetic framework that may underlie some of the phenotypic variation we observe.

A second approach uses genetic controls to obtain a sharper appraisal of environmental effects, as illustrated by Schepank (this volume). An example is the study of the structure of personality differences within MZ twin pairs (Loehlin and Nichols 1976): differences that must be environmental in origin because there is no genetic variation within MZ pairs—if one excludes rare genetic mutation or chromosomal accident.

A third approach to understanding the mechanisms of genetic and environmental influence is to focus on changes over the course of individual development, an approach represented by Pedersen (this volume). Clearly, if we find two traits that are equally heritable at, say, age 21, but one of them starts out much lower in heritability and increases to that level, whereas the other starts out much higher in heritability and decreases, we are receiving clues to differences between the mechanisms underlying the development of these two traits. An example of a developmental behavior genetic approach would be the finding of McCartney and colleagues (McCartney et al. 1990) of twins becoming less similar in personality over time, suggesting a shift of environmental influences from the familial to the nonfamilial with age.

Some day in the not too distant future, we may add to the three approaches mentioned a direct approach to the genetics of personality via the genome itself, based on the correlation of genetic sequences measured in individuals with personality differences among them. In the interim, however, or indeed continuing in parallel with molecular genetic strategies, broader approaches such as the three just mentioned should continue to be of value.

EMPIRICAL INVESTIGATIONS

Let us return to the first two levels of Table 7.1, where most of the available empirical work on the behavioral genetics of personality has taken place. As an illustration, Figure 7.1 shows correlations that have been obtained between pairs of individuals in various relationships in a number of studies of MZ and DZ twins, adoptive and biological families, the families of MZ twins, and so on. These correlations involve two traits that are ubiquitous as major dimensions in personality inventories, traits which I will refer to as *extraversion* and *emotional stability* (the latter is reversed in direction and called *neuroticism* by those in the Eysenckian tradition).

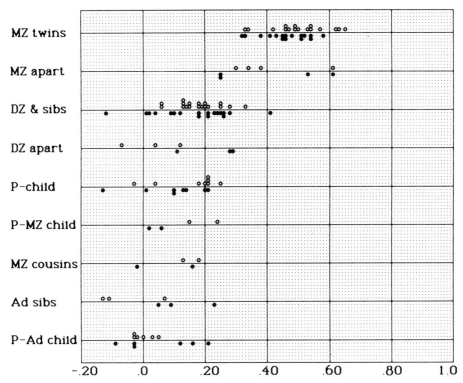

Figure 7.1 Observed correlations for extraversion (open circles) and emotional stability (solid circles) for various relationships in twin, adoption, twin-family, and separated twins studies. MZ = monozygotic, DZ = dizygotic, P = parent, MZ Ch = child of an MZ twin, MZ cousin = cousins via a pair of MZ twins, Ad = adoptive. Twin studies from Eaves et al. 1989; Floderus-Myrhed et al. 1980; Loehlin and Nichols 1976; Martin and Jardine 1986; Rose et al. 1988. Adoption studies from Eaves et al. 1989; Loehlin et al. 1985; Scarr et al. 1981. Twin-family studies from Loehlin 1986; Price et al. 1982. Separated twin studies from Langinvainio et al. 1984; Pedersen et al. 1988; Shields 1962; Tellegen et al. 1988.

Table 7.2 provides equations that constitute one simple model of the way genetic and environmental influences might account for the correlations shown in Figure 7.1. We can apply these equations to the data in either a piecemeal or an overall fashion. As a first step, however, we should ask if the observed sample-to-sample differences for given correlations are of a magnitude attributable to chance, making it reasonable to pool them. Most are, but a few are not. Three of the exceptions come from the twin studies that involve quite large samples, in several cases in the thousands of pairs. With large N's, sample-to-sample differences in sex, age, particular questionnaire scales used, etc., can lead to statistically significant chi-squares. I will go ahead and analyze the correlations jointly; however, it should be kept in mind that in all the

Table 7.2 Some relationships used in behavioral genetic studies of personality traits.

Relationship	Equation	Typical correlation E	S
MZ twins, reared together	$h^2 + i^2 + c^2$	0.51*	0.48*
MZ twins, reared apart	$h^2 + i^2$	0.39	0.39
DZ twins or sibs, reared together	$1/2\ h^2 + c^2$	0.18	0.20*
DZ twins, reared apart	$1/2\ h^2$	0.05	0.24
Parent and biological child	$1/2\ rh^2 + pc$	0.18	0.14
Parent and MZ twin's child	$1/2\ rh^2$	0.21	0.03
Cousins via MZ twins	$1/4\ h^2$	0.16	0.05
Adoptive siblings	c^2	−0.07	0.11
Parent and adopted child	pc	0.01	0.05*

Note: E = extraversion, S = emotional stability, MZ = monozygotic, DZ = dizygotic, h = additive genetic effect, i = nonadditive genetic effect (assumed due to multiple-gene epistasis), c = effect of shared family environment on children, r = correlation between child and adult genes, p = influence of parent's trait on children's environment. Adapted from Loehlin and Rowe (1992); typical correlations from model-fitting to data of Figure 7.1.
* $p < 0.05$ for chi-square test of homogeneity within set of correlations in Figure 7.1.

correlations, a certain amount of underlying heterogeneity of the kinds mentioned is most likely present. It should also be borne in mind that the measurement in these studies is of varying quality, and that all are based on questionnaires filled out by late adolescents or adults.

The "typical correlations" given in Table 7.2 represent the best maximum-likelihood fit to the observations for each relationship, by a model-fitting procedure that weights each correlation according to its sampling variance (reflecting the magnitude of the correlation and the size of sample upon which it is based). The heterogeneity tests came from these same analyses.

One possible strategy for estimating variance components would be to solve the Table 7.2 equations in various subsets, based on these typical correlations. Thus, for the trait of *extraversion* one could get estimates of c^2, the shared family environment, in several different ways: as 0.12 from the comparison of MZ twins reared together and apart, as 0.13 for the same comparison for DZ twins, or as −0.07 from adoptive siblings. Similarly, one could get estimates of the additive genetic variance, h^2, in several ways: as 0.10 from DZ twins reared apart, as 0.64 from cousins via MZ twins, and as > 0.42 from parent and MZ twin's child (because r, a correlation, has a maximum value of 1.0). These last estimates differ pretty widely, but several of the samples on which they are based are quite small.

If one is willing to assume that the various studies reflect the same underlying parameters, one can solve for single values of these parameters that best fit the entire data set of 62 correlations, and thereby get estimates which should be more stable.

Overall Model-fitting

The Table 7.2 equations were fit to the total set of 62 correlations shown in Figure 7.1 to yield approximate maximum-likelihood estimates of the parameters *h, c,* and so on, and an overall chi-square test of the goodness-of-fit of the whole model to the data (Loehlin and Rowe 1992). The chi-squares of 93.18 for *extraversion* and 167.27 for *emotional stability*, each with 57 degrees of freedom, were both statistically significant, meaning that the model did not fit the data. This lack of fit, however, appeared to be due to the heterogeneity of the same correlations across samples rather than to poorness-of-fit of the model across the different kinds of correlations. The summed heterogeneity chi-squares for *extraversion* and *emotional stability* were 84.90 and 162.99, respectively, each based on 53 *df*. Neither of the differences of these from the overall fits, differences of 8.28 and 4.28 respectively, is statistically significant (as a chi-square with 4 *df*). Further chi-square tests suggested that the parameter *p*, for the effect of parents' trait on the children's shared environment for that trait, could be set to zero without significantly worsening the fit of the model, and the correlation *r*, for the cross-age genetic correlation within an individual, could be set to 1.0. The shared environment parameter, *c*, could be dispensed with for *extraversion* but not for *emotional stability*. The best overall estimates of h^2, i^2, and c^2 were, respectively, 0.32, 0.17, and 0.02 for *extraversion*, and 0.27, 0.14, and 0.07 for *emotional stability*. The difference between the two traits appears to be real: solving for one common set of parameters for both traits led to a highly significant increase in chi-square (23.41 for 3 *df*, p < 0.001). However, in further analyses, the difference could not be determined to be due to any particular one of the three parameters.

A rather striking feature of these results, given the emphasis that many theories of personality have placed on early family experiences, is that c^2, the contribution of shared family environment to personality, is so modest. This is by now a well-replicated finding of behavioral genetic studies of personality (Plomin and Daniels 1987), and it suggests an important principle to personality theorists: If you wish to emphasize environmental factors in personality development, look for factors that can be expected to operate differentially among the children in a family.

Two additional points should be mentioned about these results. First, the fact that one model fits the data does not exclude the possibility that others might also. In particular, a model that hypothesizes a different degree of environmental resemblance for MZ twins than for the other groups, but excludes nonadditive genetic variance, will also fit these data (Loehlin 1992). Second, the models leave approximately half of the total trait variance unaccounted for: 49% in the case of *extraversion*, 52% for *emotional stability*. Probably about half of this half represents errors of measurement. The rest lumps together environmental effects unique to the individual, genotype–environment interactions, various nonlinearities, and so on.

Genotype–Environment Correlation and Interaction

The above examples illustrate behavioral genetic analysis at levels 1 and 2. What, if anything, can we say about level 3, correlation and interaction? First, we need to be clear about the sense in which we are using these terms. Genotype–environment (GE) correlation refers to correlations between genotypes and the environments to which they are exposed. GE interaction, on the other hand, refers to distinctive differences in outcomes for different GE combinations. Thus if extraverts are exposed to a more social environment than are introverts, this would be GE correlation. However, if extraverts are made more sociable by large families and introverts are made more sociable by small families, this would be GE interaction. If one imagines an analysis of variance with genotypes and environments as the two factors, GE correlation refers to nonorthogonality of the design and GE interaction to an interactive effect of the factors on the trait.

GE correlations have been classified by Plomin et al. (1977) into three types: passive, reactive, and active. A passive GE correlation is one that does not require the active participation of the individual concerned. If a child receives from emotionally unstable parents both genes and an environment that predispose him to emotional instability, this is an example of passive GE correlation. A reactive GE correlation is based on the response of others to the gene-influenced trait. An example of this would be if an emotionally unstable child is teased by playmates when he cries. An active GE correlation occurs when a gene-influenced trait results in behavior that selects particular environments. An example of a (negative) active GE correlation would be if an emotionally unstable child seeks quiet surroundings.

Passive GE correlation can be studied by comparing ordinary and adoptive families. In ordinary families, a single set of parents supplies both genes and environment so that passive GE correlation can occur. In adoptive families, the parents who supply the environment do not supply the genes so that, in the absence of the selective matching of children with adoptive families by adoption agencies, passive GE correlation should not be present. Selective placement, although it occurs for intelligence, seems empirically to be weak or nonexistent for most personality traits, so that adoption studies should allow a fairly good estimate of the contribution of passive GE correlation to personality. The existing evidence (reviewed in Loehlin 1992) suggests that passive GE correlation plays only a small role in the personality domain. (Not so for IQ, where the contribution of passive GE correlation appears to be appreciable.)

Active and reactive forms of GE correlation have been less studied. In principle, it should be easy: just measure the genotypes and environments, and see if they are correlated; if they are, then determine whether the correlation results primarily from the reactions of others or from the behavior of the person in question (there may, of course, also be mixed or ambiguous cases).

One example of an approach to GE active and reactive correlation for personality traits is provided by the work of Plomin and his colleagues in the Colorado Adoption Project (e.g., Plomin et al. 1988). They correlated personality traits of the birth parents of adopted children with aspects of the treatment of the child in the adoptive home, reasoning that (if selective placement could be ruled out) any such correlations would reflect a response of the adoptive environment to the gene-based characteristics of the child. They did not find much evidence of such correlations. However, this is not an extremely powerful test, in that additive heritabilities for personality traits are fairly low to begin with, measures are usually available for only one birth parent, and there is always room for doubt as to whether the most relevant aspects of the environment are being measured. Continued investigation of active and reactive GE correlation for personality is certainly warranted.

What about the measurement of the GE interaction? Again, if one could measure both genes and environments for individuals, it would be straightforward. Short of this, there are still some possibilities. For example, Bergeman and colleagues, in the Swedish Adoption/Twin Study of Aging (Bergeman et al. 1988), tried an interesting tactic. They used MZ twins who had been reared apart in childhood and predicted one twin's score on a personality trait by multiple regression from a measure of his childhood environment, the other twin's score on the trait, and the interaction of these. The idea was that the co-twin's phenotype provided a measure of the twins' joint genotype that was experimentally independent of the environment in which the first twin had been reared.

They did find some apparent interactions. One example: When an individual had a strong genetic predisposition toward extraversion, it did not matter whether the family environment was controlling or not—the individual was extraverted in either case. With a weaker predisposition toward extraversion, however, it did matter. These individuals were more extraverted in families exerting little control than in families exerting much control. Thus, a controlling environment affected the level of extraversion for some genotypes (weak disposition to extraversion) but not for others (strong disposition to extraversion). Complex relationships like these probably should be replicated before they are taken seriously as substantive findings; nevertheless, they illustrate the possibilities of the method.

Learning

It should be made clear that the presence of appreciable genetic variation for a given trait does not preclude the operation of learning processes in the development of that trait. It does, however, place some constraints on the arbitrariness of such learning and may say something about its locus (a familial or nonfamilial setting, at what point in the life span, etc.). Learning may also play an important role in the development of GE correlations, especially those of the active kind.

CONCLUSION

So where do we go in the future? A good deal of work is still needed at levels 1 and 2: very few personality traits have been studied as extensively as *extraversion* and *emotional stability*. Level 3 is interesting theoretically, yet relatively little has been done with it empirically. Effects at this level are complex, and if they tend to be modest in size, as some of the evidence suggests, very large samples may be required to deal with them convincingly. Finally, the most exciting developments should take place at level 4. Whether these will mostly be via the broad population-genetic approaches of the three strategies mentioned earlier, or via the narrow molecular-genetic approaches that seem to lie just around the corner, is an interesting question. Its answer should be more evident twenty years from now.

REFERENCES

Allport, G.W., and H.S. Odbert. 1936. Trait-names: A psycho-lexical study. *Psychol. Mon.* **47**:1–211.

Bergeman, C.S., R. Plomin, G.E. McClearn, N.L. Pedersen, and L.T. Friberg. 1988. Genotype–environment interaction in personality development: Identical twins reared apart. *Psychol. Aging* **3**:399–406.

Eaves, L.J., H.J. Eysenck, and N.G. Martin. 1989. Genes, Culture and Personality. London: Academic.

Floderus-Myrhed, B., N. Pedersen, and I. Rasmuson. 1980. Assessment of heritability for personality, based on a short-form of the Eysenck Personality Inventory. *Behav. Genet.* **10**:153–162.

John, O.P. 1990. The "Big Five" factor taxonomy: Dimensions of personality in the natural languages and in questionnaires. In: Handbook of Personality: Theory and Research, ed. L.A. Pervin, pp. 66–100. New York: Guilford.

Langinvainio, H., J. Kaprio, M. Koskenvuo, and J. Lönnqvist. 1984. Finnish twins reared apart. III: Personality factors. *Acta Genet. Med. Gem.* **33**:259–264.

Loehlin, J.C. 1986. Heredity, environment, and the Thurstone Temperament Schedule. *Behav. Genet.* **16**:61–73.

Loehlin, J.C. 1992. Genes and Environment in Personality Development. Newbury Park, CA: Sage Publications.

Loehlin, J.C., and R.C. Nichols. 1976. Heredity, Environment, and Personality. Austin: Univ. of Texas Press.

Loehlin, J.C., and D.C. Rowe. 1992. Genes, environment, and personality. In: Modern Personality Psychology, ed. G.V. Caprara and G.L. Van Heck. Hemel Hempstead, U.K.: Harvester Wheatsheaf Press.

Loehlin, J.C., L. Willerman, and J.M. Horn. 1985. Personality resemblances in adoptive families when the children are late-adolescent or adult. *J. Pers. Soc. Psychol.* **48**:376–392.

Martin, N., and R. Jardine. 1986. Eysenck's contributions to behaviour genetics. In: Hans Eysenck: Consensus and Controversy, ed. S. Modgil and C. Modgil, pp. 13–47. Philadelphia: Falmer.

McCartney, K., M.J. Harris, and F. Bernieri. 1990. Growing up and growing apart: A developmental meta-analysis of twin studies. *Psychol. Bull.* **107**:226–237.

Pedersen, N.L., R. Plomin, G.E. McClearn, and L. Friberg. 1988. Neuroticism, extraversion, and related traits in adult twins reared apart and reared together. *J. Pers. Soc. Psychol.* **55**:950–957.

Plomin, R., and D. Daniels. 1987. Why are children in the same family so different from one another? *Behav. Brain Sci.* **10**:1–16.

Plomin, R., J.C. DeFries, and D.W. Fulker. 1988. Nature and Nurture During Infancy and Early Childhood. Cambridge: Cambridge Univ. Press.

Plomin, R., J.C. DeFries, and J.C. Loehlin. 1977. Genotype–environment interaction and correlation in the analysis of human behavior. *Psychol. Bull.* **84**:309–322.

Price, R.A., S.G. Vandenberg, H. Iyer, and J.S. Williams. 1982. Components of variation in normal personality. *J. Pers. Soc. Psychol.* **43**:328–340.

Rose, R.J., M. Koskenvuo, J. Kaprio, S. Sarna, and H. Langinvainio. 1988. Shared genes, shared experiences, and similarity of personality. *J. Pers. Soc. Psychol.* **54**:161–171.

Scarr, S., P.L. Webber, R.A. Weinberg, and M.A. Wittig. 1981. Personality resemblance among adolescents and their parents in biologically related and adoptive families. *J. Pers. Soc. Psychol.* **40**: 885–898.

Shields, J. 1962. Monozygotic Twins: Brought Up Apart and Brought Up Together. London: Oxford Univ. Press.

Tellegen, A., D.T. Lykken, T.J. Bouchard, Jr., K.J. Wilcox, N.L. Segal, and S. Rich. 1988. Personality similarity in twins reared apart and together. *J. Pers. Soc. Psychol.* **54**:1031–1039.

8

Strategic Individual Differences: The Evolutionary Psychology of Selection, Evocation, and Manipulation

D.M. BUSS

Department of Psychology, University of Michigan,
Ann Arbor, MI 48109–1346, U.S.A.

ABSTRACT

The study of individual differences historically has been separated from the study of species-typical psychological mechanisms. This chapter proposes a model of strategic individual differences that integrates the two. The core of the model contains two essential propositions: (a) many species-typical psychological mechanisms have evolved to evaluate and act upon individually different input, both heritable and environmental in origin; (b) individually variable behavioral strategies described by the processes of selection, evocation, and manipulation represent strategic output designed to solve specific adaptive problems. The model is illustrated with an analysis of the evolutionary psychology of emotional stability and sexual jealousy. Discussion focuses on the proposition that evolved psychological mechanisms constitute the core mechanisms mediating heritable and environmentally variable determinants of behavior.

INTRODUCTION

Historical partitioning in psychology has separated the study of individual differences from the study of species-typical psychological functioning (cf. Cronbach 1957). These separate lines may be traced back within "biological" approaches to Francis Galton (1865), who focused on individual differences and their possible heritable basis, and to Charles Darwin (1859), who concentrated on species-typical characteristics. The modern forms of this split include the fields of behavioral genetics and personality psychology, which concentrate heavily on individual variation, and

Twins as a Tool of Behavioral Genetics
Edited by T.J. Bouchard, Jr. and P. Propping © 1993 John Wiley & Sons Ltd.

evolutionary psychology, social psychology, and cognitive psychology, which tend to focus on species-typical mechanisms. These divisions have had two unfortunate consequences: (a) conceptually isolating the study of individual differences from theories of basic psychological functioning, and (b) isolating theories of psychological functioning from understanding the important role played by individual differences.

In this chapter, I argue that a proper understanding of individual differences requires careful integration with evolved psychological mechanisms, most or all of which are species-typical (or sex-typical) in nature. This workshop focused on the key question: "What are the mechanisms mediating the genetic and environmental determinants of behavior?" I believe that evolutionary psychology provides a theoretically coherent answer: The key mechanisms are *evolved psychological mechanisms*, most of which have been "designed" to receive, as input, specific forms of internal and external information, to perform operations on the information, and to produce as output solutions to specific adaptive problems. I argue that important individual differences—including those that are stable over time and those that are more transient, those that originate in genetic differences, and those that originate in environmental differences—are most usefully understood within the context of our evolved psychological mechanisms.

WHY OUR BASIC PSYCHOLOGICAL MECHANISMS ARE LIKELY TO BE SPECIES-TYPICAL

Tooby and Cosmides (1990) articulate compelling arguments for why our basic psychological mechanisms are likely to be species-typical, shared by most or all humans. Essentially, all complex mechanisms require dozens, hundreds, or thousands of genes for their development. Sexual recombination, by shuffling genes with each new generation, makes it exceedingly unlikely that complex mechanisms could be maintained if genes coding for complex adaptations varied substantially between individuals. Selection and sexual recombination tend to impose relative uniformity in complex adaptive designs. This is readily apparent at the level of physiology and anatomy—all people have two eyes, a heart, a larynx, and a liver. Individuals can vary *quantitatively* in the strength of their heart or in the efficiency of their liver but do not vary in their possession of the basic physiological mechanisms themselves (except by unusual genetic or environmental accident). This suggests that individual differences, including heritable individual differences, are unlikely to represent differences in the presence or absence of complex adaptive mechanisms. Individual differences cannot be understood, apart from human nature mechanisms, any more than differences in the turning radius and stopping ability of cars can be understood, apart from the basic car-nature mechanisms, such as steering wheels and brakes.

THE NATURE OF HUMAN PSYCHOLOGICAL MECHANISMS

A long-standing dogma in this century's social science has been that the nature of humans is that they have no nature. Evidence has been accumulating over the past decade that makes this view empirically untenable (Brown 1991; Buss 1991). Conceptual analyses by scientists in artificial intelligence, psycholinguistics, cognitive psychology, and evolutionary psychology are showing why such a view is untenable theoretically, even in principle (Tooby and Cosmides 1992). Humans could not possibly perform the numerous, complex, situationally contingent tasks they do routinely without considerable intricate and domain-dedicated psychological machinery. These psychological mechanisms, coupled with the adaptive problems they were "designed" to solve, coupled with the social, cultural, ecological, and internal inputs that reliably activate them, provide a starting point for a description of human nature.

Although determining exactly which couplings are part of human nature must be determined empirically, possible candidates that have emerged empirically over the past several decades of research include childhood fears of loud noises, darkness, snakes, spiders, and strangers; characteristic emotions such as anger, envy, passion, and love; characteristic facial expressions such as happiness and disgust; competition for limited resources; competition for desirable mates; specific mate preferences; classification of kin; love of kin; preferential altruism directed towards kin; play; deceit; concepts of property; enduring reciprocal alliances or friendships; enduring mateships; temporary sexual relationships; retaliation and revenge for perceived personal violations; sanctions for crimes against the group or its members; rites of passage; concepts of self; concepts of intentions, beliefs, and desires as part of a theory of mind; status differentiation; status striving; prestige criteria; psychological pain upon loss of status or reputation; humor; gender terminology; division of labor by gender; sexual attraction; standards of sexual attractiveness; sexual jealousy; sexual modesty; tool making; tool use; tools for making tools; weapon making; weapon use; coalitions that use weapons for warfare; collective identities; cooking and fire use; and probably hundreds more (for an extended list of possibilities, see Brown 1991).

Since the cognitive revolution, psychologists have become increasingly aware of the necessity for understanding decision-making rules and other information processing devices inside people's heads. Although psychologists have largely jettisoned behaviorism's unworkable black box anti-mentalism, many have retained (perhaps inadvertently) the behavioristic assumption of equipotentiality: they assume that cognitive mechanisms are general-purpose and free of content-specialized procedures (Tooby and Cosmides 1992).

Evolutionary psychologists, in contrast to the dominant social science dogma, argue that evolved psychological mechanisms cannot be solely general-purpose, must be saturated with content, and must operate differently in response to contextual input signaling different adaptive problems. Just as the body contains a large number of specific and dedicated physiological mechanisms (taste buds, sweat glands, lungs, heart, kidneys, larynx, pituitary gland), so according to evolutionary psychologists the mind must contain a large number of specialized psychological mechanisms, each "designed" to solve a different adaptive problem. Because what constitutes a "successful solution" to adaptive problems differs across domains—criteria for successful food selection, for example, differ from criteria for successful mate selection—the requisite psychological solution mechanisms are likely to be special-purpose and domain-dedicated.

DETECTING EVOLVED PSYCHOLOGICAL MECHANISMS: IDENTIFYING THE KEY FOR THE LOCK

Psychological mechanisms are usefully regarded as evolved solutions to adaptive problems. Analogy to the human body is useful. We have sweat glands and shivering mechanisms that solve problems of thermal regulation; callous-producing mechanisms that solve the problem of repeated friction to the skin; taste preferences that solve the problem of what substances to ingest. Standards for inferring that these mechanisms are solutions to adaptive problems include economy, efficiency, complexity, precision, specialization, and reliability (Tooby and Cosmides 1992; Williams 1966). Mechanisms that solve adaptive problems are like keys that fit particular locks. The efficiency, detail, and complex structure of the key must mesh precisely with the inner "problem" posed by the lock.

Evolutionary analysis of psychological mechanisms proceeds in two directions: form-to-function and function-to-form (Tooby and Cosmides 1992). Imagine that one found a key but did not know which of the thousands of possible locks it might fit. Its size, shape, and details might suggest tentative hypotheses and rule out others. It may be too large to fit some locks, yet too small to fit others. The shape of its tines must have a corresponding mirror-image shape in the internal workings of the lock. Eventually, through an iterated process of hypothesis generation and empirical testing, we might eventually discover the exact lock that the key was designed to fit. The precision, reliability, and specialization with which a particular key fits a particular lock provide the researcher with reasonable standards for inferring that a particular key was *designed* to fit a particular lock.

Alternatively, one might identify a lock (adaptive problem) and then search for a key that might fit it. Here, the same standards would apply: precision, efficiency, complexity of design. The "bottom line" is whether the key one discovers (adaptive mechanism proposed) actually fits the lock (solves the adaptive problem with reasonable precision, efficiency, and reliability), and that alternative hypotheses about its origin (e.g., incidental byproduct of some other adaptation) and function (other adaptive problems the mechanism might solve, or other mechanisms capable of solving the adaptive problem) can be reasonably ruled out.

Evolutionary psychologists proceed in both directions, form-to-function and function-to-form. Sometimes a phenomenon or form is discovered—fever, fear of snakes, male sexual jealousy, mate preferences for "kindness"—and researchers generate and test hypotheses about its function. Often, there are competing functional theories about the same phenomenon, and these may be pitted against each other in critical empirical tests (for an example of an alternative evolutionary hypothesis for the female orgasm, see Buss 1990). This method is sometimes derided as telling "just-so stories," but it is an essential process of science. The discovery of three-degree black body radiation sent astronomers scrambling for cosmological theories or "stories" to explain it. The discovery of continental drift sent geologists scrambling for a theory such as plate tectonics, that could explain it. The power of a theory rests with its ability to explain known facts and to generate new predictions, which are then subjected to empirical test. Specific evolutionary psychological theories should be evaluated by these rigorous scientific standards. Some will pan out. Others will be jettisoned on conceptual or empirical grounds.

Evolutionary analysis provides psychologists with a powerful heuristic, guiding them to important domains of adaptive problems and guiding the development of hypotheses about adaptive mechanisms heretofore unobserved. Because fertilization and gestation occur internally within women, for example, an adaptive problem for ancestral men would have been ensuring confidence in their paternity. Men who were indifferent to this adaptive problem are less likely to be our ancestors. Identifying this adaptive problem has led evolutionary psychologists to search for adaptive solutions in psychological mechanisms such as mate preferences for chastity, fidelity, and faithfulness (Buss 1989; Buss and Schmitt 1993) and mechanisms involved in male sexual proprietariness such as sexual jealousy (Symons 1979; Daly et al. 1982; Buss et al. 1992). Function-to-form and form-to-function are both viable methods for discovering our evolved psychological mechanisms. These evolved psychological mechanisms form the foundation for the analysis of individual differences.

SEX DIFFERENCES AS ONE CLASS OF STRATEGIC INDIVIDUAL DIFFERENCES

To get from human nature psychological mechanisms to the analysis of individual differences, it is useful to go through an intermediate step: the analysis of sex differences, which may be regarded as one class of individual differences. Evolutionary psychology provides a unique meta-theory for predicting when we should and should not expect sex differences. Men and women are expected to differ only in the delimited domains where they have faced recurrently different adaptive problems (a) over human evolutionary history, (b) during their development, or (c) over different current environments inhabited. In domains where the sexes have faced the same adaptive problems, no sex differences are expected.

Men and women historically have faced many adaptive problems that are highly similar. Both sexes needed to maintain body temperature, so both sexes have sweat glands and shivering mechanisms. Repeated friction to certain areas of the skin was a "hostile force of nature" to both sexes in ancestral environments, so men and women have evolved callous-producing mechanisms. Both sexes needed to solve the adaptive problem of identifying a good cooperator for strategic confluence when seeking a long-term mate, and this may be one reason why both sexes value "kindness" in a partner so highly across all cultures whose partner preferences have been studied (Buss 1989).

In several domains, however, the sexes have faced different adaptive problems. For 99% of human evolutionary history, men faced the adaptive problem of hunting and women of gathering, possible selective reasons for men's greater upper body strength and spatial rotation ability and for women's greater spatial location memory ability (Silverman and Eals 1992). Internal female fertilization and gestation created the adaptive problem of uncertainty of paternity for men, but not uncertainty of maternity for women. Cryptic ovulation created the adaptive problem for men of knowing when a women was ovulating (Alexander and Noonan 1979). The dual male mating strategy of seeking both short-term sexual partners and long-term marriage partners created an adaptive problem for women of having to discern whether particular men saw them as temporary sex partners or as potential spouses (Buss and Schmitt 1993). Sex differences in mate preferences (Buss 1989), courting strategies (Buss 1988; Tooke and Camire 1991), jealousy (Buss et al. 1992), and sexual fantasies (Ellis and Symons 1990) correspond remarkably well to these sex-linked adaptive problems. Evolutionary psychology offers the promise of providing a coherent theory of strategic sexual differences as well as strategic sexual similarities.

STRATEGIC INDIVIDUAL DIFFERENCES CAUSED BY INDIVIDUALS CONFRONTING DIFFERENT "ENVIRONMENTALLY INDUCED" ADAPTIVE PROBLEMS

The construction workers who are laboring on the building next door have thick callouses on their hands. My academic colleagues down the hall do not. These individual differences in callous thickness are highly stable over time. At one level of analysis, the variance can be traced solely to variance in the reliably recurring experiences of the two groups. At another level of analysis, the existence of the species-typical callous-producing mechanism is a central and necessary element in the causal explanation of observed individual differences. *Just as men and women differ in the adaptive problems they confront, different individuals within each sex face different adaptive problems over time.* Some manifest individual differences are the strategic products of species-typical mechanisms responding to recurrently different adaptive problems across individuals.

In this example of callouses, the individual differences in skin-friction experiences are, in some sense, "environmental." If my academic colleagues were to trade places with the construction workers, then the manifest individual differences would reverse. Nonetheless, we cannot rule out the genotype–environmental correlation processes proposed by Plomin et al. (1977) and Scarr and McCartney (1983). Some individuals, because of heritable skills, interests, or proclivities, may preferentially select academic work or construction work as occupations. These selections, in turn, may create repeated exposure to friction-free versus friction-prevalent environments, which then differentially activate the species-typical callous-producing mechanism. I return to the genotype–environment correlation issue below.

Thus far there are three central points to my argument:

1. Stable manifest individual differences can be caused by differences in the *recurrent adaptive problems* to which different individuals are exposed.
2. The complex *species-typical mechanisms* are necessary and central ingredients in the causal explanation of individual differences because without them, the observed individual differences could not occur.
3. The manifest individual differences are *strategic outcomes* of recurrently different input into species-typical mechanisms.

There are undoubtedly many recurrent environmental individual differences of precisely this sort. Firstborn children probably face recurrently different adaptive problems compared with later-born children. These problems apparently trigger, in a firstborn, greater identification with the status quo, the parents, and the established scientific theories, and in later-born greater rebellion and identification with revolutionary scientific theories (Sulloway, pers. comm). Later-born children apparently have less to gain by identifying with a niche that is already occupied by an older sibling.

Individuals who grow up in environments where resources are unpredictable, such as inner city ghettos, may adopt a more impulsive personality style where it would be adaptively foolish to delay gratification (Buss 1990). In contrast, those growing up in middle class suburbs where resources and future prospects are more predictable may adopt a personality strategy involving greater delay of gratification. The resulting individual differences represent strategic solutions to the different adaptive problems encountered. Recurrently different environmental input into species-typical mechanisms can produce stable strategically patterned individual differences.

STRATEGIC INDIVIDUAL DIFFERENCES CAUSED BY INDIVIDUALS CONFRONTING DIFFERENT "HERITABLY INDUCED" ADAPTIVE PROBLEMS

Recurrently different input into species-typical psychological mechanisms, of course, may come from heritable individual differences, whatever their ultimate origin (i.e., whether they originated from selection for alternative genetically based strategies, frequency-dependent selection, genetic noise, pathogen-driven selection for genetic uniqueness, or assortative mating). Individuals with an ectomorphic body type, for example, confront different adaptive problems from those who are from mesomorphic. Ectomorphs may risk being at the receiving end of greater aggression than their more muscular peers, an adaptive problem that typically must be solved by means other than physical aggression. Genetic differences, in other words, pose different adaptive problems for different individuals.

In addition to facing different adaptive problems, some individuals experience *greater success* at pursuing certain strategies rather than others: "Selection operates through the achievement of adaptive goal states and any feature of the world—either of the environment, *or of one's own individual characteristics*—that influences the achievement of the relevant goal state may be assessed by an adaptively designed system" (Tooby and Cosmides 1990, p. 59; emphasis added). Individuals who are mesomorphic, for example, typically will experience far greater success at enacting an aggressive strategy than individuals who are ectomorphic (Tooby and Cosmides [1990] call this phenomenon "reactive heritability").

Consider individual differences in physical attractiveness. There is evidence that physically attractive men are better able to pursue successfully a "short-term" mating strategy involving many sexual partners (Gangestad and Simpson 1990). Physically attractive women are better able to pursue a long-term strategy of seeking and actually obtaining higher-status higher-income marriage partners (Taylor and Glenn 1976). Heritable differences in physical attractiveness affect the success of pursuing different mating strategies. The *manifest* strategy differences are in some sense "heritable," but only indirectly and reactively. Relative physical attractiveness functions as "input" into species-typical or sex-typical psychological mechanisms, which then canalize the strategies of different individuals in different directions.

Heritable dimensions of individuals—such as differences in body type, keenness of vision, oratory skills, physical attractiveness, and spatial ability—provide important input into species-typical mechanisms. These individually different inputs tell the organism about the adaptive problem it is facing and the strategic solutions likely to be successful. The resulting product consists of *strategic individual differences* that are stable over time. The observed strategic differences are *correlated* with genetic variance but cannot be understood apart from the central role played by our species-typical psychological mechanisms that were "designed" to receive input—both environmentally and heritably based—about the adaptive problems confronted and the strategic solutions likely to be successful.

A THEORY OF STRATEGIC INDIVIDUAL DIFFERENCES: GENOTYPE–ENVIRONMENT PROCESSES AS BEHAVIORAL STRATEGIES

Plomin et al. (1977) and Scarr and McCartney (1983) have proposed useful frameworks for describing the different ways in which correlations and interactions between genotypes and environments can be achieved. They describe an *active* process, whereby individuals with certain genotypes choose or select some environments over others (e.g., a verbally gifted person may expose herself to more books); an *evocative* (or reactive) process, whereby different environments are elicited by individuals with different genotypes (physically attractive people may evoke more warmth); and a *passive* process, whereby individuals experience different environments that happen to correlate with their genotypes, independent of any action on their part or reaction from others (e.g., verbally gifted children may be exposed to more books because their parents happen to read more).

Building on the work of Plomin et al. (1977) and Scarr and McCartney (1983), I broadened the concept (Buss 1984a) of genotype–environment (GE) correlation to person–environment (PE) correlation (for more recent treatments, see Plomin and Bergeman 1991; Scarr 1992). This expansion serves two purposes. First, it highlights the fact that the causal processes subsumed by the framework can occur regardless of the developmental origins of the individual differences. The linkage between physical beauty and interpersonal warmth received, for example, can occur whether individual differences in attractiveness are caused by heritable differences or environmental differences (e.g., facial injuries, scars, or cosmetic plastic surgery). Second, the term "active" is viewed as too broad to describe the causal processes involved. I have proposed (Buss 1984b) that "active" be partitioned into (a) *selection*, referring to the preferential choice of some environments over others, and (b) *manipulation*, referring to altering, influencing, or changing environments after they have been selected.

These GE and PE models provide useful frameworks for describing important causal routes by which different genotypes become linked with different environments. By themselves, however, they offer no theory about *why* such links are established. Why, for example, do people respond to attractive individuals with greater warmth? Why do the verbally gifted expose themselves to more word stimulation? Why do mesomorphs more often surround themselves with delinquent peers more than ectomorphs do? The proposal here is that the content-free processes described by Plomin et al. (1977), Scarr and McCartney (1983), and Buss (1984b) can be usefully conjoined with (a) specific theories of species-typical psychological mechanisms, (b) specific theories of the different adaptive problems confronted by different individuals, and (c) specific theories about the differential success of strategies, depending upon individually different abilities and attributes.

I believe that the theoretical and empirical work subsumed by this general framework will have to be explored domain by domain and adaptive problem by adaptive problem (Tooby and Cosmides 1992). A theory about how individuals differing in physical attractiveness encounter different adaptive problems, evoke different reactions in others, select different social niches to inhabit, and have different success with the enactment of some strategies, may tell us little about the strategic consequences of individual variation in body type, oratory skills, or spatial ability. Ultimately, of course, it will be useful to integrate domain-specific theories into a more general theory of strategic individual differences.

In this brief paper, all I can do is provide one illustration of how this form of analysis might proceed. The example I use is jealousy and the broader personality construct of emotional instability—psychological phenomena that have been treated variously as individual difference personality variables, as reactive situationally induced states, as sex-differentiated states, and as signs of pathology or immaturity. How might the evolutionary psychological analysis of individual strategic differences proceed?

THE EVOLUTIONARY PSYCHOLOGY OF JEALOUSY: STRATEGIC SEX DIFFERENCES

Jealousy is neither a peripheral nor trivial emotion, for it is experienced in all known cultures and is the leading cause of spousal battering and homicide worldwide (Daly and Wilson 1988). Why do humans experience jealousy? Do the sexes differ in the events that elicit jealousy? What contexts activate jealousy? And how can individual differences in jealousy *within sex* be accounted for?

Jealousy is a cognitive-emotional-motivational complex that is activated by threat to a valued relationship. It is considered "sexual jealousy" if the relevant relationship is a sexual one, but there are types of jealousy that do not involve sexual threat. Jealousy is often activated by cues to the apparent loss of key resources provided by a relationship: cues such as eye contact between one's partner and a rival, decreased sexual interest on the part of one's partner, and an increase in a partner's flirting with same-sex competitors. Jealousy channels attention, calls up relevant memories, and activates strategic cognitions. It may motivate actions designed to reduce or eliminate the threat, retain the valued relationship, and hence retain the valued resources it provides.

Because both men and women over evolutionary history have been damaged by relationship loss, both sexes have faced adaptive problems to which jealousy may have evolved as one solution. Several evolutionary psychologists have predicted that the sexes will differ in the weighting given to events that activate jealousy (Daly et al. 1982; Symons 1979). Because fertilization and gestation occur internally within women and not men, over evolutionary history men have faced an adaptive problem not shared by women—paternity uncertainty. The reproductive threat to a man comes from the possibility of *sexual* infidelity by his partner.

In species such as ours, a woman's certainty in genetic parenthood would not have been compromised if her partner had sex with other women. Women, however, may have risked the loss of their partner's time, attention, commitment, protection, investment, and resources. This would come as a double blow, because her loss would also be an intrasexual competitor's gain if the resources were diverted from her and her children toward another woman and unrelated children. For these reasons, evolutionary psychologists have predicted that the inputs that activate jealousy for men will be biased toward cues that relate to the sex act per se, whereas for women they will be more biased by cues to the loss of commitment and investment from a man.

Consider this question: What would upset or distress you more: (a) imagining your mate having sexual intercourse with someone else, or (b) imagining your mate forming a deep emotional attachment to someone else? In a series of studies, we found that the overwhelming majority (85%) of women to whom this dilemma was posed found emotional infidelity to be more distressing; the majority of men (60%) reported that sexual infidelity would be more distressing (Buss et al. 1992). These sex differences were also observed in physiological arousal in response to imagining the two different scenarios. In measures of heart rate, electrodermal activity, and electromyographically recorded frowning, men showed greater physiological arousal to imagined sexual infidelity than to emotional infidelity. Women, by contrast, tended to become more physiologically aroused by imagined emotional infidelity than sexual infidelity (Buss et al. 1992). These results support the hypothesis that men's and women's psychological and physiological mechanisms are tailored to differences in adaptive problems.

STRATEGIC INDIVIDUAL DIFFERENCES IN JEALOUSY: SELECTION, EVOCATION, AND MANIPULATION

If sex differences in jealousy are strategically patterned, are individual differences in jealousy within sex also strategically patterned? Given the space limitations, I restrict my attention to strategic individual differences among men and focus on a variable of key theoretical importance in evolutionary psychology—the concept of "mate value" (Buss and Barnes 1986; Symons 1987). The hypotheses are presented in abbreviated form with minimal rationale to conserve space.

Hypothesis 1: Men who are lower in "mate value" (Symons 1987) will become more upset by imagining partner infidelity due to their greater difficulty retaining their mates, their greater difficulty replacing a lost mate, and hence the greater activation of their mate retention mechanisms (cf. Buss 1988).

Hypothesis 2: Men lower in mate value, because of the greater activation of their mate retention mechanisms, will *select* different environments to inhabit, such as those with less chance of encountering intrasexual competition.

Hypothesis 3: The partners of men lower in mate value will display more frequent acts of mate defection and cues to potential infidelity.

Hypothesis 4: Men lower in mate value will *evoke* complaints from their spouses that they are too possessive, jealous, and dependent.

Hypothesis 5: Men lower in mate value will *manipulate* their spouses by displaying a variety of mate retention tactics.

To test these hypotheses, we need a good measure of mate value. Unfortunately, given the recency of the concept and the large number of components that contribute to it, no good measures have been developed. Thus, we are left with markers of mate value that are necessarily imperfect and incomplete. One way to bootstrap a marker of mate value is to identify the characteristics that are consensually the most highly valued in a potential mate. In a study of 10,047 subjects in 37 cultures, I found that

emotional stability was consistently rated as among the three most desirable characteristics out of 18 listed in a mate (the other two highly rated characteristics were
"mutual attraction—love," which represents a state rather than a trait, and "dependable
character"). On a rating scale with "0" being irrelevant and "3" being "indispensable
in a mate," women worldwide rated emotional stability as 2.68 and men rated it 2.47.
Clearly, both sexes view emotional stability as extremely important in a marriage
partner. Furthermore, women rated emotional stability as more important in a partner
than men did in the worldwide sample ($p < 0.0001$), a difference that was statistically
significant in 23 out of the 37 individual cultures. This suggests that emotional
stability is a better marker of men's mate value than of women's mate value (Buss
1989b). Perhaps not coincidentally, emotional stability is also widely regarded as one
of the most important dimensions of individual differences by personality psychologists (Goldberg 1981; Norman 1963; John 1990; McCrae 1990).

If we use individual differences in emotional stability as a marker of men's mate
value, admittedly incomplete and imperfect, how do they relate to jealousy and to
selections, evocations, and manipulations? Measures of emotional stability correlated
negatively with self-reports of jealousy to imagined sexual infidelity ($r = -0.27$, $p <
0.05$) and even more negatively with self-reports of imagined emotional infidelity ($r
= -0.41$, $p < 0.01$) (Buss and Larsen, unpublished). Furthermore, men low on
emotional stability exhibit greater heart-rate elevation in response to imagined sexual
infidelity (-0.32, $p < 0.05$) and to imagined emotional infidelity (-0.38, $p < 0.05$) (Buss
and Larsen, unpublished). These findings provide at least circumstantial support for
the hypothesis that those low in mate value, as indexed by low emotional stability,
display markedly activated mate retention mechanisms such as jealousy.

Selection

In a separate study of newlywed couples, we assessed emotional stability with three
data sources: self-report, spouse report, and independent interviewer reports (Buss
1991a). We correlated a composite index of emotional stability with reported tactics
of mate retention (Buss 1988). Men low on emotional stability tend to *select* environments that do not contain intrasexual competitors: he did not take her to the party
where other males would be present; he refused to introduce her to same-sex friends;
and he took her away from the gathering where other males were present.

Evocation

Men low on emotional stability also apparently occupy an environment in which their
partners flirt with others and threaten infidelity: he became angry when she flirted too
much; he threatened to break up if she ever cheated on him; he yelled at her after she
showed an interest in other men; he said that he would never talk to her after she
showed an interest in other men; he hit her when he caught her flirting with someone
else; he became jealous when she went out without him; he cried when she said she

might go out with someone else; he made her feel guilty about talking to other men; he told her he would "die" if she ever left him. A related finding is that men low on emotional stability evoke considerable complaints in their spouses that center on their being possessive-dependent-jealous: he was too possessive of me; he acted too dependent on me; he demanded too much attention; he demanded too much of my time; he acted jealously (Buss 1991a). Wives of men low on emotional stability apparently feel hemmed in, constricted, and excessively cloistered by their husbands.

Manipulation

Do men low on emotional stability manipulate their partners in predictable ways? We found that men low on emotional stability tended to report using tactics of mate retention marked by *abuse* (he slapped me; he spit on me; he hit me; he called me nasty names; he verbally abused me), *derogation of competitors* (he cut down the appearance of other males; he started a bad rumor about another male; he cut down the other guy's strength; he pointed out to her the other guy's flaws; he told her that the other guy she was interested in has slept with nearly everyone), *derogation of mate to competitors* (he told other guys terrible things about her so that they would not like her; he told other guys that she might have a social disease), and *intrasexual threats* (he yelled at the other guys who looked at her; he stared coldly at the other guy who was looking at her; he threatened to hit the guy who was making moves on her; he confronted the guy who had made a pass at her).

Because these findings are correlational, obviously no firm conclusions can be drawn about causality. Empirically, however, we know the following: (a) women worldwide place great value on emotional stability in a mate; (b) men low on emotional stability become more jealous psychologically and physiologically to imagined sexual and especially emotional infidelity; and (c) men low on emotional stability preferentially select environments devoid of intrasexual rivals, manipulate their mates through numerous acts of mate retention, and occupy a social environment marked by wives who give cues to involvement with other men and who complain about their partner's possessiveness. These PE correlations are established, whatever the eventual causal explanation.

One causal possibility is that initial hypotheses about emotional instability signaling low mate value are correct. Emotional instability in men is a powerful cue to low mate value, and so such men face the adaptive problem of increased odds of mate defection. As a consequence, their jealousy mechanisms are activated; they select and manipulate environments to minimize partner defection; and their activated mate retention mechanisms evoke complaints from their mates about possessiveness, dependency, and jealousy.

An alternative causal possibility is that men who occupy a recurrent environment marked by signs of mate defection (e.g., partner flirting with other men) become emotionally unstable as a result, and hence become jealous and select, evoke, and manipulate their social environments as a consequence. In both of these alternative causal accounts, different individuals confront recurrently different adaptive problems, either because of individual differences in mate value or because their partners give off different frequencies of cues to defection. The PE links reveal strategic individual differences, regardless of which causal account eventually turns out to be correct.

The causal relationships illustrated by this example are shown in Figure 8.1. Individuals differ in both psychological propensities and in the social environments they recurrently inhabit. These differences provide input into species-typical psychological mechanisms such as assessments of one's own mate value and sexual jealousy. The output consists of behavioral strategies in which individuals differentially select, evoke, and manipulate their environments. The product of these behavioral strategies is altered states: changed psychological propensities (e.g., a shift in emotional stability) and changed recurrent environments experienced (e.g., fewer intrasexual rivals). These altered products then comprise new input into psychological mechanisms as the processes are iterated over time.

CONCLUSIONS

Taken together over these half a dozen studies, the evidence shows promise for an analysis of strategic individual differences that conjoins (a) species-typical psychological mechanisms as solutions to adaptive problems (e.g., mechanisms of psychological jealousy adapted to the problems of infidelity and relationship loss); (b) individually different input into these mechanisms (e.g., differences in mate value or differences in degree to which mates recurrently show cues to defection); and (c) strategic consequences observed in the ways in which individuals *select* their social environments, *evoke* reactions in others, and *manipulate* others in predictable ways. We have the beginnings of a model of strategic individual differences in one domain and one class of adaptive problems.

Five general conclusions follow from this analysis of strategic individual differences:

1. *The mechanisms posited by PE and GE models can be placed in functional context.* It is important to know *why* people select the environments they do, *why* people evoke predictable reactions from others, *why* people manipulate their environments in predictable ways.
2. *The functional contexts in which these mechanisms become articulated will differ across different adaptive domains.* Different PE links will be forged in the domain of mate retention, for example, than in the domain of coalition-building or hierarchy negotiation.
3. *Human nature psychological mechanisms will be central to explanations of individual differences.* Without these mechanisms, observed individual differences could not occur, just as without callous-producing mechanisms, individual differences in manifest callous-thickness could not occur.

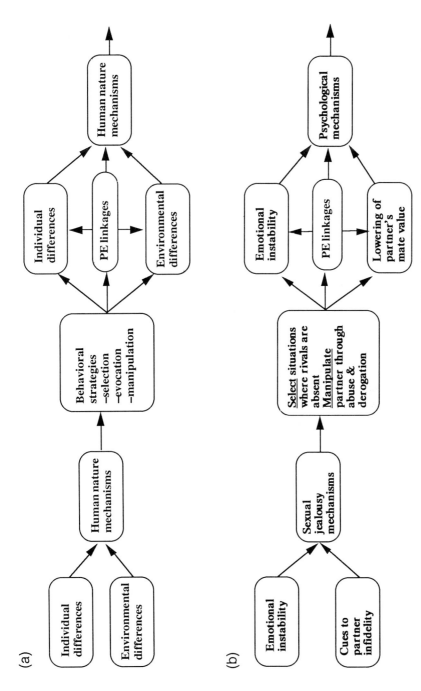

Figure 8.1 (a) A functional model of strategic individual differences. (b) An illustration with the adaptive problem of mate retention.

4. *Human nature mechanisms often cannot be fully understood without identifying the important individual differences that provide input.* Many evolved psychological mechanisms were *designed* to take individual differences as input, both heritably and environmentally generated. Therefore, understanding these differences is central to understanding the mechanisms.

5. *Content-free models of GE processes emerging from behavioral genetics can be usefully integrated with content-saturated models from evolutionary psychology to provide coherent theories of strategic individual differences.* Perhaps through this integration, we can start bridging the traditions that historically have isolated the study of individual differences from the study of human nature since the days of Galton and Darwin.

REFERENCES

Alexander, R.D., and K. Noonan. 1979. Concealment of ovulation, parental care, and human social evolution. In: Evolutionary Biology and Human Social Behavior, ed. N.A. Chagnon and W. Irons, pp. 402–435. North Scituate, MA: Duxbury Press.

Brown, D.E. 1991. Human Universals. Philadelphia, PA: Temple Univ. Press.

Buss, D.M. 1984a. Evolutionary biology and personality psychology: Toward a conception of human nature and individual differences. *Am. Psychol.* **39**:1135–1147.

Buss, D.M. 1984b. Toward a psychology of person–environment (PE) correlation: The role of spouse selection. *J. Pers. Soc. Psychol.* **47**:361–377.

Buss, D.M. 1988. From vigilance to violence: Mate retention tactics. *Ethol. Soc.* **9**:291–317.

Buss, D.M. 1989. Sex differences in mate preferences: Evolutionary hypotheses tested in 37 cultures. *Behav. Brain Sci.* **12**:1–49.

Buss, D.M. 1990. Evolutionary social psychology: Prospects and pitfalls. *Motiv. Emot.* **14**:265–286.

Buss, D.M. 1991a. Conflict in married couples: Personality predictors of anger and upset. *J. Pers.* **59**:663–688.

Buss, D.M. 1991b. Evolutionary personality psychology. *Ann. Rev. Psychol.* **42**:459–491.

Buss, D.M., and M.F. Barnes. 1986. Preferences in human mate selection. *J. Pers. Soc. Psychol.* **50**:559–570.

Buss, D.M., R. Larsen, D. Westen, and J. Semmelroth. 1992. Sex differences in jealousy: Evolution, physiology, and psychology. *Psychol. Sci.* **3**:251–255.

Buss, D.M., and D.Schmitt. 1993. Sexual strategies theory: The evolutionary psychology of human mating. *Psychol. Rev.* **100**:204–232.

Cronbach, L.J. 1957. The two disciplines of scientific psychology. *Am. Psychol.* **12**:671–684.

Daly, M., and M. Wilson. 1988. Homicide. New York: Aldine.

Daly, M., M. Wilson, and S.J. Weghorst. 1982. Male sexual jealousy. *Ethol. Soc.* **3**:11–27.

Darwin, C. 1859. The Origin of Species. London: Murray.

Ellis, B.J., and D. Symons. 1990. Sex differences in sexual fantasy: An evolutionary psychological approach. *J. Sex Res.* **27**:527–556.

Galton, F. 1865. Hereditary talents and character. *Macmillan's* **12**:157–166, 318–327.

Gangestad, S.W., and J.A. Simpson. 1990. Toward an evolutionary history of female sociosexual variation. *J. Pers.* **58**: 69–96.

Goldberg, L.R. 1981. Language and individual differences: The search for universals in personality lexicons. In: Review of Personality and Social Psychology, ed. L. Wheeler, pp. 141–165. Beverly Hills: Sage.

John, O.P. 1990. Toward a taxonomy of personality descriptors. In: Personality Psychology: Recent Trends and Emerging Directions, ed. D.M. Buss and N. Cantor, pp. 261–274. New York: Springer.

McCrae, R.R. 1990. Why I advocate the five-factor model: Joint factor analyses of the NEO-PI with other instruments. In: Personality Psychology: Recent Trends and Emerging Directions, ed. D.M. Buss and N. Cantor, pp. 237–245. New York: Springer.

Norman, W.T. 1963. Toward an adequate taxonomy of personality attributes: Replicated factor structure in peer nomination personality ratings. *J. Abn. Pers. Psych.* **66**: 574–583.

Plomin, R., and C.S. Bergeman. 1991. The nature of nurture: Genetic influence on "environmental" measures. *Behav. Brain Sci.* **14**:373–427.

Plomin, R., J.C. DeFries, and J. Loehlin. 1977. Genotype–environment interaction and correlation in the analysis of human behavior. *Psychol. Bull.* **84**:309–322.

Scarr, S. 1992. Developmental theories in the 1990's: Developmental and individual differences. *Child Devel.* **63**:1–19.

Scarr, S., and K. McCartney. 1983. How people make their own environment: A theory of genotype-environment effects. *Child Devel.* **54**:424–435.

Silverman, I., and M. Eals. 1992. Sex differences in spatial abilities: Evolutionary theory and data. In: The Adapted Mind, ed. J. Barkow, L. Cosmides, and J. Tooby. New York: Oxford Univ. Press.

Symons, D. 1979. The Evolution of Human Sexuality. New York: Oxford Univ. Press.

Symons, D. 1987. An evolutionary approach: Can Darwin's view of life shed light on human sexuality. In: Theories of Human Sexuality, ed. J.H. Geer and W. O'Donahue. New York: Plenum.

Taylor, P.A., and N.D. Glenn. 1976. The utility of education and attractiveness for females' status attainment through marriage. *Am. Sociol. Rev.* **41**:484–498.

Tooby, J., and L. Cosmides. 1990. On the universality of human nature and the uniqueness of the individual: The role of genetics and adaptation. *J. Pers.* **58**:17–68.

Tooby, J., and L. Cosmides. 1992. Psychological foundations of culture. In: The Adapted Mind: Evolutionary Psychology and the Generation of Culture, ed. J. Barkow, L. Cosmides, and J. Tooby. New York: Oxford Univ. Press.

Tooke, W., and L. Camire. 1991. Patterns of deception in intersexual and intrasexual mating strategies. *Ethol. Soc.* **12**:345–364.

Williams, G.C. 1966. Adaptation and Natural Selection. Princeton: Princeton Univ. Press.

9

Why Are Monozygotic Twins So Different in Personality?

H. SCHEPANK

Central Institute of Mental Health, Psychosomatic Hospital, Quadrat J5,
P.O. Box 122 120, 68072 Mannheim 1, F.R. Germany

INTRODUCTION

The topic of this paper, designed by the program advisory committee, is provocatively stimulating since most monozygotic (MZ) twins are actually *not* very different in their personalities. When they do exhibit differences, it is often difficult to pinpoint the causes—a statement with which most twin research experts would agree.

For the purposes of this chapter, I will use the term "personality" in the *strict* sense of the word. I will not address cognitive factors of personality, intelligence, etc. since they are the focus of another discussion group. I will address, however, normal features of personality as well as personality disorders and some psychopathological phenomena.

My professional experience includes twin research (Schepank 1974, 1975; Heigl-Evers and Schepank 1980/1981; Muhs et al. 1990; Muhs and Schepank 1991, 1993; Schepank and Muhs 1993) and practice as a physician (M.D.) in the area of psychosomatics and psychotherapy. My point of view has been decisively shaped by my education and practice as well as by the experience I have gained as director of a university clinic (a 48-bed, in- and outpatient psychotherapeutic unit) and my scientific work, which has focused on epidemiology (Schepank 1987), etiopathogenesis of the so-called psychogenic disorders (Schepank 1983, 1991), transcultural psychosomatic, and psychotherapeutic health-care research.

To rephrase the focus of this chapter more precisely, I believe the issue to be: if genetically identical monozygotic twins differ in personality, what are the reasons?

The classic twin method cannot contribute much in the way of answering this question. Instead, it investigates (on a phenomenological level, i.e., independent of metabolic processes, chromosomes, or genes) the question: to what extent are *genetic* factors included in the manifestation of a feature? This is verified by the significant

Twins as a Tool of Behavioral Genetics
Edited by T.J. Bouchard, Jr. and P. Propping © 1993 John Wiley & Sons Ltd.

difference of similarities/concordance rates of MZ twins compared to the intrapair similarity of dizygotic (DZ) twins. The topic of the *degree* of heritability will be discussed when it is applicable. To the extent that MZ twins are not identical, an environmental factor must be involved; however, the classic twin method does not provide any information regarding the nature of such factors.

All classic twin methods maintain, as a constant, one of two possible variables: the hereditary or the environmental variable. The respective intrapair environment is presumed equally constant for each pair whereas the hereditary variable fluctuates: 100% equal in MZ twins and 50% identical in dizygotic twins. If one focuses exclusively on MZ twins, the hereditary factor is held constant and only the environment varies.

Ignoring genealogical family investigations and adoption studies (neither of which utilize twins), five different techniques using MZ twins may yield further information on the effect of environmental factors:

1. An experimental design, the *co-twin control method* (Gesell 1953), is useful for short-term experiments in developmental psychology and learning experiments. However, for long-term experiments, in the case of personality (normal or pathological) development, it cannot be allowed for ethical reasons.
2. The co-twin control method utilizes MZ twins raised in different environments and has been employed, e.g., by Newman et al. (1937) and the Minneapolis group (Bouchard et al. 1990). A subgroup of this technique is the comparison of separately grown-up vs. jointly grown-up MZ pairs (nonshared versus shared environment). There are disadvantages to this method, namely, MZ twins raised in nonshared environments are extremely rare. In addition, the environmental constellations of these twin pairs differ fundamentally from normal pairs: an average bio/psycho/socially healthy family is not going to give away its twins and let them grow up separately. Institutional care and foster care are very exceptional situations for children. In terms of personality development, behavior, or psychoneurotic disorders, twin pairs from nonshared environments come from high-risk situations and therefore represent a highly selected sample group as compared to children/twins who have grown up in their original family environment. For this reason, in spite of elaborate mathematical-statistical methods, this approach is hardly suitable.
3. The *one-twin-trauma method* as applied to twins (e.g., by Dencker 1958) analyzes an MZ partner known to have had a severe infantile (somatic) trauma (brain injury).
4. In the one-twin trauma method, MZ twins report on (nontraumatic) systematic differences in their experiences while growing up (Baker and Daniels 1991).

Consistent throughout these first four methods is the fact that the *independent environmental variable*, i.e., the possible cause, is known and definable. One must search, therefore, for the dependent variables, i.e., the possible sequels in the shape of later potential differences that result between MZ twins.

5. The last method, known as the *discordance analysis of MZ twins* (Propping 1989; Vogel and Motulsky 1979), contrasts methods 1–4 inasmuch as it does not start with a known *initial state*, i.e., with a known independent variable, and then investigate its sequels; rather it begins with the phenomenon of a *discordant final state* in MZ twins. Furthermore, concerning a given feature in question (e.g., personality), it attempts to elucidate retrospectively those factors which are discordant in adults and which contributed to this very different yet final state.

The basic consideration for this method is simple: due to their identical hereditary disposition, all characteristics in which MZ twins differ phenotypically must be caused by environmental influences. These influences need to be discovered.

The disadvantage to this method is that it is based on a meticulously kept retrospective history (obtained from the probands themselves, their co-twins, parents, or hospital data). Statistical calculations and group comparisons, by definition, are hardly possible because of the small number of cases. Therefore, results are not absolute. Nevertheless, discordance analysis is capable of formulating important heuristic hypotheses and discovering potentially relevant environmental influences.

Premises upon which the application of the discordance analysis can be carried out include:

1. There must be expert *zygosity diagnostics.*
2. Both twins must still be alive and available for examination.
3. There has to be a real discordance, as great as possible, concerning a certain feature (or, in our case, globally concerning "personality").
4. The risk age for the manifestation of the characteristic in question must be reached, e.g., mature personality. (I suggest a mean age of at least 30 to 40 years, as the investigation involving 17-year-old college students (Loehlin and Nichols 1976) or even 13-year-old children (Eysenck 1944) does not satisfy these requirements.
5. Insofar as adolescents/children are examined, the source of the information gathered, e.g., from parents, must be considered in the assessment.
6. Finally, long-term observation of individuals is desirable in order to distinguish trait and state characteristics.

SAMPLING

Our investigation was based on 100 twin pairs (Heigl-Evers and Schepank 1980/1981). Fifty pairs were from an inpatient psychotherapeutic institution (State

Hospital of Lower Saxony at Tiefenbrunn near Göttingen; Heigl-Evers); the other 50 pairs (Schepank 1974) came from help-seeking clients of a large outpatient psychotherapeutic institution (Institute for Psychogenic Disorders of the Berlin General Sickness Fund). At this large outpatient clinic, children and adults with psychogenic disorders (psychoneuroses, character neuroses, and functional psychosomatic disorders) are diagnosed and psychotherapeutically treated if necessary. The overall population (of the Berlin sample) consisted of help-seeking clients (1950–1969) of 26,799 patients. Among these patients, 240 were twins; between 1963–1969 I examined 50 of them and their twin siblings (between 2.3 and 66 years of age). A necessary requirement, however, was that both partners were alive and living in Berlin. Zygosity diagnostics (Schepank 1974, p. 81) were carried out by P.E. Becker (1980), at the University of Göttingen, using human genetic anthropological markers (bloodgroups and subgroups, fingerprints, comparison of anthropological similarity according to Siemens). All twins were referred for zygosity determination *after* the psychological part of the investigation was completed. The results were: 21 MZ pairs, 16 same-sex twin pairs, and 13 opposite-sex twin pairs; 32 were adult pairs and 18 adolescent pairs. In addition, we have continuous results for the following years for all subjects and are currently conducting a 20-year follow-up, cross-sectional investigation from 1985/1986 up to the present time (Muhs et al. 1990; Muhs and Schepank 1991).

Moreover, in the outpatient department of the Psychosomatic University Hospital of Heidelberg (1971–1975), we supervised some discordance analyses in adults undergoing psychoanalytic therapy (Janus et al. 1981). We did not often find a notable discordance of personality or social neurotic symptoms (severe inability to make contacts, a rather severe compulsive impairment, suicidal tendencies, alcoholism, etc. of only one sibling). We described nine such cases (Schepank 1974, 1975; Heigl-Evers and Schepank 1980/1981; Muhs and Schepank 1993). The diagnoses were done by means of personal, psychoanalytically oriented interviews, by metric test methods (FPI), and by intelligence tests (IST, Raven, WISC, WISA).

RESULTS

The following statements summarize our results and are based on the compiled biographies of 9 discordant MZ pairs:

1. Our attention was drawn to the fact that parents of discordantly developed pairs were from a lower social class, seemed to be less intelligent, and appeared rigid in their own personality structure. (We investigated the parents of all infantile twins and, when possible, the parents of the adult twin pairs.) Obviously such constellations may have influenced a diverging development in the twins, apart from the genetic relationship to the twins themselves. These parents react in a more rigid way to the genetically granted "substance" in their children. They

can less skillfully compensate for offensive behavior, preferences, and casual discrimination.

2. Concerning neurotic character traits or psychoneurotic symptoms, the "preferred" child of the stronger or dominant parent (regardless of whether this was the mother or father) usually developed more favorably.

3. The more favorably developed twin received a Christian name connected with a positive emotional experience to the parents. For example, a mother gave the preferred twin the name of a child she once had loved or cared for, or a father tried to immortalize a positive family figure (e.g., his adored brother) in his son. Special terms of delegation can also give a positive impulse.

Further factors must be assessed differently for the male and the female sex:[1]

4. We found *male* discordant MZ twins to come from families with a strict educational upbringing and from parents who did not resolve their own marital conflicts through divorce but endured them in different manners or dismissed them.

5. The mental health of a *male MZ twin* was always better when he identified himself with the more expansive and dominating parent. It did not matter whether this parent was the mother or father. For the development of a young male, it seems to be globally beneficial for him to be capable of entering risk situations, the success of which depends on having been given the opportunity during childhood to take risks and to express expansive tendencies rather than have them restricted. Personal emotional bonds to parental figures and siblings, the so-called object relations, are perhaps more superficial, if present at all. In other words, an unfavorable (psychoneurotic/character neurotic) development of personality will result in a male if, during his (especially early) childhood, under the pretense of excessive care, he is emotionally tied down, strongly linked with the family, or if he is manipulated towards strongly disallowing or inhibiting his own impulses.

6. The risk constellation in *female MZ twins* is different. This is not surprising as the infantile libidinous object relations between the same-sex and opposite-sex parent require a more complicated shift for females than for males, inasmuch as the care is usually in the hands of mothers. The parents of several female discordant MZ twins were divorced and obviously this constellation prompted discordant developments.

---

[1] The discovered sex differences correspond to the parental investment hypothesis (Bischoff et al. 1980). A certain adaptation to this leading ethological constant is evidently followed by a healthier development of personality.

7. For a female, it seems generally to be riskier to become sociable, markedly outgoing, or "phallic" (in the psychoanalytical jargon) and for changing, unsteady object relations to prevail over partner relations, even in adulthood.
8. The *choice of a partner in later life* acts as a sort of ultimatum in that it is not easily correctable because of the legalities involved (e.g., marriage and/or births). In several female MZ twins who formed similar partner constellations (e.g., first a relationship with a much older man, then a markedly younger partner, and eventually a foreign, remote, and/or otherwise dependent partner), we observed that the healthier twin did not hold fast to neurotic bonds but instead broke away from the situation (e.g., interrupted a pregnancy, separated from the partner, etc.). By contrast, the twin sister with the more severe personality disorder cemented the neurotic relationship once established and did not correct her decision, even after she manifested a severe compulsive neurosis and a neurotic depression thereafter; suicide finally resulted.

    In a study involving another female pair of MZ twins, the healthier sibling married a much older partner. The other twin gave birth to a child, married, and later divorced her partner. Afterwards, she had numerous and frequently changing relationships with socially neglected men; she became a severe alcoholic herself, had a desolate social existence, and depended predominantly on external governmental assistance.
9. It also seems important for the sound development of a female's personality to be occupationally stable and thus to become more independent. Sublimation by occupation, linked with a substitutional homoerotic affiliation, can provide a major stabilization of personality.

FURTHER STATEMENTS

The statements below, concerning the differential psychology of twins, are based on results obtained from Schepank (1974, 1975, 1976), Heigl-Evers and Schepank (1980/1981), Muhs et al. (1990), and Muhs and Schepank (1991, 1993)

First, there is no feature of personality and, correspondingly, no diagnostic method or criterion that *basically differentiates a twin from a singleton.* I have never diagnosed someone spontaneously as a twin in my environment.

The interactions within a twin dyad subtly described by some authors (von Bracken 1938) are only observable in the twins' deportment with each other (relations of dominance, role division in "secretary" of the exterior and the interior, functional divisions as, for example, in school homework). This statement is methodically important concerning the possible generalization of twin results to all human beings. Grown-up twins are human beings just like singleton children and are by no means a species of their own. (Twin particularities only apply to certain intrauterine and perinatal risk factors, a mostly recoverable developmental retardation in childhood).

Next, concerning prevalence, it is epidemiologically true that twins do not suffer from neurotic symptoms or personality changes any more or less than singletons. The psychological twin constellation is thus neither generally stressful in the sense of being a risk factor nor is it a protective factor. This is an important result of our investigation.

A male pair of our sample grew up separately in nonshared institutional care from the first to the eighteenth year of life without knowing anything about the other twin's existence. These separately raised male MZ twins (now 45 years old) are highly concordant; they exhibit mild but continuous kinds of character neuroses, lack a profession and a stable partnership, and live in a socially parasitic way.

Finally, there is no genetic linkage between *neuroses* and *psychoses*! My reasoning: None of our 32 adult neurotic index twins had a psychotic co-twin. Among the 100 parents of the 50 twin pairs we found a psychosis in one of the mothers, which corresponds to the normal prevalence of psychosis among the general population. I believe, therefore, that the concept of "psychoticism" as a sliding dimensional continuum within reality is wrong, or is at least a very unfortunate term. Psychosis is a clearly distinguishable and genetically different disease.

As to *psychosis*, only one infantile MZ twin pair out of our 50 Berlin samples turned out to be discordant in the course of 20 years: a male twin, now 32 years old, developed a severe paranoid hallucinatory productive schizophrenia. His twin partner, however, is (still!) healthy. Any reasons for the discordance could not be pinpointed (Muhs and Schepank 1993; Schepank and Muhs 1993).

REFERENCES

Baker, L.A., and D. Daniels. 1990. Nonshared environmental influences and personality differences in adult twins. *J. Pers. Soc. Psychol.* **58**(1):103–110.

Becker, P.E. 1980. Persönlichkeit und Neurosen in der Zwillingsforschung. Ein historischer Überblick. In: Ursprünge seelisch bedingter Krankheiten, vol. 1, ed. A. Heigl-Evers and H. Schepank. Göttingen: Vandenhoeck & Ruprecht.

Bischoff, N., and H. Preuschoft, eds. 1980. Geschlechtsunterschiede: Entstehung und Entwicklung, Mann und Frau in biologischer Sicht. München: C.H. Beck.

Bouchard, T.J., Jr., D.T. Lykken, N. Matthew-McGue, L. Segall, and A. Tellegen. 1990. Sources of human psychological differences: The Minnesota Study of twins reared apart. *Science* **250**:223–250.

Bracken, H. von. 1936. Verbundenheit und Ordnung im Binnenleben von Zwillingspaaren. Z. *Pädagog. Psychol.* **37**:65–81.

Dencker, S.J. 1958. A follow-up study of 128 closed head injuries in twins using co-twins as controls. Kopenhagen: Munksgaard.

Eysenck, H.J. 1944. Types of personality: A factorial study of 700 neurotics. *J. Ment. Sci.* **97**:441–456.

Eysenck, H.J., and D.B. Prell. 1951. The inheritance of neuroticism: An experimental study. *J. Ment. Sci.* **97**:441–465.

Gesell, A. 1953. Säugling und Kleinkind in der Kultur der Gegenwart. Bad Nauheim: Christian.

Heigl-Evers, A., and H. Schepank, eds. 1980/81. Ursprünge seelisch bedingter Krankheiten: Eine Untersuchung an 100 + 9 Zwillingspaaren mit Neurosen und psychosomatischen Erkrankungen. 2 vols. Göttingen: Vandenhoeck & Ruprecht.

Janus, L., B. Janus-Stanek, and H. Schepank. 1981. Diskordanzanalyse bei 3 Zwillingspaaren, von denen jeweils ein Zwilling analytisch behandelt wurde. In: Ursprünge seelisch bedingter Krankheiten, ed. A. Heigl-Evers and H. Schepank. Göttingen: Vandenhoeck & Ruprecht.

Loehlin, J.C., and R.C. Nichols. 1976. Heredity, environment, and personality. A study of 850 sets of twins. Austin: Univ. of Texas Press.

Muhs, A., H. Schepank, and R. Manz. 1990. 20-Jahre-Follow-up-Studie von 50 Zwillingspaaren. *Z. psychosom. Med.* **36**:1–20.

Muhs, A., and H. Schepank. 1991. Aspekte des Verlaufs und der geschlechtsspezifischen Prävalenz psychogener Erkrankungen bei Kindern und Erwachsenen unter dem Einfluß von Erb- und Umweltfaktoren. *Z. psychosom. Med.* **37**:194–206.

Muhs, A., and H. Schepank. 1993. Diskordanzanalyse eineiiger Zwillinge. *Z. psychosom. Med.* **39**:174–190.

Newman, H., F.N. Freeman, and K.J. Holzinger. 1937. Twins: A study of Heredity and Environment. Chicago: Chicago Univ. Press.

Propping, P. 1989. Psychiatrische Genetik: Befunde und Konzepte. Berlin: Springer.

Schepank, H. 1974. Erb- und Umweltfaktoren bei Neurosen: Tiefenpsychologische Untersuchungen an 50 Zwillingspaaren. Monograph: Gesamtgebiete der Psychiatrie, vol. 11. Berlin: Springer.

Schepank, H. 1975. Diskordanzanalyse eineiiger Zwillinge. *Z. psychosom. Med.* **21**:215–242.

Schepank, H. 1976. Heredity and environmental factors in the development of psychogenic diseases. *Acta Genet. Med. Gem.* **25**:237–239.

Schepank, H. 1983. Anorexia nervosa in twins: Is the etiology psychotic or psychogenic? In: Pychosomatic Medicine, ed. A.J. Krakowski and C.P. Kimball, pp. 161–169. New York: Plenum.

Schepank, H. 1987. Epidemiology of Psychogenic Disorders: The Mannheim Study—Results of a Field Survey in the Federal Republic of Germany. Berlin: Springer.

Schepank, H. 1991. Erbdeterminanten bei der *Anorexia nervosa*. *Z. psychosom. Med.* **3**:215–219.

Schepank, H. 1992. Zwillingsschicksale: 20-Jahres-Follow-Up eines Samples von 50 Zwillingspaaren mit psychogenen Erkrankungen. Zürich: Burghoelzli.

Schepank, H., and A. Muhs. 1993. Zwillingsschicksale. Stuttgart: Enke.

Vogel, F., and A.G. Motulsky. 1979. Human Genetics: Problems and Approaches. Berlin: Springer.

10

Genetic and Environmental Continuity and Change in Personality

N.L. PEDERSEN

Division of Epidemiology, Institute of Environmental Medicine,
The Karolinska Institute, Box 210,
171 77 Stockholm, Sweden, and
Center for Developmental and Health Genetics,
College of Health and Human Development,
The Pennsylvania State University,
University Park, PA 16802, U.S.A.

ABSTRACT

The interface of behavioral genetic methodology with issues concerning continuity and change in personality is the focus of this chapter. Five aspects of genetic and environmental involvement for continuity and change in personality will be discussed: (1) the differential relative importance of genetic and environmental effects across the life span; (2) the importance of genetic and environmental effects for phenotypic stability and change during development and aging; (3) stability and change in genetic and environmental effects, i.e., whether the same genetic and environmental effects are operating across time; (4) the involvement of genetic and environmental effects in structural stability and change; and (5) factors influencing continuity vs. those influencing change. There are a number of misconceptions in the field of developmental psychology (e.g., that a demonstration of heritable variation for a trait implies that the trait will be stable or that the effects of genes should be stable because we are born with a full complement of genes) that should be dispelled when quantitative genetic techniques are applied to longitudinal data in genetically informative populations. Unfortunately, advances made in analytic methodology are at present offset by a paucity of longitudinal behavioral genetic studies with more than 2 points of measure. Great advancements will be made as soon as better longitudinal data are collected, and longitudinal data already collected are analyzed in a rigorous manner using the recent methodological advancements in the field.

Twins as a Tool of Behavioral Genetics
Edited by T.J. Bouchard, Jr. and P. Propping © 1993 John Wiley & Sons Ltd.

INTRODUCTION

Stability and change for personality can be characterized in a number of ways; the issues relevant to this conference, however, are concerned with stability and change in the context of genetically informative populations. In this chapter, I review evidence of phenotypic stability for personality, followed by a discussion of stability and change from a developmental quantitative genetic perspective. Contributions of behavioral genetic studies to the literature concerning stability and change in personality are then described.

PHENOTYPIC STABILITY

There are at least three perspectives from which most of the research on development and aging in personality has been generated: developmental-stage models, contextual models, and trait models. Much like the Piagetian tradition for cognitive development, developmental-stage models for personality focus on age-linked qualitative shifts in personality. Although such models, exemplified by Erikson's (1950) theory of eight ego-developmental stages, have been in existence for many years, there has been limited empirical investigation that can be evaluated psychometrically. Relatively more recent contextual models represent efforts to relate personality to age-graded roles and to historical change (for a review, see Kogan 1990). Quantitative genetic analyses have been applied almost exclusively to data stemming from trait models of personality development, and reflect the extensive influence of Eysenck and colleagues (Eaves et al. 1989). In this condensed review, the focus will be on evidence for stability and change stemming from trait models.

Three types of stability for personality have been assessed: interindividual stability, mean level stability with age, and structural invariance. Interindividual stability is the maintenance of the rank order among individuals, often measured by occasion-to-occasion correlation coefficients. Another aspect of stability is characterized by the lack of change in mean values in different age groups and is often assessed by correlations of the phenotypic scores with age. Finally, continuity in structural composition of personality traits, indicated by factor structure similarity across cohorts and times of measurements, may be considered an aspect of stability.

Interindividual Stability

Evidence from studies of varying longitudinal follow-up intervals (from 3 to 50 years) and an array of measures converge on the conclusion that there is considerable interindividual stability for personality in adulthood and old age (Kogan 1990). Stability coefficients are substantial but tend to decline as the testing interval increases. There is less agreement as to whether stability differs during various portions of the life span and, more specifically, whether the stability of personality is greater among older than younger adults (Finn 1986; Costa and McCrae 1988).

Mean Level Stability

Reliance on cross-sectional data has resulted in considerably less accord concerning stability of mean levels. Even when conclusions are based on longitudinal data, change in mean levels has been described as "trivial and insignificant" at the one extreme, and "substantial and meaningful" at the other (Costa et al. 1987; Hann et al. 1986). As in the case of interindividual stability, there may be support for differential change dependent on the stage of the life span, i.e., that the rate of change declines with increasing chronological age (Finn 1986). Evidence of cohort differences in mean level stability (Schaie and Willis 1991) suggests that caution is warranted in drawing generalizations concerning the nature of mean level stability and change with age.

Structural Invariance

Longitudinal evidence concerning structural invariance in personality with age comes largely from four studies, each of which employed different instruments. Kogan (1990) concludes:

> Thus, the evidence is highly suggestive of a consensual five-factor structure within the personality trait domain across projects and over time within projects....In sum, the pattern of similarity in factor structure across different instruments, types of measurement, cohorts, and times of measurement is surely a powerful tribute to the structural invariance principle. (Kogan 1990, p. 333).

Support for this conclusion comes in part from the Oakland Growth Study (Hann et al. 1986), which included periodic measurements from childhood through late adulthood. Considering the controversy regarding whether personality is comprised of five factors and, if so, which factors should be included, Kogan's interpretation may be very optimistic.

BEHAVIORAL GENETIC CONCEPTS OF STABILITY AND CHANGE

The concepts presented in this section are by no means new to students of behavioral genetics, nor is the list necessarily comprehensive. However, for the reader with a life-span developmental perspective, this section may provide an insight into the many facets of genetic influences on development.

Concept 1: Differential Heritability across the Life Span

Perhaps the most commonly found representation of genetic contributions to development is the assessment of potential differential heritability across the life span. This is typically characterized by the question: Does the relative importance of genetics

(G) and environment (E) differ in different age groups? Differential heritability is most often assessed in cross-sectional studies with subjects from different cohorts or age groups, or by comparisons of heritabilities from studies with differing age compositions. In the absence of longitudinal data, estimates of age differences in heritability tell us little about the etiology of stability and change for personality, and yet (see below) such studies constitute the bulk of the information gathered to date.

Concept 2: Stability and Change in Genetic and Environmental Effects

Rather than focusing on longitudinal phenotypic stability, genetic and environmental correlations describe the extent to which genetic and environmental effects, respectively, are temporally stable. Loosely interpreted, the genetic correlation indicates the extent to which there is "overlap" in genetic effects operating at two or more occasions; considerable caution must be exercised, however, in the interpretation of such pleiotropic effects (for a rigorous discussion of the interpretation of genetic correlations, see Carey 1988). Genetic effects may be highly stable, regardless of their relative importance at any two time points. Similarly, new or "innovative" genetic effects may come into play at selected developmental stages. Perhaps one of the most common misconceptions concerning the nature of genetic effects is that because we are born with a full complement of genes, their influence must be stable and invariant throughout the life span. As discussed below, this is not the case, but rather a phenomenon which should be evaluated empirically for each phenotype of interest.

Concept 3: Importance of Genetic and Environmental Effects for Stability and Change

Questions concerning the importance of G and E for stability and change are more central to the developmental perspective concerning interindividual stability. Regardless of the relative importance of genes and environments at any series of time points (Concept 1), the emphasis is now on determining the extent to which phenotypic stability reflects stability of genetic and environmental effects. Phenotypic stability is a result of the influence of genes and environments at the two measurement occasions as well as the stability of the genetic and environmental effects between those occasions. High heritability at two times does not necessarily imply substantial phenotypic stability nor does it imply that the genetic component of stability is substantial. If the genetic correlation across occasions is low, the genetic contribution to stability may be low despite high heritabilities at the two time points.

There are a number of elegant models available for assessing the genetic architecture of development (for a summary, see Pedersen 1991). Molenaar and colleagues have developed and elaborated on the "genetic simplex" (Molenaar et al. 1991) while Eaves, Hewitt and colleagues present a combined common factor and transmission (simplex) model (Eaves et al. 1986; Hewitt et al. 1988). Application of these models allows us to evaluate not only the concepts described above, but also to identify specific events such as change due to distinct subsets of genes turning off or on.

Thus far, the concepts and models discussed refer primarily to interindividual stability. Models are also available for addressing issues concerning mean level stability or average growth curves (e.g., McArdle 1986; Molenaar et al. 1991). Simultaneous modeling of means and covariances allows for hypothesis testing concerning whether these are influenced by common underlying processes reflecting stability. Perturbations in the developmental system can certainly be assessed in all three of these major groups of models.

Concept 4: Genetic Architecture of Structural Stability

In developmental terms, structural invariance or stability can be assessed by comparing factor structures at different measurement occasions. This same issue can be examined within a quantitative genetic design, with the advantage of gaining information on the role of genetic and environmental factors for structural invariance or change. At any single occasion, multivariate genetic analyses of the covariances among items on a scale or covariances between traits provide an insight into, for example, the importance of common genetic and environmental factors for the structure of higher-order traits, such as *extraversion* (for further examples, see Eaves et al. 1989). How stable is the genetic covariance structure? To what extent does phenotypic structural invariance reflect stability in the genetic or environmental structure? Answers to these questions may illuminate the controversies mentioned above concerning the five-factor theory of personality or lower stability in cohorts of young, college-age adults.

Concept 5: Etiology of Stability as Distinct from Etiology of Change

Stability and change are not necessarily antonyms, at least from a quantitative genetic perspective. The etiology of change may be very different from the etiology of stability (Plomin and Nesselroade 1990), as the heritability of change is a function not only of the heritability of the trait, but also the genetic correlation between occasions and the phenotypic stability. Although studies of change are intuitively appealing, power is considerably lower in analysis of change score, and the likelihood of finding significant genetic variance for change is reduced.

THE EVIDENCE

Differential Heritability across the Life Span

Infancy and Childhood

The majority of twin studies concerning stability and change in personality are cross-sectional and thus provide little more than evidence for evaluating differential heritability during the various phases of the life span. Plomin (1986) has summarized twin and adoption studies of temperament during infancy and childhood, concentrating primarily on the EAS (emotionality, activity, and sociability) temperament traits and similar measures. Although temperament may be classified by other schema, the EAS distinction is convenient for comparisons with the adult personality domains of *neuroticism* and *extraversion*. Differences in instruments and assessment techniques (observer reports, parental ratings, and "objective" measurements of behavior) make comparisons among studies and age groups difficult. Nevertheless, Plomin concludes that, because the intraclass correlations for monozygotic (MZ) twins are consistently greater than dizygotic (DZ) correlations, there is evidence for the importance of genetic effects on temperament during both infancy and childhood. Furthermore, because DZ correlations are often less than half the magnitude of the MZ correlations, nonadditive effects *or* sibling interaction effects (contrast) are suggested. The studies provide little evidence of differential heritability across these two developmental periods. Finally, Plomin considers that the EAS temperament traits are among the most heritable during infancy and childhood.

The data appear to be less conclusive and consistent than Plomin suggests. For many of the variables, a classic model including additive genetic variance and within family environments cannot be fitted, either because of negative DZ correlations or, in a few cases, MZ correlations that do not exceed DZ correlations. Despite claims that the EAS traits are among the most heritable during childhood, Plomin (1986) points out that there are problems with the patterns of correlations from several early childhood twin studies that used parental ratings on EAS questionnaires: DZ correlations for both *activity* and *sociability* are negative. Recomputation of intraclass correlations from twin studies using other parental rating measures (summarized in Plomin 1986) suggests that *activity* in the latter group of studies may be "better behaved," i.e., that the ratio of MZ to DZ correlations is closer to 2, than in those based on the EAS (Table 10.1). Goldsmith (1989) suggests that some parental rating scales may be more susceptible to rating biases in which mothers of fraternal twins tend to exaggerate their differences. As he points out, the issue of high identical twin similarity accompanied by minimal fraternal similarity "is unresolved, but it is clear that the resolution of substantive behavior-genetic issues interacts strongly with psychometric properties of temperament assessment instruments" (Goldsmith 1989, p. 121).

On the whole, there appears to be considerable heterogeneity in heritability estimates (if one simply manipulates the intraclass correlations reported) across measures and studies. However, sample sizes are generally small (fewer than 100 pairs each of MZ and DZ), hence standard errors would be large and power too low to find significant differences in estimates. The hazards of comparing numerous cross-sectional studies with relatively small sample sizes for the purpose of understanding development are painfully apparent.

There are some bright spots in the data, notably the consistent results from the Louisville Twin Study (LTS; Matheny 1990), and the work of Goldsmith and

Table 10.1　Summary of selected cross-sectional findings for temperament in childhood (Plomin 1986).

	Pattern of Intraclass Correlations					
	Emotionality		Activity		Sociability	
	MZ	DZ	MZ	DZ	MZ	DZ
EAS	0.63	0.12	0.62	−0.13	0.53	−0.03
Other measures	0.61	0.22	0.80	0.57	0.67	0.25

EAS = emotionality, activity, sociability.

colleagues (Goldsmith and Gottesman 1981; Goldsmith and McArdle, submitted), both of which have, fortunately, included longitudinal assessment. These two studies will be discussed below in the section concerning longitudinal results.

Adolescence

Although there are fewer twin studies of personality in adolescence than at earlier ages, these studies are considerably larger, but again provide only cross-sectional data at present. Heritabilities for *extraversion* and *neuroticism* in the juvenile subset (262 pairs, 7–17 years old) of the London Twin Study are 0.54 and 0.44, respectively (Eaves et al. 1989). Reanalyses of the NMSQT data (850 pairs of American high school juniors [Loehlin and Nichols 1976]) indicate heritabilities of 0.61 and 0.52 for *extraversion* and *neuroticism*, respectively (Eaves et al. 1989) (Table 10.2).

Adulthood

There have been numerous twin studies of personality during adulthood (for one review, see Eaves et al. 1989), many of which were reported prior to 1976. Rather than repeat attempts to summarize these data, I will concentrate on the results from extremely large data sets from twin registries in Scandinavia and Australia.

The Australian Twin Registry contains information on *extraversion* and *neuroticism* from 3,810 twin pairs aged 18–88, average age 34 (Martin and Jardin 1986). Correlations of *extraversion* and *neuroticism* scores with age ranged from −0.13 to −0.16 in men and women, indicating small mean level changes with age. To test whether twin similarity differed as a function of age, Martin and Jardin correlated absolute within-pair differences with age and found a significant association only for *neuroticism* in DZ females. This result was interpreted as indicating increasing genetic differences with age for *neuroticism*. The data were not corrected for age, nor analyzed separately by age group in the model-fitting analyses. Heritability was estimated at 0.52 for *extraversion*, 0.45 for *neuroticism* in men, and 0.51 for *neuroticism* in females (Table 10.2).

Table 10.2 Summary of cross-sectional findings for personality in adolescence and adulthood.

		Heritability Estimates			
		Neuroticism		Extraversion	
Adolescence	Age				
London Twin Study	7–17	0.44		0.54	
NMSQT Twin Study	18	0.52		0.61	
Adulthood		Men	Women	Men	Women
Australian Twin Registry	$\bar{X} = 35$	0.45	0.51	0.52	0.52
Finnish Twin Registry[1]	25–43	0.31	0.44	0.42	0.44
"New" Swedish Twin Registry	19–28	0.53	0.61	0.51	0.27
	29–38	0.39	0.51	0.35	0.49
	29–48	0.38	0.40	0.45	0.50
SATSA	$\bar{X} = 59$	0.37 (0.31^2)		0.45 (0.41^2)	

[1] Kaprio, Rose, Viken, and Koskenvuo, pers. comm.; from first measurement occasion, 1975.
[2] Estimate when effects of shared rearing environments included in the model.
NMSQT = National Merit Scholarship Test.
SATSA = Swedish Adoption/Twin Study of Aging.

Personality data have been collected at two time points from twins in the Finnish Twin Registry (J. Kaprio, R. Rose, R. Viken, and M. Koskenvuo, pers. comm.). Cross-sectional results from the first measurement occasion have not been evaluated with respect to age differences in heritability for personality during adulthood; however, inspection of intraclass correlations and heritabilities suggests slight decreases in heritability with increasing age (Koskenvuo et al. 1979). A more complete description of results from the Finnish data will be discussed below (see section concerning longitudinal studies).

In the large Swedish study of 12,898 pairs (Floderus-Myrhed et al. 1980), mean squares between and within pairs were reported separately by birth cohort, allowing reanalyses which specifically addressed the question of heterogeneity of heritability by age group. (Correlations with age and between intrapair differences and age were not performed.) There were significant age and gender interactions for both *neuroticism* and *extraversion*, such that models allowing parameters to be estimated separately for the six groups fit the data significantly better than those constraining estimates to be equal across either gender or age (summarized in Eaves et al. 1989). Without further analysis of differences in variances across age and gender, only tentative conclusions about trends in age differences in heritability can be made. From Table 10.2, one can see that heritability for *neuroticism* appears to decrease with increasing age for both genders. The pattern is less clear for *extraversion*.

In one of the few twin studies specifically designed to examine questions concerning individual differences in aging, my colleagues and I have exploited a unique subsample of Swedish twins separated at an early age and reared apart, and a matched subsample of twins reared together. Data on 25 self-reported personality measures have been collected on three occasions in the Swedish Adoption/Twin Study of Aging (SATSA). The distribution of age at the first time point (1984) is highly skewed; mean age = 58.8, range 26–87. Results from the first measurement occasion indicate that genetic effects are significant (average heritability 0.30) for all but 3 of the 25 measures (Pedersen et al. 1991). Results for *extraversion* and *neuroticism* (Pedersen et al. 1988) are summarized in Table 10.2. Hierarchical multiple regressions were applied to all of the scales to determine whether twin resemblance and/or heritability differed as a function of age. Only two of the measures, the mental health scale of *depression* and the *life direction* subscale of *locus of control*, show differences in heritability across four age bands (Pedersen et al. 1991). The latter measure shows a decline in heritability in the oldest age band and the former shows a variable pattern. On the whole, cross-sectional analyses of the SATSA data set indicate no reliable age differences in heritability for self-report personality measures.

It is interesting that Eaves et al. (1989) found age group differences in the younger cohort Swedish Twin Registry data (known as the "New" Swedish Twin Registry), whereas we found no similar differences in the SATSA data, which is in part a subset of the larger registry. This may reflect the different age composition of the two samples (the youngest SATSA cohort is the same age as the oldest in the New Swedish Twin Registry). On the other hand, the sample size is sufficiently large in the registry to find significant differences with very small effect sizes while SATSA has less power for comparisons of this type. Furthermore, it is not clear whether the age differences demonstrated in the registry data reflect age differences in total variance or in genetic variance.

Finally, the developmental meta-analysis of twin studies by McCartney et al. (1990) deserves mention. In a very ambitious effort to summarize whether twin similarity differs as a function of age, all cross-sectional twin studies published between 1967 and 1985 with results in the form of intraclass correlations were reviewed. Intraclass correlations were then correlated with age; a negative correlation indicated that twin similarity decreased as a function of age (which for personality ranged from 1–50 years). Unfortunately, the results from the London Twin Study, the Australian Twin Registry, and SATSA were not included in this analysis as they postdate the studies covered in the meta-analysis. It is not clear whether the New Swedish Twin Registry data were treated as three separate age groups or whether the summary statistics pooled over the age range were used. Nevertheless, McCartney et al. (1990) report MZ correlations of − 0.11 for *emotionality* (*neuroticism*) and − 0.24 for *sociability* (*extraversion*) and DZ correlations of 0.26 and 0.30, respectively. None of these correlations are significant, although they are larger than those reported within the Australian study (which primarily included adults). This convergent pattern would suggest that across cohorts and studies, heritability is somewhat lower in the older age groups by some, as yet undetermined, amount.

In summary, the cross-sectional data indicate that personality traits are heritable throughout the life span. It is not possible to determine whether there are differences in the relative importance of genetic effects across infancy, childhood, adolescence, and adulthood. However, large sample studies of adult twins do suggest that there are detectible changes in heritability in adulthood. If any generalization is to be made, it would be to suggest that heritability is greater in young adults than in older adults (at least for *neuroticism*). Furthermore, there appear to be no differences in the (lack of) importance of shared environmental influences for familial resemblance across the life span. Before sweeping generalizations about age changes in heritability are made, we should look at the longitudinal data. It is important to bear in mind that the age differences reported above may reflect cohort effects and have no parallel in true aging trends.

Longitudinal Evidence from Twin Studies

If the volume of twin study literature was distributed proportionately to the five concepts presented in the beginning of this chapter, the following section should be about four times as long as the section on cross-sectional studies. The reverse is more nearly accurate. To the best of my knowledge, there are only two twin studies (LTS and SATSA) with temperament or personality data from three or more occasions, and five with information from two occasions with at least a 2-year inter-test interval.

Infancy and Childhood

Longitudinal data on temperament from up to 6 measurement occasions between 3 and 24 months of age have been reported from the LTS (Matheny 1990). Age-to-age stabilities averaged 0.44 for measures of temperament based on lab, questionnaire, and infant behavior record ratings. These stabilities are somewhat lower than those found for self-reported personality scores in adults. Besides cross-sectional reports of intraclass correlations, LTS has reported "trend correlations for age-to-age change" for MZ and DZ pairs which "take into account the concordance for overall level of scores as well as change in scores" (Matheny 1990, p. 32). MZ trend correlations were significantly greater than DZ correlations for ages 6–12–18 months for *task orientation* and *test affect-extraversion*, and for 12–18–24 months for all three measures (including *activity*). Thus, age to age stability for temperament during infancy is influenced in part by genetic effects.

Goldsmith and McArdle (submitted) have applied several of the biometric models described earlier to longitudinal twin data concerning temperament collected in conjunction with the Collaborative Perinatal Project (Goldsmith and Gottesman 1981). Autoregressive, growth curve (difference score), and structural invariance approaches were applied to assessments of *reactivity* and *persistence* at ages 4 and 7 in 85 MZ and 65 DZ pairs of twins. From autoregressive models, Goldsmith and McArdle demonstrated that heritability was greater at age 7 than at age 4 for *reactivity*, but greater at 4 than age 7 for *persistence*. Although genetic and environmental correlations were not reported, the authors conclude that genetic influences accounted for most of the phenotypic stability in both *reactivity* and *persistence*.

Goldsmith and McArdle also addressed issues of structural invariance across measurement occasions (Concept 4) within both autoregressive and difference score approaches. Models specifying structural invariance across the 3-year interval were found to fit the data adequately; however, specific biometric effects acting over time were found.

Adulthood

Information on *neuroticism* and *extraversion* from two occasions is available in the Australian Twin Registry, the Finnish Twin Registry, and SATSA (Table 10.3). These data provide preliminary evidence relevant to Concepts 2 and 3 above. Preliminary analyses (provided as a personal communication by A. Heath and N. Martin) from an 8-year follow-up of the Australian Twin Registry indicate high longitudinal stability for *extraversion* (0.70–0.81) and *neuroticism* (0.61–0.74) for respondents aged 25 and over. Stability was moderately reduced for those aged 18–25 at initial assessment. A very similar pattern of phenotypic stabilities across a 6-year interval was found in the Finnish Twin Registry (Kaprio, Rose, Viken and Koskenvuo, pers. comm.). For both men and women, stability was 0.70 for *extraversion* and 0.60–0.62 for *neuroticism*, again with somewhat lower stabilities for those under age 25 at initial assessment. Two longitudinal follow-ups at 3-year intervals have been completed for SATSA (Pedersen et al., submitted). Analyses of the second period are currently underway. Three-year stabilities for *extraversion* and *neuroticism* were very similar to the Australian and Finnish results: 0.71 for *extraversion* and 0.68 for *neuroticism*. (Stabilities for all 20 personality scales for which there were longitudinal data ranged from 0.48 to 0.71, with an average of 0.65.)

For all three samples, there was considerable stability in genetic effects across the intervals (Concept 2): genetic correlations ranged from 0.81 to 1.0 (Table 10.3). Nonshared environmental factors were less stable: environmental correlations ranged from 0.26 (for the youngest Finnish twins) to 0.58 (for the older SATSA twins), indicating the appearance of considerable "innovative" environmental variance at the second time point. Despite the high stability of genetic effects, they accounted for only about half of the phenotypic stability, owing primarily to the modest heritabilities at each of the measurement occasions (Concept 3).

One of the contributions from SATSA is the potential to delineate the relative importance of shared rearing environments (SRE) through the comparison of twins reared together and twins reared apart. Very few of the personality measures in SATSA show significant shared rearing environmental variance. When present, these effects seldom account for more than 10% of the variation in personality late in life (Pedersen et al. 1991). However, rearing effects appear to be quite stable. When present, occasion-to-occasion correlations for SRE are over 0.90.

N.L. Pedersen

Table 10.3 Summary of longitudinal personality results from the Australian and Finnish Twin Registers and SATSA.

	Men			Women		
	Time 1	r_E or r_G	Time 2	Time 1	r_E or r_G	Time 2
Neuroticism						
Australian sample[1]						
V_E	0.59	0.43	0.54	0.55	0.48	0.55
V_A	0.00	1.00	0.05	0.26	1.00	0.42
V_{NA}	0.41	1.00	0.41	0.20	1.00	0.03
Finnish sample[2]						
Born 1951–1957						
V_E	0.49	0.26	0.68	0.48	0.27	0.57
V_G	0.51	0.85	0.32	0.52	0.81	0.43
Born 1932–1950						
V_E	0.69	0.42	0.71	0.56	0.35	0.56
V_G	0.31	0.99	0.29	0.44	0.96	0.42
SATSA[3]						
V_G	0.37	1.00	0.37			
V_E	0.58	0.50	0.58			
V_{ES}	0.05	1.00	0.05			

[1] Heath and Martin, pers. comm.
[2] Kaprio, Rose, Viken, and Koskenvuo, pers. comm.
[3] Genders combined.
SATSA = Swedish Adoption/Twin Study of Aging.

In summary, the three studies that focused on interindividual stability (the Australian study, the Finnish study, and SATSA) indicate substantial genetic and phenotypic stability. The relative importance of genetic and environmental effects was the same at the two occasions. Because of the modest heritabilities, genetic effects contributed slightly more to phenotypic stability than environmental effects in the Australian sample and about equally in the SATSA and Finnish material.

The Analysis of Change

Two of the four longitudinal studies that highlighted the analysis of change were based on the MMPI (Dworkin et al. 1976; Pogue-Geile and Rose 1985). None of these studies found reliable evidence of genetic influence on directional change (the tendency of both members of a twin pair to change in the same direction across time). Likewise, analyses of absolute change scores indicated that nonshared environmental influences were of predominant importance for change. As mentioned earlier, these

Table 10.3 *continued*

	Men			Women		
	Time 1	r_E or r_G	Time 2	Time 1	r_E or r_G	Time 2
Extraversion						
Australian sample[1]						
V_E	0.49	0.46	0.51	0.48	0.53	0.52
V_A	0.23	1.00	0.37	0.24	1.00	0.35
V_{NA}	0.28	1.00	0.13	0.28	1.00	0.13
Finnish sample[2]						
Born 1951–1957						
V_E	0.54	0.35	0.53	0.48	0.36	0.58
V_G	0.46	0.88	0.47	0.52	0.88	0.42
Born 1932–1950						
V_E	0.58	0.51	0.61	0.56	0.46	0.52
V_G	0.42	0.99	0.39	0.44	0.97	0.48
SATSA[3]						
V_G	0.42	0.86	0.37			
V_E	0.49	0.58	0.58			
V_{ES}	0.09	0.83	0.05			

V_E = proportion of variance due to nonshared environmental effects.
V_A = proportion of variance due to additive genetic effects.
V_{NA} = proportion of variance due to nonadditive genetic effects, probably additive–additive epistasis.
V_G = variance proportion due to additive genetic effects.
V_{ES} = variance proportion reflecting shared rearing environmental effects.

findings are not surprising given the lack of power for finding significant heritability for change.

Goldsmith and McArdle (submitted) found that whereas biometric methodologies focusing on continuity versus change represent different conceptualizations of developmental processes, analyses of the Collaborative Perinatal Project data suggest only minor differences in substantive interpretation from autoregressive and difference score approaches.

THE CHALLENGE

There is clearly a need for longitudinal follow-up of the already existing data sets so that at least four occasions of measure are available. Until such data are available, many of the elegant methodological advances recently elaborated will not be utilized as they should and important developmental questions will remain unanswered. If longitudinal data were collected during periods for which stability is lower, e.g., early adulthood, one might find interesting relationships between genetic and environmental effects. Perhaps interindividual instability is due to differential synchronization of developmental events and thus to differential age of onset of changes in level. Developmental synchronization and age of onset may in turn be differentially affected by G and E. If personality scores show the same sort of change prior to death that has been demonstrated with cognitive data ("terminal decline"), then longitudinal genetic analyses of the young–old and old–old will be particularly interesting.

None of the studies have included p-technique types of measurement nor have they focused on intraindividual variability with extremely frequent measurement occasions (Plomin and Nesselroade 1990). If such data were available, individual genetic and environmental growth curves could be derived (Molenaar et al. 1991). It also appears that no ipsative data have been collected in behavioral genetic studies of personality, although ipsative approaches have been important in classic studies of stability for personality.

In the quest for providing generalizations, I have ignored the issue of differential heritability for personality traits, which is a hot topic in behavioral genetic circles and is related to developmental concerns of structural invariance. This is yet another area in which the interface of developmental psychology and behavioral genetics may provide fruitful results.

Many of the issues concerning the inability to fit genetic models to data from infancy and childhood can be resolved by studies of siblings, parents and offspring, and adoptees. There are, of course, problems inherent in cross-generational studies of development. Nevertheless, results from twin studies can best be interpreted when complemented by studies with multiple configurations of genetic and environmental relationships. Fortunately, there are well-known adoption and twin-family studies of development currently in progress which include data on personality.

Research on stability and change for personality has been characterized by developmental studies with too little attention paid to genetic effects on individual differences and by behavioral genetic studies that incorporate only a smattering of developmental theory and methodology. The challenge is to find a better synergism of behavioral genetic and life-span developmental perspectives.

ACKNOWLEDGEMENTS

The research and writing of this chapter were supported by grants from the Research Network on Successful Aging of the John D. and Catherine T. MacArthur Foundation and the National Institute on Aging (AG 04563). SATSA is an ongoing study conducted at the Division of Epidemiology of the Institute of Environmental Medicine at the Karolinska Institute in Stockholm in collaboration with the Center for Developmental and Health Genetics at The Pennsylvania State University.

REFERENCES

Carey, G. 1988. Inference about genetic correlations. *Behav. Genet.* **18**:329–338.

Costa, P., and R. McCrae. 1988. Personality in adulthood: A six-year longitudinal study of self-reports and spouse ratings on the NEO personality inventory. *J. Pers. Soc. Psychol.* **54**:853–863.

Costa, P., A. Zonderman, R. McCrae, J. Cornoni-Huntly, B. Locke, and H. Barbano. 1987. Longitudinal analyses of psychological well-being in a national sample: Stability of mean levels. *J. Gerontol.* **42**:50–55.

Dworkin, R., B. Burke, B. Maher, and I. Gottesman. 1976. A longitudinal study of the genetics of personality. *J. Pers. Soc. Psychol.* **34**:510–518.

Eaves, L., H. Eysenck, and N. Martin. 1989. Genes, Culture and Personalty: An Empirical Approach. London: Academic.

Eaves, L., J. Long, and A. Heath. 1986. A theory of developmental change in quantitative phenotypes applied to cognitive development. *Behav. Genet.* **16**:143–162.

Erikson, E. 1950. Childhood and Society. New York: Norton.

Finn, S. 1986. Stability of personality self-ratings over 30 years: Evidence for an age/cohort interaction. *J. Pers. Soc. Psychol.* **50**:813–818.

Floderus-Myrhed, B., N. Pedersen, and I. Rasmusson. 1980. Assessment of heritability for personality, based on a short form of the Eysenck Personality Inventory: A study of 12,898 twin pairs. *Behav. Genet.* **10**:153–162.

Goldsmith, H.H. 1989. Behavior-genetic approaches to temperament. In: Temperament in Childhood, ed. G.A. Kohnstamm, J.E. Bates, and M.K. Rothbart, pp. 111-132. Chichester: Wiley.

Goldsmith, H.H., and I.I. Gottesman. 1981. Origins of variation in behavioral style: A longitudinal study of temperament in young twins. *Child Devel.* **52**:91–103.

Hann, N., R. Millsap, and E. Hartka. 1986. As time goes by: Change and stability in personality over fifty years. *Psychol. Aging* **1**:220–232.

Hewitt, J., L. Eaves, M. Neale, and J. Meyer. 1988. Resolving causes of developmental continuity or "tracking." I. Longitudinal twin studies during growth. *Behav. Genet.* **18**:133–151.

Kogan, N. 1990. Personality and aging. In: Handbook of the Psychology of Aging, ed. J. Birren and K. Schaie, 3rd ed., pp. 330–346. San Diego: Academic.

Koskenvuo, M. 1979. The Finnish Twin Registry: Baseline Characteristics. Section III. Occupational and Psychosocial Factors, vol. 49. Helsinki: Univ. of Helsinki Press.

Loehlin, J., and R. Nichols. 1976. Heredity, Environment and Personality: A Study of 850 Sets of Twins. Austin: Univ. of Texas Press.

Martin, N., and R. Jardin. 1986. Eysenck's contribution to behavior genetics. In: Hans Eysenck: Consensus and Controversy, ed. S. Modgil and C. Modgil, pp. 13–62. Lewes, Sussex: Falmer.

Matheny, A. 1990. Developmental behavior genetics: Contributions from the Louisville Twin Study. In: Developmental Behavior Genetics: Neural, Biometrical, and Evolutionary Approaches, ed. M. Hahn, J. Hewitt, N. Henderson, and R. Benne, pp. 25–34. Oxford: Oxford Univ. Press.

McArdle, J. 1986. Latent-variable growth within behavior genetic models. *Behav. Genet.* **16**:79–95.

McCartney, K., M. Harris, and F. Bernieri. 1990. Growing up and growing apart: A developmental meta-analysis of twin studies. *Psychol. Bull.* **107**:226–237.

Molenaar, P., D. Boomsma, and C. Dolan. 1991. Genetic and environmental factors in a developmental perspective. In: Problems and Methods in Longitudinal Research: Stability and Change, ed. D. Magnusson, L. Bergman, G. Rudinger, and B. Törestad, pp. 250–273. Cambridge: Cambridge Univ. Press.

Pedersen, N.L. 1991. Behavioral genetic concepts in longitudinal analyses. In: Problems and Methods in Longitudinal Research: Stability and Change, ed. D. Magnusson, L. Bergman, G. Rudinger, and B. Törestad, pp. 236–249. Cambridge: Cambridge Univ. Press.

Pedersen, N., G. McClearn, R. Plomin, J. Nesselroade, S. Berg, and U. de Faire. 1991. The Swedish Adoption/Twin Study of Aging: An update. *Acta Genet. Med. Gem.* **40**:7–20.

Pedersen, N., R. Plomin, J. Nesselroade, G. McClearn, and L. Friberg. 1988. Neuroticism, extraversion and related traits in adult twins reared apart and reared together. *J. Pers. Soc. Psychol.* **55**:950–957.

Plomin, R. 1986. Development, Genetics, and Psychology. Hillsdale, NJ: Erlbaum.

Plomin, R., and J. Nesselroade. 1990. Behavioral genetics and personality change. *J. Pers.* **58**:191–220.

Pogue-Geile, M., and R. Rose. 1985. Developmental genetic studies of adult personality. *Devel. Psychol.* **21**:547–557.

Schaie, K., and S. Willis. 1991. Adult personality and psychomotor performance: Cross-sectional and longitudinal analyses. *J. Gerontol. Psychol. Sci.* **46**:275–284.

Standing (top), left to right:
J.C. Loehlin, N. Waller, M. Nöthen, P. Borkenau, D. Buss, W. Maier, S. Biryukov
Standing (middle), left to right:
N. Pedersen, P. Kline
Seated, left to right:
H.H. Goldsmith, H. Schepank, K. McCartney

11

Group Report: What Can Twin Studies Contribute to the Understanding of Personality?

K. McCARTNEY, Rapporteur

S.D. BIRYUKOV, P. BORKENAU, D.M. BUSS,
H.H. GOLDSMITH, P. KLINE, J.C. LOEHLIN,
W. MAIER, M. NÖTHEN, N.L. PEDERSEN,
H. SCHEPANK, N. WALLER

CURRENT STATUS ON TWIN RESEARCH FOR PERSONALITY

The current landscape of twin studies of personality is remarkably different from that of two decades ago. The field once consisted largely of studies of small samples, assessed with one or more self-report inventories. The evidence for heritable effects on personality variation from these pioneering studies encouraged new studies that expanded the field in at least five ways. While older studies tended to focus on students and young adults, newer studies collectively encompass the lifespan, from the neonatal period (Riese 1990) to older adults (Pedersen et al. 1991). With access to population registers and other large population bases, sample sizes have reached several thousand twin pairs in some projects. Another change is a trend toward longitudinal designs, with a dual focus on continuity and change. Studies now also tend to incorporate nonpersonality variables in a broader assessment strategy. Newer studies tend to include assessment batteries consisting of cognitive and physiological measures, in addition to personality measures. Finally, current assessment methodologies also include observational techniques, both with respect to the assessment of personality and the assessment of the environment.

There are a number of recent reviews of behavioral genetics research, including twin research, with respect to personality (see Eaves et al. 1989; Loehlin 1992; McCartney et al. 1990). Two points are worth summarizing. The first point concerns what we know generally about genetic and environmental contributions to personality.

Twins as a Tool of Behavioral Genetics
Edited by T.J. Bouchard and P. Propping © 1993 John Wiley & Sons Ltd.

Loehlin's examination of the *five-factor model* of personality (1992), using scales of moderate reliability, led to the following conclusion:

> Additive genetic effects ... fell generally in the moderate range, 22% to 46%; shared family environment effects were small, 0% to 11%; and the ambiguous third factor—nonadditive genetic or special monozygotic twin environments—was intermediate at 11% to 19%, except for culture, where it was a lesser 2% to 5%. The remaining variation, 44% to 55%, presumably represents some combination of environmental effects unique to the individual, genotype-environment interaction, and measurement error. (Loehlin 1992, p. 68).

The second point concerns developmental differences in twin similarity. McCartney et al.'s (1990) meta-analysis demonstrated an interesting developmental trend in the intraclass correlations of monozygotic (MZ) and dizygotic (DZ) twins, namely that they decrease with the age of twins. This change appears to be due to a shift toward individual environments that are not shared by twin pairs.

The current twin literature yields a reasonably coherent set of conclusions. These conclusions are largely consonant with the broader behavioral genetic literature. There are, however, a few features of twin study results that differ from most family and adoption results. In addition, there are many more complex questions about genetic and environmental architecture and, especially, developmental genetics that require extensive study. Furthermore, the traditional variance partitioning approach of twin studies needs to be combined with rich theoretical perspectives from both biology and psychology to address questions of process and mechanism. These perspectives and questions occupied most of our group discussion, which focused on six diverse topics. Our prescriptions for future directions in twin research with respect to these topics follow.

THE STRUCTURE OF PERSONALITY AND TWIN STUDIES

Behavioral Genetics and the Structure of Personality

Behavioral genetic studies have typically relied on the assessment of dispositions with respect to emotional experience (e.g., anxiety) and social interaction (e.g., extraversion). Historically, the field of personality is viewed as broader, e.g., it also encompasses aspects of self development and defense mechanisms. Personality as a field does not have rigid boundaries; there is typically a distinction between the personality and cognitive domains, although even this boundary is fuzzy.

What is the extent of the personality domain and how is it best understood? After 60 years of empirical research, personologists have yet to agree on the number or nature of necessary and sufficient dimensions of personality (Eysenck 1975; Cattell 1981; Waller et al. 1991). Perhaps the most popular model currently is the so-called *five-factor model* (John 1990). These traits—(1) surgency or extroversion, (2) agreeableness, (3) dependability or responsibility, (4) emotional stability, and (5) intellectual orientation, culture, or openness—are considered as minimally sufficient for characterizing the domain of lexical terms for personality traits, as limited by somewhat arbitrary decision rules for inclusion in the domain. Loehlin (1992) recently summarized the behavioral genetic literature for these five factors and found moderate heritabilities for all five of these higher-order dimensions. The group noted that these five dimensions are not exhaustive of all that is interesting to personality researchers.

Some additional content domains have yet to be investigated from a behavioral genetic perspective. For example, future twin studies might include dimensions of sexuality, self and other evaluation, and adult manifestations of temperament. In the realm of sexuality are characteristics such as short- versus long-term sexual strategies, mate preferences, and adolescent sexual drive (see Buss, this volume). Explicitly evaluative terms were omitted from the original dictionary searches that yielded the *five-factor model*, and these may be of genetic interest.

Temperament is a construct that, in the United States, is chiefly associated with early personality (for a review of the behavioral genetic literature, see Goldsmith 1989). However, there are also concepts of adult temperament that have been little investigated in the West. The group considered in detail the model of temperament by Rusalov (1987, 1989), which addresses the formal dynamic properties of behavior. This research originated in the work of Pavlov, Teplov, and Nebylitsyn. The formal aspects of behavior (temperament) are distinguished from personality, which is understood as involving sociopsychological content. In brief, this conceptualization involves three facets of activity (ergonicity, flexibility, and tempo) and emotional sensitivity. Each of these four aspects of temperament is divided into reactivity in the world of objects versus the social world. Research in this tradition has been both neurophysiological and, more recently, questionnaire-based.

Cross-cultural and family studies have been undertaken, and a twin study in Russia is being contemplated. Integration of this research tradition into the mainstream of behavioral genetics is an interesting possibility. It offers a conceptual vehicle for jointly investigating psychometric personality structure and neurophysiology. Surely, evidence from twin studies that a genetic basis existed for the covariation among neurophysiological variables and the formal aspects of personality would hasten acceptance of these models as well as stimulate integration of the fields of personality and human neuroscience.

Types vs. Dimensions

We discussed the utility of categorical models of personality and noted that, from a behavioral genetic perspective, the search for personality types can be focused at three levels of analysis: genotype, phenotype, and environment. Tooby and Cosmides (1989) have viewed the question from an evolutionary perspective and note the infeasibility of a genetic basis for personality types. They argue that personality types require a suite of complex psychological mechanisms that are likely to be broken up across the generations because of sexual recombination. "However, different genomic constellations could be activated by single gene or hormonal 'switches,' and these could serve as a heritable basis for types" (Tooby and Cosmides 1989). The group discussed the likelihood that features which appear to constitute a type at the level of overt behavior might be dimensional at the level of underlying physiology, or vice versa.

Much of the discussion focused on genetic and environmental effects of phenotypically derived dimensions, e.g., the heritability of factor scores from the *five-factor model* vs. other models of personality. It is clear that in the future, large sample studies with a wide range of items and measures can be used to assess whether there are distinct phenotypic, genetic, or environmental factor structures. Just as Fulker and Cardon (this volume) demonstrated a differential genetic vs. environmental structure for cognitive abilities, a fruitful direction for future research (indeed, one which may serve to clarify issues concerning the five-factor vs. other N-factor models of personality) may be the application of biometrics factoring to personality.

EVOLUTIONARY PSYCHOLOGY OF PERSONALITY

The Human Nature of Personality

Buss (this volume) proposes that all behavior is a function of internal mechanisms, combined with input that activates those mechanisms. Psychological mechanisms are information processing devices that take in certain input, operate on that input with decision rules, and produce some form of output such as manifest behavior or information to other mechanisms. All manifest behavior requires the existence of such mechanisms. If the same environmental event affects men and women differently, or extroverts and introverts differently, it is because the two comparison groups differ in some aspects of their psychological mechanisms. The only known causal process capable of producing complex organic mechanisms is evolution by natural or sexual selection.

If mechanisms produce behavior, and evolutionary processes produce mechanisms, then the key issue concerns the nature of the mechanisms evolution has produced, and where, when, and how mechanisms are relevant. Evolutionary psychologists believe that the mechanisms are likely to be large in number, complex in nature, and domain-specific in the sense of being uniquely tailored to solving specific adaptive problems. The mechanisms are also likely to be species-typical (see Buss, this volume). The group debated the usefulness of such a broad term as psychological mechanism and the likelihood that many specific mechanisms could have evolved. In constructing a more precise definition, it will be necessary to distinguish between manifest behavioral responses and the mechanisms that are responsible for producing those responses (Tooby and Cosmides 1989).

Compelling Form-to-Function Arguments

Evolutionary biologists have discussed the dangers inherent in relying on form-to-function arguments (Gould and Lewontin 1978). Contemporary evolutionary psychologists recognize the need to formulate evolutionary-informed hypotheses in a falsifiable manner and to test evolutionary hypotheses using rigorous empirical standards. The following example illustrates how twin studies can supplement form-to-function arguments in evolutionary personality psychology.

More than fifty years ago, R.A. Fisher (1930) predicted that fitness-related traits (such as brood size) would have low, rather than high, narrow-sense heritabilities if there had been strong directional selection operating on the trait. Furthermore, the genetic variance that remains should be primarily of the nonadditive type. This idea, known as *Fisher's fundamental theorem of natural selection*, has recently been tested in a wide range of species and found to be essentially correct (Mousseau and Roff 1987). Preliminary results from a biometrics analysis of mate preference data in twins are also consistent with Fisher's theorem (Waller and Zavala, in preparation).

Data from twins cannot prove that directional selection has operated on a trait. In the absence of inbreeding data, however, twin analyses provide the best means of testing evolutionary-informed hypotheses in the interface between evolutionary psychology and human behavioral genetics.

CONTINUITY AND CHANGE IN PERSONALITY DEVELOPMENT

Continuity and change in personality development can be approached in a variety of ways using a behavioral genetic methodology. Aspects of stability and change that can be considered include (a) stability in the relative importance of genetic and environmental effects; (b) the contribution of genetic and environmental influences to phenotypic stability and change; (c) stability in genetic and environmental effects (i.e., the magnitude of genetic and environmental correlations across time); and (d) the importance of these influences for structural stability.

Thus far, most of the developmental behavioral genetic studies of personality and temperament have been cross-sectional. The only tentative conclusion that can be drawn from this body of literature is that from childhood through early adulthood, the importance of nonshared environmental effects increases (McCartney et al. 1990). Clearly, longitudinal designs, with at least four waves of data collection, are necessary to delineate unambiguously cohort effects from age effects and true developmental changes.

Depending on the purpose of a given study, investigators might consider several desirable features of longitudinal designs. If change is of primary interest (rather than stability), studies should focus on known or suspected periods of transition, e.g., the transition to puberty or the period of "terminal decline." Transitions at midlife have been neglected by developmentalists; this would be a fruitful time to assess the effects of various life events on personality change. Adjacent periods of assessment should include overlapping measures to allow for inter-interval test equating. This is particularly important for studies spanning several developmental periods. Because longitudinal studies require repeated use of the same instrument over extended periods of time, multiple indicators of latent constructs should be included such that factors estimated are not instrument specific. Consideration should also be given to potential correlates of sample attrition. For example, obsessive-compulsives are more likely to continue in a longitudinal study. Finally, studies should include direct assessment of the environment. Questions concerning environmental correlates of differential rates of change (i.e., slope change) can then be examined.

There are a number of elegant models available for assessing the genetic architecture of development (see Pederson, this volume; Meyer et al. 1991). These methodological advancements as well as survival analyses and growth curve models (McArdle 1986) are valuable tools for further research strategies. However, their value will only be realized as longitudinal data become available.

CONCEPTUALIZING THE ENVIRONMENT IN BEHAVIORAL GENETICS DESIGNS

Some have criticized early twin studies for estimating the effect of the environment indirectly (e.g., Wachs 1983). Critics fail to note that the effect of the genetic contribution is estimated indirectly as well. Behavioral genetic techniques work by apportioning variance to environmental and genetic components based on associations between family members with varying genetic resemblance. With structural modeling techniques, it is now possible to include direct assessments of the environment in the model (see Morton 1974; Rao et al. 1976). Conceptualizations of the environment have expanded to include a differentiation between shared and nonshared environments, a consideration of how genotypes and environments are correlated, and a question of whether genotypes and environments interact. These advances have made the field of behavioral genetics more central to psychology. This trend is likely to continue if behavioral genetics studies include the assessment of various aspects of the environment.

Shared and Nonshared Environments

As previously mentioned, there is converging evidence that little variance in personality is accounted for by the shared environment. The nonshared environment appears more important, although its estimation is typically confounded with measurement error. It is important to differentiate the nonshared environment from the nonfamilial environment; the two terms are not equivalent, although they have been confused by some. Nonshared environments may be either familial (e.g., differential parental treatment) or nonfamilial (e.g., different peer groups). Thus, the failure to identify large effects for the shared environment does not mean that parents do not influence their children's development through environmental mechanisms.

There are a number of reasons why the effect of the shared environment is generally small. It is possible that contradictory shared processes are at work, such that the resultant impact is small; however, the probability of this occurring uniformly across a wide variety of personality traits would seem to be small. Another possible explanation is that shared environments of siblings do not result in shared experiences. McCall (1983) offers an anecdote involving two siblings visiting a museum, presumably a shared event. One child is bored while the other is engaged by the exhibits. The environment, the museum, is the same for both children; the experience (boredom vs. intellectual stimulation) is not. Consideration of contradictory processes and the assessment of experience (as opposed to environment) have implications for future assessments of the environment.

Developmentalists and behavioral geneticists alike have been taken with the finding that the nonshared environment is important. It is time to identify specific aspects of the environment that are important for the development of personality characteristics. Some environmental mechanisms will operate within families, while others will not.

One of the few examples of intrafamily mechanisms that has been examined is differential parental treatment of children (see Schepank, this volume). Initial attempts to assess siblings' perceptions of differential parental treatment involved the use of a questionnaire (see Baker and Daniels 1990). More recent attempts that involve nonipsative, independent assessments of siblings' experience are underway (Baker, pers. comm.).

In general, the group was in consensus that future research should include direct assessment of the environment. Four classes of environmental variables are likely to be important for personality: (a) those assessing prenatal and perinatal trauma (see Bryan and Rutter, both this volume), (b) those assessing parent-child interaction, especially with respect to the emotional connection, (c) those assessing the mother-father-child triad, and (d) those assessing peer group characteristics.

Researchers who study twins reared both together and apart should examine intrapair differences as a function of differences in environmental experiences (although we are all aware of the caveats implied by the identification of genetic effects for measures of environment). The search for environmental differences between discordant identical twins, which has been so central to psychiatric genetics (Schepank 1974, 1981; see also this volume), would also be a powerful technique when applied to normal-range personality variation, given that adequate measures of the environment are available.

The issue gets complicated when we consider that measures of the environment may indeed reflect genetic variance (Plomin and Bergman 1992; Pedersen et al. 1991). For example, there is heritability for a life event such as divorce, which may be influenced by a heritable personality trait, such as neuroticism or assertiveness.

Genotype–Environment Interaction and Correlation

The concepts of genotype-environment correlation and interaction, and suggestions that have been made for detecting them empirically, are covered by Loehlin (this volume). Examples of genotype-environment interaction appear to be rare in the empirical literature. Behavioral geneticists may be more successful in documenting such interactions by examining theoretically determined subgroups. It is less clear whether genotype-environment correlations have been demonstrated empirically, because there is disagreement concerning whether it is profitable to treat the active and reactive varieties as distinct from "direct" effects of the genes, i.e., whether genotype-environment correlation of these types is distinct from genetic effects. Insofar as all effects of genes on personality are indirect, is it appropriate to distinguish some of them by giving them a separate label of genotype-environment correlation? One argument for doing this is that it may be helpful to focus attention on the mechanisms by which such correlations arise, and the relative contributions of such mechanisms to overall genetic variation.

By distinguishing among such sources as reactive, or evocative, and active genotype-environment correlation, one points to two different ways in which genetic and environmental influences may become positively (or negatively) associated. By further distinguishing forms of active genotype-environment correlation into those derived from the manipulation of environments and those derived from the selection of environments (see Buss, this volume), one can facilitate research into mechanisms. Obviously, this process of subdivision can be continued further, e.g., one might distinguish manipulations of the environment into those that affect the nonsocial and social environment, and the latter into those that occur via persuasion and force.

In the long run, such theoretical elaborations must be evaluated by their heuristic contributions: Do they indeed extend our understanding of the relevant mechanisms of personality development and change? Do they assist rather than hamper our ability to communicate with colleagues in related fields?

TWIN ECOLOGY, ASSUMPTIONS OF THE TWIN METHOD, AND SPECIAL TWIN ENVIRONMENTS

The nature of twinning, twin development, and twin relationships is a key issue for behavioral geneticists, for both substantive and methodological reasons. Substantively, twins are of interest to behavioral geneticists in their own right, and some psychologists focus on this aspect (Hay 1985). More relevant to behavioral geneticists are the implications of twin ecology for the two key assumptions of the traditional twin design: the assumption that twins are representative of the general population and the "equal environments" assumption. Twins clearly differ from singletons in some regards (e.g., birth complications and the condition that they always have a sibling—their co-twin); identical co-twins are clearly treated more alike than fraternal co-twins in some regards (e.g., being dressed alike, being mistaken for one another). However, the crucial question is whether such factors are associated with personality differences.

Available evidence suggests that inferences from twin studies of personality are not substantially biased by differences in how similarly identical versus fraternal pairs are treated (see Plomin et al. 1989). However, the group discussed the need for more extensive validational evidence, and considered the desirability of including in each study existing methodological checks. These checks include examining the distribution of twin differences, correlating differences in measured, imposed, and environmental factors with differences in outcome, and comparing twins whose parents are mistaken about zygosity with those who are not. The group also considered the need to continually search for other twin-specific factors and relate them to outcome personality variables. Of course, the crucial test for the validity of twin designs is consistency with family and adoption data, as tested in comprehensive models.

The Possibility of Special Twin Environments for Personality

It is possible that twins are forced into different social niches in family contexts or that they select different niches. This process is sometimes called a contrast effect. The process can be considered a bias in twin studies, especially if the niche placement is independent of phenotype. However, it seems likely that niche placement is likely to be niche selection and that it would be conditioned on genotype. Assimilation effects are the opposite of contrast effects. It seems plausible that assimilation effects would occur more frequently in identical pairs. The question is whether there is actual evidence for either assimilation or contrast effects in personality data. Twin data are consistent with the possibility of identical twin assimilation effects, but they cannot be distinguished from the alternative possibility of nonadditive genetic variance (Loehlin 1992). Because fraternal co-twins are certainly more alike than ordinary siblings, one might at first dismiss the possibility of fraternal contrast effects. Of course, the same sort of contrast effects might occur in ordinary siblings.

The group reached the conclusion that twinship, in all likelihood, does not have a global or uniform effect on personality. Behavioral geneticists need to assess and consider the quality of twin relationships. With a multivariate view of the relationships that twins experience, the issue could be evaluated. For example, twin relationships could be assessed as they relate to variables such as communication style, role differentiation, and perception of the twins as a pair vs. as individuals.

PERSONALITY VARIATION AND PSYCHOPATHOLOGY

There is inadequate research on the association between the sphere of normal personality variation and the realm of personality disorders and severe psychoses. The relationship between these two areas has been viewed differently by clinicians and by personality psychologists. For clinicians, psychopathological disorders identified in clinical samples serve as the starting point. The study of personality tends to focus on patterns associated with a particular disorder, as evidenced by assessments of premorbid functioning. For personality psychologists, personality traits identified in the general population serve as the starting point. In this approach, the study of personality tends to focus on risk factors, or protective factors, for the subsequent development of a disorder.

Not surprisingly, the clinical approach has been used in the study of severe and rare disorders (e.g., psychoses), while the personality approach has been used in the study of more common disorders (e.g., personality disorders and depression). The group considered the association between personality variation and psychopathology primarily with respect to schizotypy and schizophrenia, psychoanalytic constructs, and neuroticism and depression.

Schizotypy and Schizophrenia

Two types of studies have been used to identify personality and behavior patterns related to schizophrenia: retrospective assessments of premorbid functioning and prospective assessments of high risk families. These studies demonstrate that schizophrenia is associated with a history of social withdrawal or schizoidia, perceptual aberrations, affective flattening, anhedonia, and bizarre behavior and speech. Non-schizophrenics who exhibit these traits and behaviors are considered to be at risk for developing the disorder. This pattern of abnormal behavior might be conceptualized as a personality *type*; for example the schizotype (Meehl 1989) or the schizotypal personality disorder (American Psychiatric Association 1987). Alternatively, this pattern of behavior might be conceptualized as a personality *dimension*; for example, the schizotypy trait (Claridge 1981, 1987) or proneness to psychoses (Chapman et al. 1980). The personality type-schizophrenia link forms a schizophrenia spectrum; the validity of this spectrum is based on family studies (see Kaprio, this volume).

Twin research has been used to examine heritabilities for schizophrenia and schizotypal personality disorder. There is contradictory research concerning which is more heritable: schizophrenia or the schizophrenia spectrum (see Farmer et al. 1987; Torgeresen et al. 1980). Some personality patterns seem to be particularly heritable, for example affective flattening.

Neuroticism and Depression

Neuroticism is a frequently studied personality trait; depression disorders are the most common disorders in psychiatry; and research has to link the two. For example, neuroticism is a risk factor for the development of depression (see Hirschfield et al. 1990) and neuroticism is elevated in the healthy relatives of depressed prohands (see Maier et al. 1992). Twin studies have demonstrated a strong genetic component for some varieties of neuroticism and for depression; however, it is not clear whether these two disorders are linked genetically.

Psychoanalytic Constructs

Psychoanalytic constructs have been conceptualized on a continuum ranging from healthy, well-adapted states to severe neurotic and/or functional disturbances. Three major constructs discussed are oral-depressive, obsessive-compulsive, and hysterical structure. Two twin studies are available in this area (Schepank 1974; Heigl-Evers and Schepank 1980, 1981; Torgeresen 1980). Both studies found evidence for a genetic determination; however, environment was at least equally important. Yet it remains to be determined whether these results hold up in twin samples from the general population, with a follow-up component.

Beyond these theoretically derived constructs, less specific dimensional measures of disability and impairment should be additionally included in future twin studies. One example for this kind of measure would be to focus on psychogenic impairment, split up into three dimensions: somatic, psychic, and socio-communicative impairment (Schepank 1987).

Unresolved Issues

There are many unresolved issues concerning associations between personality variation and psychopathology. It may be that twin research can help address some of these issues. Future twin studies on psychiatric disorders should incorporate standard personality assessments; this would be particularly important in gene linkage and association studies. It is less clear how twin studies could be used to decide whether personality disorders would provide one sort of bridge between research on personality and that on psychopathology. How or whether psychiatric studies can inform research on personality is also not clear.

The spectrum concept appears to be fertile ground for integrating the study of psychopathology and normal range personality variation. However, there are some impediments to research in this area. For example, the base rates of personality extremes are unknown, partially because the diagnostic criteria for personality disorders have undergone repeated revision. Another point emphasized by Gottesman is that we may need a "double standard" for what constitutes relevant personality dysfunction within twin pairs, where one member is an index case, versus the use of such criteria in the general population. For example, fidgeting in the offspring of one of two offspring of a Huntington's Disease patient has different diagnostic significance than fidgeting in an individual from the general population.

CONCLUSIONS

The group's prescriptions for new directions in twin research on personality abound in this chapter. Some of the suggestions are familiar ones, for example, the need for comprehensive, longitudinal study of large samples. Such designs require substantial resources, yet are worth the investment. Another familiar suggestion concerns the need for direct assessments of the environment in biometric models and the need for better conceptualizations of environment and experience. Again, such research efforts require resources, especially since it is not clear which aspects of the environment should be measured. What does seem clear is that it is not a trivial matter to assess an individual's environment. Important components of the environment appear not to be shared by siblings. Studies of differential parental treatment of twins appear to be a fruitful place to begin.

A central topic concerns how twins can be used to understand the structure of personality. Twins can be very useful in the identification of distinct phenotypic, genetic, and environmental factor structures of personality. Links between normal personality and psychopathology can also be assessed with family studies, including twin studies. Yet it is clear that further work needs to be done concerning potential biases in twin studies. The field has a great deal to learn about contrast and assimilation effects. Such studies will yield important methodological information concerning the equal environment hypothesis and important theoretical information concerning sibling effect more generally.

Other suggestions are less familiar, even novel. For example, the group discussed the fact that methods will soon be available to analyze the genetic background of personality variables. Twin studies will be useful to identify where such molecular investigations should begin, i.e., with respect to traits with high heritability; however, twins are no more valuable than other pairs of family members for molecular studies, except when age spacing is an issue. The group also discussed how evolutionary theory might suggest new personality variables for investigations. Twin studies will be useful here, as with other personality variables, in assessing components of variance and in assessing genetic effects on the environment.

The study of twins has facilitated the development of the field of behavioral genetics, as researchers consider whether personality, as well as other traits, are heritable. As the field moves beyond such considerations, twin samples will continue to offer one means for studying questions of process and mechanism.

REFERENCES

American Psychiatric Association (APA). 1987. DSM-III-R: Diagnostic and Statistical Manual of Mental Disorders (3rd edition). Washington, D.C.: APA.

Baker, L.A., and D. Daniels. 1990. Nonshared environmental influences and personality differences in adult twins. *J. Pers. Soc. Psychol.* **58(1)**:103–110.

Cattell, R.B. 1981. Personality and Learning Theory. New York: Springer.

Chapman, L.F., W.S. Edell, and F.P. Chapman. 1980. Physical anhedonia, perceptual aberration, psychosis proneness. *Schizo. Bull.* **6**:639–653.

Claridge, G.S. 1981. Psychotocism. In: Dimensions of Personality, ed. R. Lynn, pp. 79–110. Oxford: Pergamon.

Claridge, G.S. 1987. The schizophrenias as nervous types revisited. *Br. J. Psychiat.* **151**:735–743.

Eaves, L.J., H.J. Eysenck, and N.G. Martin. 1989. Genes, Culture and Personality: An Empirical Approach. New York: Academic.

Eysenck, H.J. 1975. The EPQ. London: Hodder and Stoughton.

Farmer, A.E., P. McGuffin, and I.I. Gottesman. 1987. Twin concordance for DSM-III schizophrenia: Scrutinizing the validity of the definition. *Arch. Gen. Psychiat.* **44**:634–641.

Fisher, R.A. 1930. The Genetical Theory of Natural Selection. Oxford: Clarendon.

Goldsmith, H.H. 1989. Behavior-genetic approaches to temperament. In: Temperament in Childhood, ed. G.A. Kohnstamm, J.E. Bates, and M.K. Rothbart, pp. 111–132. Chichester: Wiley.

Gould, S.J., and R.C. Lewontin. 1978. The spandrels of San Marco and the Panglossian paradigm: A critique of the adaptationist programme. *Proc. Roy. Soc. Lond.* **205**:581–598.

Hay, D.A. 1985. Essentials of Behavioral Genetics. Melbourne: Blackwell.

Heigl-Evers, A., and H. Schepank. 1980. Ursprünge seelisch bedingter Krankheiten. Göttingen: Vandenhoeck & Ruprecht.

Hirschfeld, R.M.A., G.L. Klerman, P. Lavori, M.B. Keller, P. Griffith, and W. Coryell. 1989. Premorbid personality assessments of first onset of major depression. *Arch. Gen. Psychiat.* **46**:345–354.

John, O.P. 1990. The Big Five factor taxonomy: Dimensions of personality in the natural language and in questionnaires. In: Handbook of Personality Theory and Research, ed. L.A. Pervin, pp. 66–100. New York: Guilford.

Loehlin, J.C. 1992. Genes and Environment in Personality Development. Newbury Park, CA: Sage.

Maier, W., D. Lichternmann, J. Minges, and R. Heun. 1991. Personality traits in subjects at risk for unipolar major depression: A family study perspective. *J. Affec. Dis.* **24**:153–164.

McArdle, J.J. 1986. Latent variable growth within behavior genetic models. *Behav. Genet.* **16(1)**: 163–200.

McCall, R.B. 1983. Environmental effects on intelligence: The forgotten realm of discontinuous nonshared within-family factors. *Child Devel.* **54(2)**:408–415.

McCartney, K., M.J. Harris, and F. Bernieri. 1990. Growing up and growing apart: A developmental meta-analysis of twin studies. (Paper presented at International Conference on Psychology, Sydney, 1988). *Psychol. Bull.* **107(2)**:226–237.

Meehl, P.E. 1989. Schizotaxia revisited. *Arch. Gen. Psychiat.* **46**:935–944.

Meyer, J.M., L.J. Eaves, A.C. Heath, and N.G. Martin. 1991. Estimating genetic influence on the age-at-menarche: A survival analysis approach. *Am. J. Med. Genet.* **39(2)**:148–154.

Morton, N.E. 1974. Analysis of family resemblance. *Am. J. Hum. Genet.* **26**:318–330.

Mousseau, T.A., and D.A. Roff. 1987. Natural selection and the heritability of fitness components. *Heredity* **59**:181–197.

Pedersen, N.L., G.E. McClearn, R. Plomin, J.R. Nesselroade, et al. 1991. The Swedish Adoption/Twin Study of Aging: An update. *Acta Genet. Med. Gem.* **40(1)**:7–20.

Plomin, R., and C.S. Bergeman. 1992. The nature of nurture: Genetic influence on "environmental" measures. *Behav. Brain Sci.* **14(3)**:373–427.

Plomin, R., J.C. DeFries, and G.E. McClearn. 1989. Behavioral Genetics: A Primer, 2nd ed. New York: Freeman.

Rao, D.C., N.E. Morton, and S. Yee. 1976. Resolution of cultural and biological inheritance by path analysis. *Am. J. Hum. Genet.* **28**:228–242.

Riese, M.L. 1990. Neonatal temperament in monozygotic and dizygotic twin pairs. *Child Devel.* **61**:1230–1237.

Rusalov, V.M. 1987. Foundations of a special theory of human individuality. In: Abstracts of 8th International Congress of Logic, Methodology and Philosophy of Science, vol. 5, part 3, pp. 218–220. Moscow: Nauka.

Rusalov, V.M. 1989. Object-related and communicative aspects of human temperament. *Pers. Indiv. Diff.* **10**:817–827.

Schepank, H. 1974. Erb- und Umweltfaktoren bei Neurosen. Tiefenpsychologische Untersuchungen an 50 Zwillingspaaren. Psychiatry Series, volume 11. Berlin: Springer.

Schepank, H. 1981. Diskordanzanalytische Untersuchungen an 9 ES Paaren über spezielle neurose-pathogene Umwelteinflüsse. In: Ursprünge seelisch bedingter Krankheiten, ed. A. Heigl-Evers and H. Schepank, pp. 619-683. Göttingen: Vandenhoeck & Ruprecht.

Schepank, H. 1987. Epidemiology of Psychogenic Disorders. The Mannheim Study. Results of a Field Survey in the F.R. Germany. Berlin: Springer.

Tooby, J., and L. Cosmides. 1989. On the universality of human nature and the uniqueness of the individual: The role of genetics and adaptation. Special Issue: Biological Foundation Personality. *Evol. Behav. Genet. Psychophys.* **58(1)**:17–67.

Torgeresen, S. 1980. The oral obsessive and hysterical personality syndromes. *Arch. Gen. Psychiat.* **37**:1272-1277.

Wachs, T.D. 1983. The use and abuse of environment in behavior-genetic research. *Child Devel.* **54(2)**:396–407.

Waller, N.G., S.O. Lilienfeld, A. Tellegen, and D.T. Lykken. 1991. The tridimensional personality questionnaire: Structural validity and comparison with the multidimensional personality questionnaire. *Mult. Behav. Res.* **26(1)**:1–23.

12

How Informative Are Twin Studies of Child Psychopathology?

M. RUTTER[1], E. SIMONOFF[1], and J. SILBERG[2]

[1]MRC Child Psychiatry Unit, Institute of Psychiatry, De Crespigny Park, Denmark Hill, London SE5 8AF, U.K.
[2]Department of Human Genetics, Medical College of Virginia, Virginia Commonwealth University, Box 3, Richmond, Virginia 23298–0003, U.S.A.

ABSTRACT

There is a rich potential for twin studies of childhood psychopathology; however, if this potential is to be realized, it will be crucial to undertake further research into the assumptions of the twin design. Investigations must examine environmental similarity between MZ and DZ pairs, the comparability of twins and singletons, and the origins and effects of assortative mating; care will also be needed in looking for possible ascertainment bias. Further research is also needed into assumptions regarding psychopathology, with studies designed to use genetic data to develop and test phenotypic definitions and to examine continuities/discontinuities between normality and disorder. Twin studies may be used to assess levels of heritability but, more particularly, they are of value for investigating the nature of disorders, the meanings of comorbidity and heterotypic continuity, risk mechanisms, the testing of environmental effects, and for using twins as a natural experiment to examine the causes of disorder that differ in rate between twins and singletons.

INTRODUCTION

The basis of science lies in the experimental method and, when contrived experiments in the laboratory are not ethically or practically possible, there has to be recourse to "natural experiments" in which real life circumstances provide situations that "pull apart" variables that ordinarily go together (Rutter 1981). More than a century ago, Galton noted the utility of twins as an experiment of nature that allowed a separation of the effects of nature and nurture, because monozygotic (MZ) and dizygotic (DZ) twin pairs were similar in the extent to which they shared the same environment but differed in the extent to which they shared segregating genes. The method has proved a powerful one and has well justified its place as a key genetic design (Plomin 1986).

Twins as a Tool of Behavioral Genetics
Edited by T.J. Bouchard, Jr. and P. Propping © 1993 John Wiley & Sons Ltd.

Its application in the field of childhood psychopathology is rather more recent (Rutter et al. 1990a, b) but there, too, its findings have been informative, and it is clear that it has a rich potential yet to be realized. Nevertheless, now that twin studies of childhood psychopathology are beginning to burgeon, the time is ripe for a critical look at some of the key assumptions of twin designs and the purpose to which they may be put, in order to make some inferences on possible ways ahead. These are the purposes of this chapter.

RATIONALE OF TWIN STUDIES OF PSYCHOPATHOLOGY

Assumptions of Twin Design

Environmental Similarity

For a very long time, it has been appreciated that it was crucial to test the assumption that familial environmental effects were comparable for MZ and DZ twins. Although it is apparent that the social environment of MZ twins is more similar than that of DZ twins, such evidence as is available suggests that, for the most part, the similarity is the result (and not the cause) of their behavioral similarity; moreover, the environmental features that vary by zygosity appear not to be ones that have much effect on the degree of concordance for psychopathology (Rutter et al. 1990a). Whereas it seems unlikely that this zygosity difference in environmental similarity creates a serious bias in the use of twin designs, it has to be noted that there has been surprisingly little systematic research into the psychological development and parental treatment of twins (Rutter and Redshaw 1991), and it would be premature to conclude that the matter is necessarily wholly irrelevant for all types of childhood psychopathology.

However, there are other aspects of twin environments that have been relatively neglected up to now and which might be of greater relevance in some circumstances. Three may be highlighted. First, it is well established that there tend to be greater discrepancies in birth weight between twins in MZ than DZ pairs, largely due to the feto-fetal transfusion syndrome and it may be that this nongenetic biological discordance in MZ pairs has implications for within-pair differences in psychological attributes. It deserves further study using designs in which MZ within-pair concordance/correlations are contrasted according to the presence/absence of the feto-fetal transfusion syndrome. Second, there are several factors that influence the frequency of DZ twinning but which have little or no effect on MZ twinning. High maternal age is one (MacGillivray et al. 1988), and this might be pertinent either in terms of its association with an increased risk for congenital anomalies or with respect to possible effects on styles of parenting. However, increasingly, fertility drugs and various methods of *in vitro* fertilization are also influential and it remains unclear how far this is leading to the parents of DZ twins being selected for characteristics that may be psychologically relevant. Third, so far, very little attention has been paid to the relationships between the twins themselves with respect to either assimilation (i.e., a tendency to identify with each other and share activities) or de-identification (i.e., a tendency to accentuate their differences and emphasize their individuality) (Goodman 1991). There is some evidence that MZ twins are more likely than DZ twins to cooperate effectively (see Rutter and Redshaw 1991), but we know very little about how twin-twin relationships develop and even less about their psychological implications.

In investigating these issues further, it will be necessary to study between-twin relationships directly (rather than infer them from variance estimates) because it may well be that *both* assimilation and de-identification occur, but that they do so in different pairs. Insofar as that is the case, their effects may counterbalance each other with respect to population variance but yet still have substantial effects at the individual level. It is also important to test whether twin-twin relationships are the same as sib-sib relationships (see below).

Comparability of Twins and Singletons

A further key assumption of the twin design is that twins and singletons are comparable with respect to patterns of genetic and environmental effects on psychopathology. Unless that assumption is justified, the results of twin analyses cannot be generalized to singletons. Because the rate of many disorders of interest, such as schizophrenia, has not been found to differ between twins and singletons, reviewers have tended to dismiss this concern. However, with respect to childhood psychopathology, this dismissal is premature. First, several studies have shown that the language and reading skills of twins tend to be lower than those of singletons (Rutter and Redshaw 1991). This means that twin studies of both traits are likely to underestimate the general importance of genetic factors (because the depression of verbal skills in twins must stem from some environmental factor that has a greater impact, or frequency, in twins than in singletons). However, because both language delay and reading difficulties are associated with an increased risk of psychopathology, the bias may be more widespread in its effects. Little is known about twin-singleton differences in childhood psychopathology but there is some suggestion that conduct disorders may possibly be slightly more common in twins (Simonoff 1992). Second, it is clear that both obstetric complications and congenital anomalies are more frequent in twins than singletons (Rutter and Redshaw 1991). As both are associated with an increased psychopathological risk, this has some relevance for twin studies. Third, the mothers of twins have an increased risk of depression during the first few years after the twin-birth (Thorpe et al. 1991). Presumably this is a consequence of the many stresses involved in looking after very young twins, but it is also likely to have implications for parenting and infant development (Rutter 1989). In addition, there are suggestions that twins and singletons tend to differ in other aspects of rearing, e.g., the former tend to have more adults involved in caregiving.

It is unlikely that any of these noncomparabilities between twins and singletons compromise twin designs for the genetic study of most traits; however, it is probable that the biases may be important for some characteristics (particularly language-related ones). At the moment, genetic modeling cannot take into account possible biases due to twin-singleton noncomparabilities because neither their origins nor their effects are well understood.

There is a considerable need for more systematic research to compare the experiences and psychological development of twins and singletons using designs that can test causal hypotheses, both because the findings are needed to test twin design assumptions and because such studies provide a powerful tool for testing environmental hypotheses (see below).

Ascertainment

Twin researchers have long been aware of the biases created through nonsystematic twin samples, with their tendency to be biased in terms of overinclusion of concordant and MZ pairs. There is less appreciation, however, of the need for care in the calculation of the expected proportion of DZ pairs (because the rate of DZ twin births in many countries has fallen greatly in recent years, but is now rising again as a result of infertility treatments; see below), because the proportion of DZ pairs in samples chosen due to some form of disorder will be higher than that in the general population, when the MZ concordance is very high and DZ concordance very low (because when the rate of disorder is the same in MZ and DZ twins, the cases in DZ twins will be spread across more pairs [Rutter et al. 1990a]).[1] Also, care is needed in calculating the expected number of twins, because the use of infertility treatments has led to a marked increase in rates of multiple births (although this increase is relatively much greater in triplets and above, than in twins). Due to these important secular trends, calculations of the expected number of twins and of the MZ–DZ ratio will need to control for date of birth if the twin sample covers a wide age range.

---

[1] At first sight, it might seem that this would not apply if the ascertainment was truly systematic, but, in fact, this is not so. Suppose, in order to illustrate the point, that the MZ pairwise concordance for a disorder is 90% and the DZ concordance 10%. Suppose, also that the prevalence of a disorder is 1%, and that the general population provides a sample of 1000 MZ and 1000 same sex DZ pairs. The MZ pairs will include 10 in which there is an affected twin; in 9 pairs they will be concordant so there will be 19 affected probands (assuming perfect ascertainment). The DZ pairs provide 2000 genetically distinct individuals resulting from separate eggs and hence 20 affected individuals spread over 18 pairs including a proband (again assuming perfect ascertainment). Thus, the sample based on probands gives an MZ/DZ ratio of 0.5, although in the general population from which it was drawn the ratio was 1.0. This difference arises, of course, from the missing cell (in a two-way table of the four possible pair-combinations of affected or nonaffected status) in the sample of pairs including an affected proband; that is, those concordant for being *non*affected. For obvious reasons, these will include an excess of MZ pairs, an excess that counterbalances the excess of DZ pairs in the proband sample. The use of proband-wise concordance rates means that there is no distortion of concordance patterns but it does not "correct" the MZ/DZ ratio of pairs with probands back to the MZ/DZ ratio in the general population. This needs to be taken into account when calculating the MZ/DZ ratio that should be obtained if ascertainment is complete, a calculation that is needed to check on the adequacy and nonbiased nature of the sampling.

Assortative Mating

A fourth, much neglected, issue concerns the effect of assortative mating (which is known to take place in relation to psychiatric disorder [Merikangas 1982]), the occurrence of which tends to lower estimates of genetic influence (because by increasing the sharing of genes in DZ pairs, it will inflate concordance/correlation within DZ pairs, whereas it will have no effect in MZ pairs who already share all their genes).[2] Curiously, we have very little systematic knowledge on the extent or nature of assortative mating with respect to psychopathology. However, if proper account of it is to be taken in genetic analyses, we must understand both its causation and patterning. Such findings as are available emphasize the complexities that are involved. Thus, although, in general, people tend to marry people with similar psychological characteristics, the same may not apply in the field of psychopathology. For example, Mednick found that, in his sample, schizophrenic women tended to have children by antisocial men (Mednick 1978). Also, the tendency to exhibit assortative mating may vary greatly according to circumstances. For instance, in his long-term follow-up study of institutionally reared children, Quinton (pers. comm.) found a strong tendency for young people with conduct disorder to marry a behaviorally deviant spouse; this tendency was also associated with age at marriage, being greater in those who married in their teens. Thus, assortative mating appears to be a function of people's social circumstances as well as their behavior. It is also necessary to bear in mind that individuals living together may grow more alike as a result of their mutual influence on one another, and hence that inter-spouse likeness does not necessarily imply assortative mating. Longitudinal studies of individuals identified in childhood and followed through to adult life are greatly needed in order to understand the processes involved in assortative mating. Thus, it is necessary to determine how far spouse selection is influenced by factors such as age at marriage and the peer group applicable at the time of selection, and the extent to which these factors operate in the same way, and to the same degree, in behaviorally extreme groups as in the general population. However, it is particularly important to use twin designs with psychopathological groups to compare choice of spouse in MZ and DZ pairs to assess the extent to which this has a genetic component. For the findings to be directly applicable to childhood psychopathology, it will be necessary to use longitudinal twin designs extending from childhood to adult life.

The implication of these considerations is both that there needs to be further research into some of the most important, but little tested, assumptions underlying twin designs, and also that twin strategies need to be combined with other genetic strategies (such as adoptee designs, which have seldom been used in the study of childhood psychopathology), because each strategy has a rather different mix of strengths and limitations.

---

[2] Assortative mating will increase covariance in both MZ and DZ pairs but this will not affect heritability estimates in a single sample.

Assumptions Regarding Psychopathology

Definition of the Phenotype

There is a tendency among some research psychiatrists to assume that the solution to the problems of phenotype definition lies in the use of standardized structured measuring instruments and the laying down of precise rules for diagnosis, but it does not (Rutter and Pickles 1990). These methodological improvements may help, but they do not provide an effective answer because the best of measures tends to tap both the latent construct of interest (but only partially) and also other confounding variables. In other words, any single measure will be subject to both systematic and random error. Multiple data sources help but some form of latent trait/class method of analysis will be needed to capitalize on their advantages. However, several other considerations also need to be taken into account. To begin with, it cannot be assumed that behavioral attributes are sufficient in themselves. For example, with respect to the trait of "behavioral inhibition," it seems that it may best be defined in terms of a combination of physiological and behavioral characteristics (Kagan et al. 1989). Or, if the interest is in an antisocial tendency that is likely to persist into adult life, an early age of onset may need to form part of the definition (see Cantwell and Rutter 1993). Psychiatric nosologists in their well-justified desire to get rid of unsubstantiated theory from psychiatric classifications have sometimes seemed to want to rely only on behavioral groupings without paying attention to conceptual distinctions that derive out of well-substantiated research. That constitutes a step backwards, not forwards.

However, with respect to genetic analyses, it is also important to recognize that history teaches us that the valid phenotype and traditional diagnostic concepts rarely coincide exactly. Of course, it is necessary to start with some hypothesized definition of the phenotype in order to undertake any type of genetic analysis; however, it is equally crucial to use the genetic findings to revise the phenotypic concept and then to go on to apply the revised concept in further genetic analyses, in an iterative process. Needless to say, it is of course essential for the testing to be extended both to further independent samples and to alternative genetic designs to cross-validate the conclusion. Independent replication, not statistical significance, is the benchmark in science.

A further consideration is that it is clear that many single gene Mendelizing disorders show an astonishing degree of variable expression. Thus, neurofibromatosis may be manifest in just a few cafe au lait spots or in a mass of deforming nerve tumors. In addition, pleiotropy is likely to be as great in the case of many disorders showing oligogenic or polygenic inheritance (Plomin et al. 1991). Moreover, as shown by the probably increased risk of schizophrenia in the offspring of apparently unaffected MZ co-twins of schizophrenic probands (Gottesman and Bertelson 1989), genetically affected individuals may not manifest the phenotype at all (at least not in ways that we know how to recognize). Of course, carrier status has long been recognized in recessive disorders but the recent findings on the Fragile X anomaly (see Bolton and Holland 1993) show hitherto unexpected complexities in terms of some carriers being clinically affected, of inheritance that does not follow an entirely regular Mendelian pattern, and of change in the gene product as a result of vertical transmission across the generations.

Dimensions and Categories

Much heat, and rather little light, has been generated in disputes between nosologists who have argued for dimensional and those who have wanted categorical approaches to diagnosis (see Cantwell and Rutter 1993). Psychiatric geneticists have sometimes seemed to want to have it both ways by studying categories (such as schizophrenia) but then transforming twin concordance to tetrachoric correlations on the basis of an untested assumption of some normally distributed underlying liability, in order to apply a multifactorial model. Plomin et al. (1991) have argued for the need to put this assumption to empirical test and to replace the hypothesized "black box" liability with some measurable attribute—on an analogy with convulsive threshold and epilepsy, or glucose tolerance and diabetes. Of course, it does not necessarily follow that a liability *will* be manifest in any psychological, physiological, or biochemical feature, but it seems highly desirable to seek to translate the hypothetical construct into something measurable, and reasonable to assume that it will often be possible to do so if the appropriate research is undertaken.

Sometimes it is also assumed that the dimensional/categorical dispute can be resolved by determining if variations at the end of a dimension function in the same way as variations in the middle. Four points need to be made in that connection. First, the category may be most validly considered in terms of a *constellation* of features rather than the extreme of a single trait. Thus, the most pervasive and persistent of conduct disorders are probably best conceptualized in terms of the combination of early onset, poor peer relationships, hyperactivity, attention deficits, and aggression rather than just extreme aggression or antisocial behavior. Indeed, it seems that the predictive power of aggression with respect to adult criminality is lost once this minority subgroup of children with multiple problems is removed from the general population (Magnusson and Bergman 1988). Second, it is clear that a trait may behave dimensionally in most respects but yet be genetically *dis*continuous. This is the case with severe mental retardation and IQ variations in the normal range (Plomin et al. 1991). Third, the comparison of individual and group familiality, using the methods developed by DeFries and Fulker (1985, 1988), now provides a means of testing genetic continuity/discontinuity; however, this is much more satisfactory in ruling out continuity than in ruling out discontinuity. Fourth, the same point that was made with respect to liability applies also to thresholds in multifactorial models, i.e., the concept is clearly biologically sound (there are many examples of its operation in medicine). It is important, however, for us to seek to find out the mechanisms involved in the qualitative change in phenotypic manifestation that accompanies the quantitative change in liability at the point of crossing the threshold.

PURPOSES OF TWIN STUDIES

Levels of Heritability

Traditionally, much of behavioral genetics has been concerned with establishing the degree of heritability of various traits. Heritability estimates constitute a crucial element in research to tackle questions concerning processes and mechanisms (see below). Nevertheless, in our view, heritability per se is of very little interest in its own right, despite the large literature devoted to it. The reasons for rejecting that objective are both theoretical and practical. The theoretical reason is that measures of heritability apply only to the populations on which they are based; if either environmental circumstances or genetic variability alter, so will heritability estimates go up or down. Heritability estimates have no general validity or meaning. Moreover, they apply strictly to population variance and not to *levels* of a trait in a population (Scarr 1992), hence they cannot be used to draw inferences about the modifiability of a trait given a change of circumstances (Rutter 1991a). The practical reason is that almost all human behavior is subject to substantial genetic influence, and most psychological attributes have not been found to vary greatly in levels of heritability (Plomin 1986). We gain negligible understanding by finding a heritability of, say, 36% for trait A and 47% for trait B. It had been hoped that temperamental attributes could be differentiated from other personality features on the basis of being much more strongly heritable, but empirical findings have not supported this distinction.

Of course, extremes may be of interest. Thus, the finding (on the assumptions of a multifactorial model) that the heritability of autism is over 90% (Rutter 1991b) does make it stand out from almost all other forms of childhood psychopathology. Also, apparent changes in heritability with age raise important issues—such as the seemingly greater importance of genetic factors in adult antisocial behavior as compared with juvenile delinquency—despite strong continuity between the two at an individual level (Rutter et al. 1990b).

Mode of Genetic Transmission

It is well recognized that twin studies are not the design of choice for the testing of alternative models of genetic transmission; family genetic designs using segregation analyses are required (Rutter et al. 1990a). Twin data are useful, however, for the detection of multiplicative genetic effects. The logic is straightforward. Because MZ twins share all their genes, they necessarily also share all patterns of *combinations* of genes. Because DZ twins share on average just half their genes, it follows that they will share only a quarter of specified two-gene combinations, an eighth of three-gene combinations and so on. Thus, if the DZ correlation is less than half the MZ correlation, a multiplicative effect involving two or more genes is suggested. Until recently, the uncertainty has been how to apply this approach to concordance data for categories. If the disorder is very rare (say 2 to 4 per 10,000), even near-zero DZ concordance (say 3% or so) will translate into tetrachoric correlations approaching 0.5 (this is because the translation is strongly affected by the general population base rate, which constitutes a key element in logic and mathematics), making it virtually impossible to detect multiplicative effects for the inferred continuously distributed liability. However, developing a well-established method, Risch (1990) recently pointed out that if the falloff in rate going from genetic correlations for relationships of 1.0 (MZ twins) to 0.5 (DZ twins), or from 0.5 to 0.25 (e.g., grandparents) exceeds a half, multiple gene combinations may be inferred. This inference clearly applies in the case of autism, where the MZ concordance is some 60% but the rate in sibs only 3% (Rutter 1991b). The research implication is that there is a special value in studies that combine twin and family data.

Guide to the Nature of Disorders

Genetic data, deriving from twin studies, may also be informative on the nature of child psychiatric disorders. For example, in adulthood, the genetic component in bipolar and severe unipolar disorders appears high whereas it seems quite low in the more common milder depressive disorders seen in outpatient and community samples (Rutter et al. 1990b). Childhood depression presents an apparent paradox in that it differs from depression in adult life in biological findings (such as cortisol levels, drug response, and sleep architecture), yet it shows a strong tendency to lead on to the latter (see Harrington 1993). Twin data incorporating a follow-up into adult life would be useful in showing whether there was genetic continuity between affective disorders occurring in the two life periods, and in indicating whether the phenotypic continuity applies mainly to strongly or weakly genetic varieties of depression.

Twin studies of autism, in conjunction with family studies, suggest that the phenotype extends beyond autism as traditionally diagnosed to include a combination of cognitive and social deficits occurring in individuals of normal intelligence (Rutter 1991b). Clearly, the next step should be research, using genetic designs, that seeks to span the neuropsychological and neurobiological domains in order to determine what constitutes the basis of this broader phenotype. However, the findings of the Scandinavian twin study by Steffenburg et al. (1989) differed in suggesting a narrower phenotype more closely associated with severe mental retardation. The limited twin data available so far do *not* suggest any overall difference in the degree of heritability of autism according to IQ level, but the possibility of genetic heterogeneity (in which autism associated with profound retardation might differ from other cases of autism) warrants explanation. Once again, for this purpose, twin studies will need to be combined with other genetic designs.

The genetic findings on "dyslexia," or severe reading retardation, have been rather contradictory and difficult to interpret. The recent evidence that phonological deficits may involve a strong genetic component, that these deficits may constitute the basis of "dyslexia," and (possibly) that "dyslexia" may not represent simply the end of the normal distribution of reading skills, constitutes a further interesting example of how

genetic data may throw light on the nature of a disorder (DeFries and Gillis 1991;
Olson et al. 1989)

Meaning of Comorbidity/Heterotypic Continuity

Epidemiological, as well as clinical, data have shown the high frequency of comorbidity in child psychiatric disorders (Caron and Rutter 1991), and longitudinal data
indicate the existence of heterotypic continuity, meaning that the behavioral manifestations of the same basic latent trait may change with age (Rutter and Rutter 1993).
Genetic designs, including twin studies, may be informative in testing competing
hypotheses on the mechanisms involved. Thus, adoptee data (Cadoret et al. 1990)
suggest that antisocial behavior and depression are genetically distinct but that the
former predisposes to the latter because it is associated with (and possibly creates)
environmental risk factors. A longitudinal twin study following children with conduct
disorder and/or depression into adult life would be helpful in taking this further.
Similar issues arise with respect to the strong associations between anxiety and
depression, and between hyperactivity and conduct disorder. Research leverage is
provided by the contrasting age trends across adolescence for anxiety and depression
and by the greater heritability for hyperactivity than for conduct disorder. Longitudinal designs of samples spanning the teenage years would be particularly valuable.
Pilot data from the Virginia Twin studies indicate the promise of twin research for
analyzing both anxiety-depression (Simonoff, Heath, and Kendler, pers. comm.) and
hyperactivity-conduct disorder comorbidity (Silberg, unpublished data). However, to
study comorbidity mechanisms effectively, it will be crucial to pay careful attention
to measurement issues by using multiple data sources and latent class analyses (Rutter
and Pickles 1990).

Mechanisms of Risk

Evidence that there is a genetic component to psychopathology is of little practical
use unless it leads to an understanding of risk mechanisms. Twin studies, however,
can be used for this purpose, as may be illustrated by two examples. The possibility
that genetic factors may play a role in individual variations in exposure to risk
environments has already been noted (Plomin and Bergeman 1991; Scarr 1992).
However, the evidence that people behave in ways that shape and select their
environments does not necessarily mean that the behavior that functions in this way
is genetically determined (Rutter and Rutter 1993). Thus, longitudinal data indicate
that conduct disturbance in childhood is associated with a much increased rate of life
stress in adult life (Champion, pers. comm.) but that the genetic component in conduct
disorder is low (Rutter et al. 1990b). Longitudinal twin samples are required to
investigate the genetic and environmental mechanisms involved.
    The second example is provided by the evidence showing that certain temperamen-
tal traits are associated with an increased risk of psychiatric disorder (Maziade 1989;

Rutter 1989). There are methodological difficulties involved in the "clean" separation of temperament and psychopathology; however, these terms are conceptually distinct, and one possibility is that the genetic component in the more common varieties of emotional and conduct disturbance resides entirely in temperamental predisposition. That hypothesis could be tested with longitudinal twin data, provided that multiple data sources and latent construct analyses are used to provide a satisfactory differentiation between the hypothesized risk factor (i.e., temperament) and the outcome of interest (i.e., psychopathology). In addition, however, it is desirable to take the further step of linking the behavioral manifestation of temperament with the supposed underlying biological substrate (a step that is beginning to be taken with behavioral inhibition).

It should be added that genetic factors may also operate by increasing vulnerability to particular environmental influences. Although few replicated gene–environment (GE) interactions have been found so far in behavioral genetic studies, person–environment interactions are widespread in biology, and it is necessary that they be searched for with respect to specific environments and specific subgroups, and not just in terms of some overall statistical interaction term (Rutter and Pickles 1991). The history of medical advances suggests that this search should focus on hypothesized mechanisms in specified subgroups (usually involving extremes) using informative research strategies. There are very few (if any) examples of a genetic mechanism operating through susceptibility to an environmental feature that has been discovered through the finding of a significant $G \times E$ interaction term in a multivariate analysis.

Testing of Environmental Effects

A further use of twin designs concerns the testing of environmental effects. Behavioral genetic data have been provocative in their suggestion that, for many psychological traits, shared environmental influences are far less important than nonshared effects (Plomin and Daniels 1987). It remains uncertain how far this applies to psychopathology; it probably does not in the case of conduct disorder (Plomin et al. 1991), but our own twin data, based on the Child Behavior Checklist, raise queries regarding emotional disturbance as well (Silberg, unpublished data). The issue is both theoretically and practically important and it warrants further study paying careful attention to the conceptual and methodological issues involved (Goodman 1991).

To date, a major weakness of most behavioral genetic analyses is that the environment has been dealt with only in "black box" terms as that which is not genetic. An urgent need is to include well-designed specific environmental measures in twin studies. Kendler et al. (1992) have shown that the move from an inferred environmental effect to a specified measured environmental variable (parental loss) may reveal a significant shared environmental influence that may be absent, as judged from the traditional partitioning of the variance in the usual analysis of data.

It is, of course, important to recognize that many supposed environmental measures reflect genetic influences (Plomin and Bergeman 1991). However, the strength of the twin design is that it allows determination of the extent to which effects are truly environmental. In order for this to be done, clearly it is necessary to know the twin upon which the effect impinged. This poses a special difficulty in the case of prenatal obstetric hazards. In the past, it has often been assumed that these are likely to represent an environmental influence (see, e.g., Steffenburg et al. 1989). Yet this constitutes a most uncertain inference in view of the evidence that genetically abnormal fetuses show an increased risk of obstetric complications (Goodman 1990). In the case of autism, our own data indicating a very strong association between obstetric complications and early determined congenital anomalies, together with a lack of any role of such complications in explaining twin concordance, suggest that they reflect genetic factors (Bailey, pers. comm.). This inference is also supported by the parallel finding from a family study (Bolton, pers. comm.), which found the familial loading for autism is *increased* in relation to probands with an adverse obstetric optimality score.

In using twin designs to test whether effects are truly environmental, however, it is necessary to bear in mind that the *origins* of a variable and the *mechanisms* of its effects are not necessarily synonymous (Rutter 1991c). The point may be illustrated by considering the extreme case of what twin analyses would show if the former were entirely genetic and the latter entirely environmental. Suppose that smoking was a wholly genetic trait (which it is not) and its carcinogenic effect wholly environmental (which it may well be). If that were the case, the usual partitioning of the variance would show *no* environmental effect on lung cancer, even though in reality the mechanism was entirely environmental. Of course, real life is unlikely to include such an extreme but the warning regarding the danger of misleading inferences from multivariate genetic analyses is real. The need is to supplement twin analyses with experimental research strategies designed to test hypotheses on mechanisms.

A further caveat is that the proportion of variance explained provides an inadequate index of the strength of an effect at the individual level (Rutter 1987; Rutter and Pickles 1991). This is because the variance-explained estimate is hugely influenced by the proportion of the population subject to the influence of the risk factor. If that proportion is very small, the variance explained will also be small, even when the size of the effect in individuals subjected to the risk factor is very large. Fortunately, the issue is easily dealt with by calculating the effects by means of odds ratios and proportion of the variance explained.

An additional point with respect to nonshared environmental influences is that, necessarily, this is an inference and not an observed variable or effect. Thus, if a family-wide influence (such as family discord) impinges on different children in the same family to different degrees or in different ways, it will appear in twin analyses as a nonshared effect even though that effect derives from a family-wide influence. The implication, again, is that twin analyses need to be combined with studies designed to provide a direct test of postulated environmental mechanisms.

Twins as a "Natural Experiment"

The last purpose of twin studies we wish to mention concerns their use as an experiment of nature (Rutter 1981) with respect to the testing of hypotheses on the mechanisms involved in the etiology of disorders, such as language delay, that are more frequent in twins than singletons (Rutter and Redshaw 1991). The key point is that this difference must be due to some nongenetic factor that is either more common or more powerful in its effects among twins. Three main alternatives need to be contrasted: amount and style of family interaction, obstetric and perinatal complications, and congenital anomalies. Two designs are likely to be informative: first, twin-singleton comparisons to determine whether the features that differentiate twins and singletons also relate to language delay within the twin sample; and second, comparisons of singletons, twins reared as twins, and twins reared as singletons (because the co-twin died in infancy or because of separation deriving from adoption or fostering). The latter comparison is informative because if the key variable concerns postnatal patterns of rearing, the twins reared as singletons would not differ from singletons in their verbal skills.

Twins also provide a natural experiment for studying risk factors that are more common in twins than singletons. For example, depression in the early years of childhood is much more frequent in the mothers of twins than singletons (Thorpe et al. 1991). Because the increased risk of maternal depression seems to derive from the stress of caring for young twins, it provides a useful test of the suggestion from findings in singletons that maternal depression in the first year of life depresses the children's cognitive performance (Coghill et al. 1986). The advantage of the twin situation is that the maternal depression is likely to be less confounded with genetic risk factors and with social disadvantage than in singletons, thus providing a better opportunity for testing the causal hypothesis. Of course, it would be necessary in such a study to check that the families of twins and singletons were comparable on the relevant variables.

CONCLUSION

With the dawning of the new era of molecular genetics there is a danger of underplaying the strengths of behavioral genetics and of the twin paradigm for genetic research, as well as underestimating how much remains to be done. We have drawn attention to a series of important issues that remain to be tackled in relation to the rationale of the twin design in genetic research. These will involve studies of the psychological development of twins (with a particular interest in the relationship between them); comparisons of twins and singletons with respect to their experience of risk factors and to rates of psychopathology; and studies of the patterns of, and causal mechanisms underlying, assortative mating. There is also a need for further research into phenotypic definitions (together with the genetic validation of such definitions); the use of genetic analyses to examine continuities and discontinuities between normality and pathology; and the use of latent construct analyses that reflect operationalized theoretical concepts. In considering the purposes of twin studies, attention has been drawn to the particular value of longitudinal twin designs incorporating both transitions of key age periods (such as adolescence), in which there are major changes in rates or patterns of psychopathology, and the spanning of childhood and adulthood; the need to incorporate specific environmental measures; the advantages of combining twin, adoptee, and family research strategies; the importance of examining the mechanisms underlying person–environment interactions and correlations; the value of studying the mechanisms underlying comorbidity and heterotypic continuity; the use of genetic data for gaining an understanding of the nature of disorders; and the utility of twin designs and of twin-singleton comparisons for the testing of environmental effects. There is a rich potential for twin studies of child psychopathology; however, if that potential is to be realized, it will be important for twin designs to be used in imaginative, as well as scientifically rigorous, ways.

ACKNOWLEDGEMENTS

We are grateful to the Medical Research Council, the Wellcome Trust and NIMH (Grant No. MH45268 to J. Hewitt) for support in the preparation of this paper.

REFERENCES

Bolton, P., and A. Holland. 1993. Chromosomal abnormalities. In: Child and Adolescent Psychiatry, 3rd ed., ed. M. Rutter, E. Taylor, and L. Hersov. Oxford: Blackwell Scientific, in press.

Cadoret, R.J., E. Troughton, L.M. Merchant, and A. Whitters. 1990. Early life psychosocial events and adult affective symptoms. In: Straight and Devious Pathways from Childhood to Adulthood, ed. L. Robins and M. Rutter, pp. 300–313. Cambridge: Cambridge Univ. Press.

Cantwell, D., and M. Rutter. 1993. Classification: Conceptual issues and substantive findings. In: Child and Adolescent Psychiatry: Modern Approaches, 3rd ed., ed. M. Rutter, E. Taylor, and L. Hersov. Oxford: Blackwell Scientific, in press.

Caron, C., and M. Rutter. 1991. Comorbidity in child psychopathology: Concepts, issues and research strategies. *J. Child Psychol. Psychiat.* **32**:1063–1080.

Coghill, S., H. Caplan, H. Alexandra, K. Robson, and R. Kumar. 1986. Impact of postnatal depression on cognitive development in young children. *Br. Med. J.* **292**:1165–1167.

DeFries, J.C., and D.W. Fulker. 1985. Multiple regression analysis of twin data. *Behav. Genet.* **15**:467–473.

DeFries, J.C., and D.W. Fulker. 1988. Multiple regression analysis of twin data: Etiology of deviant scores versus individual differences. *Acta Genet. Med. Gem.* **37**:205–216.

DeFries, J.C., and J.J. Gillis. 1991. Etiology of reading deficits in learning disabilities: Quantitative genetic analysis. In: Neuropsychological Foundations of Learning Disabilities, ed. J. E. Obrzut and G.W. Hynd, pp. 29–47. New York: Academic.

Goodman, R. 1990. Technical note: Are perinatal complications causes or consequences of autism? *J. Child Psychol. Psychiat.* **31**:809–812.

Goodman, R. 1991. Growing together and growing apart: The nongenetic forces on children in the same family. In: The New Genetics of Mental Illness, ed P. McGuffin and R. Murray, pp. 212–224. Oxford: Heinemann Medical.

Gottesman, I., and A. Bertelsen. 1989. Confirming unexpressed genotypes for schizophrenia. *Arch. Gen. Psychiat.* **46**:867–872.

Harrington, R. 1993. Affective disorders. In: Child and Adolescent Psychiatry: Modern Approaches, 3rd ed., ed. M. Rutter, E. Taylor, and L. Hersov. Oxford: Blackwell Scientific, in press.

Kagan, J., J.S. Reznick, and N. Snidman. 1989. Issues in the study of temperament. In: Temperament in Childhood, ed., G.A. Kohnstamm, J.E. Bates, and M.K. Rothbart, pp. 133–144. Chichester: Wiley.

Kendler, K.S., M.C. Neale, R.C. Kessler, A.C. Heath, and L.J. Eaves. 1992. Childhood parental loss and adult psychopathology in women: A twin study perspective. *Arch. Gen. Psychiat.* **49**:109–116.

MacGillivray, I., D.M. Campbell, and B. Thompson, ed. 1988. Twinning and Twins. Chichester: Wiley.

Magnusson, D., and L. Bergman. 1988. Individual and variable-based approaches to longitudinal research on early risk factors. In: Studies of Psychosocial Risk: The Power of Longitudinal Data, ed. M. Rutter, pp. 45–61. Cambridge: Cambridge Univ. Press.

Maziade, M. 1989. Should adverse temperament matter to the clinician? An empirically based answer. In: Temperament in Childhood, ed. G.A. Kohnstamm, J.E. Bates, and M.K. Rothbart, pp. 421–435. Chichester: Wiley.

Mednick, S.A. 1978. Berkson's fallacy and high-risk research. In: The Nature of Schizophrenia: New Approaches to Research and Treatment, ed. L.C. Wynne, R.L. Cromwell, and S. Matthysese. New York: Wiley.

Merikangas, K.R. 1982. Assortative mating for psychiatric disorders and psychological traits. *Arch. Gen. Psychiat.* **39**:1173–1180.

Olson, R., B. Wise, F. Conners, J. Rack, and D. Fulker. 1989. Specific deficits in component reading and language skills: Genetic and environmental influences. *J. Learn. Dis.* **22**:339–348.

Plomin, R. 1986. Development, Genetics and Psychology. Hillsdale, NJ: Erlbaum.

Plomin, R., and C.S. Bergeman. 1991. The nature of nurture: Genetic influences on "environmental" measures. *Behav. Brain Sci.* **14**:373–386.

Plomin, R., and D. Daniels. 1987. Why are children in the same family so different from one another? *Behav. Brain Sci.* **10**:1–15.

Plomin, R., R. Rende, and M. Rutter. 1991. Quantitative genetics and developmental psychopathology. In: Internalizing and Externalizing Expressions of Dysfunction: Rochester Symposium on Developmental Psychopathology, vol. 2, ed. D. Cicchetti and S.L. Toth, pp. 155–202. Hillsdale, NJ: Erlbaum.

Risch, N. 1990. Linkage strategies for genetically complex traits. *Am. J. Hum. Genet.* **46**:222–253.

Rutter, M. 1981. Epidemiological/longitudinal strategies and causal research in child psychiatry. *J. Am. Acad. Child Adol. Psych.* **20**:513–544.

Rutter, M. 1987. Continuities and discontinuities from infancy. In: Handbook of Infant Development, 2nd ed., ed. J. Osofsky, pp. 1256–1296. New York: Wiley.

Rutter, M. 1989. Psychiatric disorder in parents as a risk factor for children. In: Prevention of Mental Disorders, Alcohol and Other Drug Use in Children and Adolescents, ed. D. Shaffer,

I. Philips, and N.B. Enzer, OSAP Prevention Monograph 2, pp. 157–189. Rockville, MD: U.S. Department of Health and Human Services, Office for Substance Abuse Prevention.

Rutter, M. 1991a. Nature, nurture, and psychopathology: A new look at an old topic. *Devel. Psychopathol.* **3**:125–136.

Rutter, M. 1991b. Autism as a genetic disorder. In: The New Genetics of Mental Illness, ed. P. McGuffin and R. Murray, pp. 225–244. Oxford: Heinemann Medical.

Rutter, M. 1991c. Origins of nurture: It's not just effects on measures and it's not just effects of nature. *Behav. Brain Sci.* **14**:402–403.

Rutter, M., P. Bolton, R. Harrington, A. Le Couteur, H. Macdonald, and E. Simonoff. 1990a. Genetic factors in child psychiatric disorders: I. A review of research strategies. *J. Child Psychol. Psychiat.* **31**:3–37.

Rutter, M., H. Macdonald, A. Le Couteur, R. Harrington, P. Bolton, and A. Bailey. 1990b. Genetic factors in child psychiatric disorders: II. Empirical findings. *J. Child Psychol. Psychiat.* **31**:39–83.

Rutter, M., and Pickles, A. 1990. Improving the quality of psychiatric data: Classification, cause and course. In: Data Quality in Longitudinal Research, ed. D. Magnusson and L.R. Bergman, pp. 32–57. Cambridge: Cambridge Univ. Press.

Rutter, M., and A. Pickles. 1991. Person–environment interactions: Concepts, mechanisms, and implications for data analysis. In: Conceptualization and Measurement of Organism–Environment Interaction, ed. T.D. Wachs and R. Plomin, pp. 105–141. Washington, D.C.: APA.

Rutter, M., and J. Redshaw. 1991. Annotation: Growing up as a twin: Twin-singleton differences in psychological development. *J. Child Psychol. Psychiat.* **32**:885–895.

Rutter, M., and M. Rutter. 1993. Developing Minds: Challenge and Continuity across the Life Span. New York: Basic Books.

Scarr, S. 1992. Developmental theories for the 1990s: Development and individual differences. *Child Devel.* **63**:1–19.

Simonoff, E. 1992. A comparison of twin and singleton child psychiatric disorders: An item sheet analysis. *J. Child Psychol. Psychiat.* **33**:1319–1332.

Steffenburg, S., C. Gillberg, L. Hellgren, L. Andersson, C. Gillberg, G. Jakobsson, and M. Bohman. 1989. A twin study of autism in Denmark, Finland, Iceland, Norway and Sweden. *J. Child Psychol. Psychiat.* **30**:405–416.

Thorpe, K.J., J. Golding, I. MacGillivray, and R. Greenwood. 1991. Comparison of prevalence of depression in mothers of twins and mothers of singletons. *Br. Med. J.* **302**:875–878.

13

Twin Studies as an Approach to Differential Manifestations of Childhood Behavioral Disorders across Age

D.L. PAULS

Child Study Center, Yale University School of Medicine,
230 S. Frontage Road, New Haven, CT 06510, U.S.A.

ABSTRACT

The purpose of this chapter is to review the difficulties associated with phenotypic variable expressivity in genetic studies. Before the genetics of psychopathology can be elucidated, it is necessary to understand the full range of symptoms associated with a particular disorder. This is particularly true for psychiatric disorders of childhood since childhood phenotypes change over time. If the genetics of these disorders are to be elucidated, it will be necessary to understand the variant expression of these conditions across development. Traditional twin studies can yield some information for the understanding of phenotypic change over time. However, twin studies that take advantage of new molecular genetic techniques and incorporate assessment strategies from developmental psychology could provide the more definitive data necessary to understand the variable expressivity of childhood behavioral phenotypes.

The single most important issue in the genetic study of psychiatric/behavioral disorders is the definition of the phenotype. In order to examine genetic hypotheses about a disorder adequately, it is necessary to know how wide the spectrum of symptoms is for the condition and whether family members or co-twins are affected. If all phenotypes were unvarying, knowing whom to include as affected in the genetic analyses (identifying affected individuals) would not be a difficult problem. However, as any student of genetics knows, there is considerable variability in the manifestation of all phenotypes (both behavioral and nonbehavioral), even in those conditions where the underlying genetic mechanisms are both necessary and sufficient. Thus, before the inheritance of a behavioral disorder can be elucidated, it is important to understand the variable expression of the trait being studied.

Twins as a Tool of Behavioral Genetics
Edited by T.J. Bouchard, Jr. and Propping ©1993 John Wiley & Sons Ltd.

Variable expressivity is defined as the range of phenotypes resulting from a specific genotype. Variable expressivity is not to be confused with co-morbidity. Co-morbidity occurs when two disorders are expressed by the same individual at the same time. These disorders may be independent of each other or one disorder may be secondary to the first; however, the two are not separate manifestations of the same underlying genetic mechanisms. Variable expressivity occurs when a range of symptoms is associated with the same underlying genotype(s). This range of symptoms does not always occur in the same way so that individuals may have considerably different phenotypes even though they carry the same susceptibility genes. On the other hand, it is possible for some individuals to have the full range of symptoms so that they may appear to have two co-morbid conditions. For example, Gilles de la Tourette (GTS) syndrome patients often exhibit obsessive-compulsive symptomatology sufficiently severe to satisfy criteria for a diagnosis of obsessive-compulsive disorder (OCD) (for a review, see Pauls 1990b). The patients with both GTS and OCD appear to have two co-morbid conditions. However, family and twin data suggest that the two are variant expressions of the same underlying genetic factors: (a) OCD by itself occurs with higher frequency in the family members of GTS patients (Pauls et al. 1986; Pauls et al. 1991); and (b) co-twins in monozygotic (MZ) pairs discordant for GTS are more likely to manifest OCD than expected by chance (Walkup et al. 1988).

Variable expressivity and co-morbidity can be confused when one co-morbid disorder is secondary to the first. Variable expressivity differs from this type of co-morbidity in the following way. When two disorders are co-morbid and one is dependent on the other, the expression of the second disorder is entirely a consequence of the manifestation of the first and occurs only in the presence of the first. On the other hand, when two disorders are variable expressions of one another, each is caused by the same set of underlying etiological factors and each can occur independently of each other in different individuals. When two heritable disorders are variable expressions of one another, specific patterns of transmission in families and concordance in twins are expected. In families, the rate of the two disorders should be the same in relatives of probands with either or both diagnoses. For example, if traits A and B are variable expressions of the same genotypes, the rate of A should be the same among relatives of probands with B as it is among relatives of probands with A and vice versa. In twins, the rate of A should be the same in co-twins of probands with B as in co-twins of probands with A and vice versa. If A and B are co-morbid conditions, with B secondary to A, then the affected relatives should only have B in the presence of A.

If the goal of a genetic study is to identify the genes responsible for the manifestation of a condition, it is essential to know the range of expression of that disorder at any given time and at any given age. This is particularly true for disorders with onset in childhood since the manifestation of childhood disorders changes over the lifespan of the individual. For example, children with an anxiety disorder may have different symptoms at 8 years of age than at 18. Thus, when undertaking genetic studies with family members of all ages, it is critical to know whether a disorder is expressed differently at different ages.

In most studies (whether twin, adoption, family, or genetic linkage) probands with unambiguous diagnoses are ascertained and their families are studied. Typically, probands are ascertained that are as phenotypically homogeneous as possible. However, when relatives are studied (either co-twins or other family members), it quickly becomes evident that the same strict classifications used for the probands are not entirely adequate to document the range of phenotypic expression among the relatives. This is particularly true when adult relatives are included in studies of childhood disorders. Unless the phenotype is invariant over time, it is often difficult to know whether adult relatives have the same condition being studied in their children. Furthermore, even when relatives in a narrow age range are included for study, there can be a considerable range of symptomatology observed, and it becomes necessary to consider additional classifications when deciding which relatives to include as affected.

Understanding the range of phenotypic expression is critical to the ultimate success of studies designed to identify and characterize the genetic factors important for the expression of behavioral disorders. Misclassification of relatives in such studies can be disastrous. For example, a change in diagnosis of only a few relatives can have a profound impact on the results of genetic linkage analyses (Kelsoe et al. 1989). On the other hand, if the intent of a study is simply to document that genetic factors are important in the expression of some trait, then less phenotypic specificity can be tolerated. In a large twin study examining the heritability of a behavioral condition, a 5% misclassification rate will not have a major effect on the conclusions of a study. That is, even with a 5% error rate, it will still be possible to establish that genetic factors are important for the expression of the disorder. However, a 5% misclassification rate in a genetic linkage study will make it impossible to localize accurately the gene(s) responsible for the disorder (Egeland et al. 1990; Kelsoe et al. 1987; Pauls 1993).

For childhood conditions a clear understanding of the phenotype over the life span is necessary to help clarify the inheritance of a specific disorder unless the psychopathology being studied is homotypic. For example, childhood OCD is diagnosed with the same criteria employed for adults. If it can be assumed that childhood onset OCD is the same disease as adult onset OCD, then it is possible to examine genetic hypotheses regarding the disorder. However, most childhood illnesses are not invariant over time, and it is unlikely that adult phenotypes will be expressed in exactly the same way as the phenotypes manifested in childhood or adolescence. Considerable research is needed to document how phenotypes change over time.

Consistent with the need to identify variable expressivity among adult family members, it is necessary to know how illness is manifested over different stages of development. Attention to development is particularly important when the goal is an understanding of the genetics of a disorder. It is vital to know how genes are expressed through all stages of development and what impact their function has on the ultimate expression of the phenotype throughout the life span. The manifestation of gene function may be observed through specific symptoms and behaviors. However, the expression of those phenotypes may be affected by interaction with family members or peers, cellular levels of neurochemicals, response to specific environmental stressors, regulator gene function at the cellular level, or some yet-to-be determined factors. Thus, it is important to understand the context in which the phenotype develops to document more completely the symptomatologic range of that phenotype over time.

It is imperative for future research to collect data that allow the examination of both genetic and environmental contributions to the variable expression of psychiatric illness. While in most studies of psychiatric illnesses it has been possible to demonstrate that genes are important in their etiology, it is clear that genetic factors are not both necessary and sufficient for the expression of any behavioral disorder. It is possible that the genetic mechanism for a specific illness will only be identified when the impact of nongenetic factors on the expression of the illness has been adequately described. It may also be the case that only with the careful documentation of environmental factors will it be possible to determine the phenotype that is inherited. Environmental factors may mask the true phenotype and make it impossible to understand the impact on genetic factors for the "core phenotype."

In order to examine the impact of environment adequately, it is important for this work to take place in a developmental context. Work in developmental psychology has demonstrated that early life experiences can have a significant impact on later mental health. For example, the quality of an infant's attachment to her or his mother has a strong predictive power for later childhood behavior. While attachment theorists attribute a majority of deviant behavior to deviant attachments and the deviant attachments to inadequacies in the caregiving environment (Bowlby 1988), it is becoming apparent that the genetic endowment of both parent and child may influence the attachment. Not all children respond in the same way to specific events. Different genotypes or temperamental predispositions may mediate response to different caregiving environments and may even elicit differences in caregiving.

Documentation of the influence of both genetic and nongenetic factors on behavior requires prospective longitudinal studies of children at genetic risk. Studies designed to document both genetics and environment need to take into account developmental stages throughout the life span. Thus, if twins are studied, they should be studied in a prospective way. Furthermore, if the aim of a study is to examine a specific disorder, then the children being followed longitudinally should be at risk for the disorder in question. So, ideally, if twins are to be studied, then the twin pair should be at risk for some illness. These studies should not only be designed to document early signs and symptoms of psychopathology (although this is a worthy goal). The design should include documentation of the experience of the individual, not only as measured by a global assessment of the home and family life, but by documentation of specific experiences and interactions, both within and outside (i.e., school experiences) of the home using sensitive observational measurements.

Studies of environmental risk are needed for children of all ages. However, it is essential that we study the effect of environment on infants. From an evolutionary point of view that recognizes the dependency of infants on caregiving at the time of rapid growth and maturational change, it is logical to assume that humans may be more sensitive to variable environmental influences. Attachment research has shown that the parental environment is critical for emotional development in young children. Thus, parent-child dyads need to be carefully studied. However, it is important to learn about sib-sib dyads and peer-peer dyads as well. This would be of particular interest for twins. The importance of unique environments for the development of behavior has been demonstrated (Plomin and Daniels 1987). Thus, it is necessary to document those environments as carefully as possible for each twin or relative pair. Furthermore, attachment paradigms should be designed to do more than just document interactions. Response to specific environments should be measured behaviorally, physiologically, and biologically. Multiple developmental domains need to be assessed by examining several different behaviors. In all assessments, particular attention should be given to critical developmental periods. It is important to document whether specific stressful events take place at a particularly critical developmental stage.

Until now, what has been proposed is not much different from traditional high-risk studies except to suggest that infant twins at risk be studied with more comprehensive observational measures of the caregiving environment. This design may be unrealistic given the paucity of at-risk twins available for study. Furthermore, it is not clear that a traditional twin design would allow the documentation that the same genes impact on different phenotypes over time. While it is possible to determine with dizygotic (DZ) twins that two conditions are not variable expressions of one another, it is not possible to determine unambiguously that different phenotypes are variant expressions over time. That is, it is not possible to determine with current twin designs that the same genes are expressed over time to result in the variant phenotypes. At best, twin studies can demonstrate that the heritability of different phenotypes is stable over time, but it is not possible to determine unequivocally that the same genes are involved. For additional discussions of this topic for cognitive traits see Fulker and Cardon, and Boomsma (both this volume). New methods are needed that allow the identification of the genes functioning to produce the variable phenotypes. Molecular genetic methods could provide a way to identify regions of the genome contributing to the expression of the variant phenotypes. By utilizing the developing linkage map in a twin study, specific regions of the genome accounting for a significant amount of the phenotypic variance observed could be identified (Carey and Williamson 1991; Fulker et al. 1991; Goldgar 1990).

Genetic linkage has long been recognized as one of the methods useful in clarifying the role of genetic and environmental factors in the expression of complex disorders. Historically, the method was limited because of the small number of sufficiently polymorphic genetic markers available for study in humans. This situation has changed dramatically in the last decade. Advances in DNA technology have made it possible to detect many highly polymorphic genetic markers, and these markers, based on DNA sequence polymorphisms, have stimulated a renewed interest in linkage approaches to the study of human disorders. Summaries from the most recent workshop on gene mapping (Kidd et al. 1989) suggest that over 5,000 genetic markers have been detected. As a result, extensive linkage mapping of all human chromosomes is underway, and a genomic map including some of these markers is nearing completion (Kidd et al. 1989). The application of these techniques will help to clarify the underlying genetic mechanism of many disorders, including psychiatric illnesses. Theoretical and empirical work suggests that, irrespective of problems of incomplete penetrance, linkage studies can identify the location and thereby verify the existence of genetic loci important in the expression of complex human disorders (Kramer et al. 1989; Kramer et al. 1990; Price et al. 1989).

The localization of a gene or genes responsible for the expression of a psychiatric illness will be a major step forward in our understanding of the genetic/biological risk factors important for the expression of behaviors. In addition, this work will allow the potential identification of nongenetic factors associated with the manifestation or the amelioration of the symptoms of psychological disorders (Leckman et al. 1990; Pauls 1990a). On the one hand, the identification of linked markers will permit the design of much more incisive studies to illuminate the physiological/biochemical etiology of behavioral disorders by examining the gene product and its impact on the manifestation of the traits. On the other hand, by controlling for genetic factors it will be possible to document more carefully the environmental/nongenetic factors important for the expression of psychiatric disorders. One of the main obstacles to the success of this work is the precise assessment of the phenotype. If it is not possible to identify affected individuals accurately, traditional genetic linkage strategies will be useless. However, if valid and reliable assessments are developed which capture variable expressivity, then genetic linkages will be quite useful in the elucidation of both genetic and nongenetic factors important for the manifestation of abnormal behavior.

Much has been written about the relative importance of genetic and environmental factors in the manifestation of specific behaviors (Plomin 1990). Current discussions are no longer debates between proponents of nature vs. nurture, but rather focus on the extent to which both contribute to the expression of behavior. Included in this discussion is the notion that genotype-environment interaction may have a major role in the development of behavioral disorders. It is difficult to assess adequately the impact of this interaction for humans because of our inability to control environments. Elegant methods have been employed in the study of plants and animals but are not directly applicable to studies of humans (for a review, see Eaves 1984). In the past, work in humans has suffered from both inadequate measurement of phenotypes as well as environments. As detailed earlier, phenotype descriptions employing comprehensive assessments to characterize developmental aspects of expression are needed. Likewise, shared and unique environmental factors need to be assessed directly. Studying genetic marker data together with data characterizing phenotypic expression in the context of specific environments should allow a more complete examination of the co-contribution of genetic and nongenetic factors to the expression of the phenotype.

Once the location of a gene or genes has been verified through genetic linkage studies, it will be possible to type the children in prospective studies and their relatives to determine with a high probability who is carrying a susceptibility gene. It will then be possible to examine in more detail the interaction between genotype and environment. Thus, linked genetic markers can serve as the basis for defining an appropriate control group for a "genetic case-control" design (Kidd 1984). Closely linked genetic markers can identify quite accurately those children who are genetically identical with their affected sibs and/or parents for the relevant loci. Genetically identical individuals constitute an ideal control group for the identification of nongenetic factors relevant to the onset of psychiatric disorders. Thus, the study of discordant MZ twins might prove useful in the identification of nongenetic risk factors. However, MZ twins reared together share their rearing environment to some degree, and MZ twins reared apart are rare. Genetically identical but unaffected siblings have the genetic susceptibility but, since they are not affected, have presumably not been exposed to necessary environmental agents, or may have been exposed to them at different times in development, or may have encountered factors that protected them from manifesting the disorder. In contrast, siblings who are not genetically identical at the relevant loci might very well have experienced the same risk factors but be unaffected because they lack the necessary genetic susceptibility. Comparing probands with their respective genetically identical siblings or children provides the basis of a genetic case-control paradigm. This particular control provides the optimum power for identifying major nongenetic risk factors for the illness. This paradigm could be enhanced by including a sample of twins at risk. For example, if age is important in the manifestation of the behaviors being studied, then twins would provide information that would be helpful in examining its effect on the phenotype.

The ability to utilize genetic linkage and other aspects of genetic studies to design and carry out a study of nongenetic etiologic factors of a psychiatric illness is a significant methodological advancement that has not been possible heretofore. Data from prospective studies as outlined will make it possible to examine individuals with specific genotypes to determine which factors protect some from manifesting the syndrome. However, it is not necessary to wait for the localization of a gene or genes for a specific disorder. Given that a genomic map is being developed and that the cost of typing individuals with highly polymorphic markers is decreasing, it is possible to use DNA markers to: (a) increase the power of twin studies, (b) provide data to examine the change in phenotype over time, and (c) identify regions of the genome containing loci of interest for a specific disorder.

One type of study that would help to elucidate the phenotypic expression over the life span would include a sample of MZ and DZ twins at risk for a particular disorder. Preferably, the twins should be selected prior to onset of the disorder. To document the nongenetic factors important for the expression of the phenotype adequately, the twins should be selected at birth or during pregnancy. Because variable expressivity could be due to different environmental exposure, assessments of the twins should include careful documentation of all relevant environments. Attention should be given to prenatal exposure to environmental events, parenting practices, attachment of the child to a caregiver, and temperament of both parents and children. Measurements should be taken at critical times in development and all symptoms should be carefully documented and described.

The assessment of the phenotype should include the use of continuous measures in contrast to dichotomous measures. While this may seem obvious to most researchers in behavioral genetics, much of psychiatric genetic research has been limited by reliance on categorical definitions of phenotype. On the other hand, most research in behavioral genetics has focused on normal behavior, and thus the measures used in those studies may not be appropriate for the study of abnormal behavior. In current psychiatric research, much effort is directed to collecting data necessary to make DSM and/or ICD diagnoses. In so doing, much of the richness of the information is lost when categories of illness are defined. Relying on categorical definitions has limited the power in many of the analyses and may have led to incomplete conclusions regarding the importance of genetic factors and their transmission within families. Given the variability in the manifestation of a psychiatric diagnosis, it would be useful to have assessments that provide continuous definitions or multidimensional definitions of the phenotype. Categorical definitions may be quite useful when considering treatment and outcome; however, in research designed to determine the extent of the involvement of genetic factors and to attempt to elucidate patterns of illness within families, it is helpful to have information about the range of symptom expression that might occur in "unaffected" relatives. Although some attempts have been made to include continuous assessment of phenotype, these have not been used extensively in psychiatric genetic research.

As noted above, the assessment of the phenotype should include a developmental focus on the manifestation of psychopathology. At the present time, there are subdisciplines in psychology and psychiatry focused on infants, children, adolescents, adults, and the elderly. This specialization may be necessary for effective treatment; however, for research focused on understanding the underlying genetics of abnormal behavior, it is necessary to reshape our thinking in terms of human psychology/psychiatry. It is also necessary to think in terms of life span developmental psychology/psychiatry. It is critical to learn more about continuity and discontinuity between childhood and adult psychopathology. It is also important to know about the possibility of sensitive periods in development and their relationship to later psychopathology.

In addition to assessing the phenotype using continuous measures, the genotypes of twins should also be examined more closely. All twins should be typed with highly polymorphic markers spanning the genome. Including more detailed information about both phenotype and genotype would increase the power to identify the behaviors that are variant expressions of the same underlying genetic factors. To cover the genome adequately, approximately 500 evenly spaced markers should be typed on all twins. This typing would allow an estimate of the degree of genetic similarity of the DZ twins. By including a more precise estimate of the number of genes shared in common by DZ twin pairs, the power of genetic analyses could be increased (Pakstis et al. 1972). By including this estimate of genetic similarity, it could be possible to estimate more accurately the proportion of the phenotype due to genetic factors. Furthermore, by including careful measures of the environment, it could be possible to estimate the impact of nongenetic factors on the expression of the phenotype. By combining all of these variables, it would then be possible to determine, with a much higher degree of certainty, which behaviors were representative of the phenotype under study. Furthermore, the data collected would also allow analyses that would provide clues as to the actual location of the genes of interest (Carey and Williamson 1991; Fulker et al. 1991; Goldgar 1990). In addition to learning about variable expressivity, it would be possible to begin the process to localize the genes of import.

As proposed, these twin studies have not been possible heretofore and may not be feasible for any one investigative team at one university to undertake. A multi-site collaborative project, however, could accomplish this work. Furthermore, this approach need not be limited to twins. Siblings in at-risk families could also be typed and followed prospectively. In fact, by including individuals of different ages in the experimental design, it would be possible to evaluate the impact of similar environments at different developmental stages—analyses not possible if the data were collected only on twins. A study of the type proposed will inform us of much more than just variable expressivity. Data collected in a prospective longitudinal study of twins and singletons should allow careful examination of the importance of genetic and environmental factors and their interaction in the manifestation of behavioral disorders.

ACKNOWLEDGEMENTS

This work was supported, in part, by a Research Scientist Development Award (MH–00508) and a grant from the National Institute of Neurological Disorders and Strokes (NS–16648).

REFERENCES

Bowlby, J. 1988. Developmental psychiatry comes of age. *Am. J. Psychiat.* **145**:1–10.

Carey, C., and J. Williamson. 1991. Linkage analysis of quantitative traits: Increased power by using selected samples. *Am. J. Hum. Genet.* **49**:786–796.

Eaves, L.J. 1984. The resolution of genotype × environment interaction in segregation analysis of nuclear families. *Genet. Epidem.* **1**:215–228.

Egeland, J.A., J.N. Sussex, J. Endicott, A.M. Hostetter, D.R. Offord, J.J. Schwab, and D.L. Pauls. 1990. The impact of diagnoses on genetic linkage study for bipolar affective disorders among the Amish. *Psychiat. Genet.* **1**:5–18.

Fulker, D.W., L.R. Cardon, J.C. DeFries, W.J. Kimberling, B.F. Pennington, and S.D. Smith. 1991. Multiple regression analysis of sib-pair data on reading to detect quantitative trait loci. *Read. Writ. J.* **3**:299–313.

Goldgar, D.E. 1990. Multipoint analysis of human quantitative genetic variation. *Am. J. Hum. Genet.* **47**:957–967.

204 *D.L. Pauls*

Kelsoe, J.R., E.I. Ginns, J.A. Egeland, D.S. Gerhard, A.M. Goldstein, S.J. Bale, D.L. Pauls, R.T. Long, K.K. Kidd, G. Conte, D.E. Housman, and S.M. Paul. 1989. Reevaluation of the linkage relationship between chromosome 11p loci and the gene for bipolar affective disorder in the Old Order Amish. *Nature* **342**:238–243.

Kidd, K.K. 1984. New genetic strategies for studying psychiatric disorders. In: Genetic Aspects of Human Behavior, ed. T. Sakai and T. Tsuboi, pp. 225–246. Tokyo: Igaku-Shoin Ltd.

Kidd, K.K., A.M. Bowcock, J. Schmidtke, R.K. Track, F. Ricciuti, G. Hutchings, A. Bale, P. Pearson, and H.F. Willard. 1989. Report of the DNA committee and catalog of cloned and mapped genes and DNA polymorphisms: Human Gene Mapping Workshop 10. *Cyto. Cell Genet.* **51**:622–947.

Kramer, P.L., D. de Leon, L. Ozelius, N. Risch, S.B. Bressman, M.F. Brin, D.E. Schuback, R.E. Burke, D.J. Kwiatkowski, H. Shale, J. Gusella, X.O. Breakefield, and S. Fahn. 1990. Dystonia gene in Ashkenazi Jewish population is located on chromosome 9q32–34. *Ann. Neur.* **27**:114–120.

Kramer, P.L., D.L. Pauls, R.A. Price, and K.K. Kidd. 1989. Estimation of segregation and linkage parameters in simulated data: I. Segregation analysis with different ascertainment schemes. *Am. J. Hum. Genet.* **45**:83–94.

Leckman, J.F., E.S. Dolnansky, M. Hardin, M. Clubb, J.T. Walkup, J. Stevenson, and D.L. Pauls. 1990. The perinatal factors in the expression of Tourette's syndrome. *J. Am. Acad. Child Adol. Psych.* **29**:220–226.

Pakstis, A., S. Scarr-Salapetek, R.C. Elston, and R. Siervogel. 1972. Genetic contributions to morphological and behavioral similarities among sibs and dizygotic twins: Linkage and allelic differences. *Soc. Bio.* **19**:185–192.

Pauls, D.L. 1990a. Emerging genetic markers and their role in potential preventive intervention strategies. In: Conceptual Research Models for Preventing Mental Disorders, ed. P. Muehrer, pp. 184–195. Rockville, MD: NIMH.

Pauls, D.L. 1990b. Gilles de la Tourette's syndrome and obsessive-compulsive disorder: Familial relationships. In: Obsessive-Compulsive Disorders: Theory and Management, 2nd Ed., ed. M.A. Jenike, L. Baer, and W.E. Minichiello, pp. 149–153. Littleton, MA: Year Book Medical Publ., Inc.

Pauls, D.L. 1993. Genetic linkage studies in psychiatry: Strengths and weaknesses. In: Genetic Studies in Affective Disorders: Overview of Basic Methods, Current Directions and Critical Research Issues. Einstein/Montefiore Monograph Series in Clinical and Experimental Psychiatry, ed. D.F. Papolos and H.M. Lachman. New York: Brunner/Mazel, in press.

Pauls, D.L., C.L. Raymond, J.F. Leckman, and J.M. Stevenson. 1991. A family study of Tourette's syndrome. *Am. J. Hum. Genet.* **48**:154–163.

Pauls, D.L., K.E. Towbin, J.F. Leckman, G.E.P. Zahner, and D.J. Cohen. 1986. Gilles de la Tourette syndrome and obsessive compulsive disorder: Evidence supporting an etiological relationship. *Arch. Gen. Psychiat.* **43**:1180–1182.

Plomin, R. 1990. The role of inheritance in behavior. *Science* **248**:183–188.

Plomin, R., and D. Daniels. 1987. Why are children in the same family so different from one another? *Behav. Brain Sci.* **10**:1–15.

Price, R.A., P.L. Kramer, D.L. Pauls, and K.K. Kidd. 1989. Estimation of segregation and linkage parameters in simulated data: II. Simultaneous estimation with one linked marker. *Am. J. Hum. Genet.* **45**:95–105.

Walkup, J.T., J.F. Leckman, R.A. Price, M. Hardin, S.I. Ort, and D.J. Cohen. 1988. The relationship between Tourette syndrome and obsessive compulsive disorder: A twin study. *Psychopharm. Bull.* **24**:375–379.

14

Can We Identify Specific Environmental Influences on Behavioral Disorders in Children?

D.C. ROWE[1] and J.L. RODGERS[2]
[1]School of Family and Consumer Resources, University of Arizona,
Tucson, AZ 85721, U.S.A.
[2]Department of Psychology, University of Oklahoma,
Norman, OK 73021, U.S.A.

ABSTRACT

In this chapter, evidence is reviewed that the major environmental influences on personality development are nonshared rather than shared by siblings. Three methods of investigating nonshared influences are proposed: (a) correlating differential treatments with behavioral differences within pairs of monozygotic (MZ) twins, (b) comparing total trait variances for only-children versus siblings, and (c) explaining residual variation with specific measures of differential treatment in DeFries/Fulker regression models. These methods will be illustrated with applications to childhood problem behavior. Our conclusion is that the etiologic role of nonshared environmental influences is difficult to demonstrate because of their similarity to reactive and active gene × environment correlations.

INTRODUCTION

Since environmental views on behavioral development became dominant in the 1920s to 1930s, variation in personality has been attributed to variation in rearing environments. Children's poor self-esteem was blamed on a lack of parental love, their poor vocabulary on deficiencies in parental vocabulary, their smoking and drinking on imitation of parental habits, and their social prejudices on parents' espousing negative views of minorities. The environmental features most prominent in theories were those that varied *between* families (such as income and years of parental education), the emotional climate of the home, parental attitudes and beliefs, and structural family features (such as father presence versus absence). Although dissenting voices noted

Twins as a Tool of Behavioral Genetics
Edited by T.J. Bouchard, Jr. and P. Propping © 1993 John Wiley & Sons Ltd.

that the statistical associations between rearing and children's traits were not strong, no one sought to challenge fundamentally the notion that rearing environments in the home affected children's personality development.

In the postwar period, this challenge came, instead, from behavioral genetic studies of personality, which used quasi-experimental research design to estimate rearing influences independently of genetic ones. In the family study—the workhorse of socialization research—genes and rearing environment can be inextricably confounded with associations between rearing practices and children's traits that are potentially the result of genes shared by parent and child.

The behavioral genetic approach to rearing influence relied on apportioning personality variation to genes, rearing environment, and nonshared environment. The three sources of variation were defined mathematically, and the research strategy was to look for effects implied by each one on the trait correlation for pairs of social or biological relatives.

The shared source refers to *any* environmental influence that is shared by siblings or by parent and child. By mathematical definition, a shared component of variation can only increase the behavioral resemblance of family members. The nonshared environment source refers to an environmental influence that makes family members unlike, i.e., the effect of such an influence on one member of a family is uncorrelated with its effect on any other member. Genes refer, of course, to variation in the population caused by the substitution of one gene for another at the genetic loci affecting a particular trait.

Behavioral geneticists reasoned that if shared environmental influences were operative, then children who were adopted and raised by genetically unrelated parents would come to resemble them. In the twin research design, they reasoned that fraternal twins would be nearly as alike as identical twins, despite the latter's greater genetic resemblance. In recent studies, mathematical model-fitting techniques have permitted the estimation of the influence of genes, shared environment, and nonshared environment on a particular trait using data from several kinds of relatives simultaneously.

Individual twin and adoption studies, as well as model-fitting approaches applied to diverse data sets, converged on a surprising conclusion: a lack of influence of shared rearing environments on personality development (Rowe and Plomin 1981; Plomin and Daniels 1987). A harbinger of the general result was Loehlin and Nichols' (1976) twin study of the California Psychological Inventory, a standardized personality test. In this large study of 850 pairs of twins, the MZ twin personality correlations were greater than those found in dizygotic (DZ) twins, usually at least twice as large. This large gap between MZ and DZ correlations was consistent, with a total absence of shared rearing influence. Although one might think that more similar treatments made the MZ twins more alike in personality, several ingenious analyses by Loehlin and Nichols rendered this interpretation implausible. Their pioneering effort finds support in more recent studies of biologically unrelated children reared together in the same home: despite a shared rearing environment, adoptees are hardly more alike in personality than children reared in different homes (Goldsmith 1983; Plomin and Daniels 1987). What is the influence of shared rearing? These empirical results, and others similar to them, have diminished scientific support for shared rearing influences. The remaining source of environmental influence, therefore, is nonshared influences that operate within families to reduce siblings' behavioral resemblance. The magnitude of nonshared influence may be estimated using the MZ twin correlation. It represents the total effect of heredity on a trait, whereas the reliability coefficient represents the total trait variation that is systematic (not due to measurement errors). Thus the difference between these numbers reflects the proportion of variation due to systematic, but nonshared, environmental influences. For personality traits, a reliability coefficient of 0.80 and a typical MZ twin correlation of 0.50 indicates that about 30% of trait variation is nonshared. For intellectual abilities, the reliability of 0.90 minus the adult MZ twin correlation of 0.70 indicates that about 20% of variation is nonshared.

WHAT IS NONSHARED ENVIRONMENTAL INFLUENCE?

A problem for social science is to identify particular nonshared environmental influences, and here, an important paradox emerges: many influences that scholars have claimed as possible *nonshared* family influences are also shared ones. Suppose we find that children who have "warmer" parents also express greater self-esteem. Parents vary in their degree of warmth among families, so that a typical child in high-warmth families should express greater self-esteem than a typical child of low-warmth parents. However, if we have already shown that the variation in children's self-esteem owes *not* to shared rearing but rather to genes and nonshared environmental influences, then between-family (shared) associations of parental warmth and self-esteem must be spurious; causally, they must reflect the genes shared by biological relatives. In a parent, we assume that these genes affect child-rearing style; in the child, they may affect self-concept and other personality traits.

To resolve this difficult dilemma, we must suppose that the new nonshared influences are different *in kind* from the discredited shared rearing influences. Yet how can they be different *in kind* from ordinary environmental influences? One possibility is that they represent influences not ordinarily considered in socialization theory. Such influences may include the biological circumstances of embryological development, particular life events of strong emotional quality, chance events, or ordinary social encounters that are essentially unreplicable because their effects are mediated through the unique traits of each individual. Nonshared influences could even mimic the process of polygenic inheritance—where each nonshared "quanta" operates in a way similar to genes at a single locus—having a relatively small influence, but cumulatively, over many nonshared "quanta," accounting for an appreciable variation in outcomes.

Another alternative, put forth cogently by Dunn and Plomin (1990), is that non-shared rearing influences are psychologically unique and therefore unlike shared rearing influences. They pose this alternative:

> ... environmental factors that create differences within families can act independently of factors that cause differences between families, even when the same general issue is involved—parental affection, for example. Children really know only their own parents. They are unlikely to know if their parents love them more or less than other parents in other families love their children ... they are likely to be painfully aware if their parents show them less affection than siblings (Dunn and Plomin 1990, p. 161).

Thus, Dunn and Plomin concluded that nonshared rearing influences depend on social comparisons made within the family. What is important is the parental favoritism shown towards a sibling, not whether a child who lives down the street is well- or mistreated.

HUNTING FOR NONSHARED INFLUENCES

Measurement of Differential Experience

Siblings differ in the kinds of peers with whom they associate, in parental treatments, and in how they treat one another. To measure siblings' differential experience, Daniels and Plomin (1985) designed the "sibling inventory of differential experience (SIDE)." For each SIDE item, a sibling makes a relative judgment of the degree to which he/she is exposed to a particular environment more strongly than his/her brother or sister. On a 1–5 scale, a middle score of "3" indicates neutrality (social influence is the same for both siblings), a score of "1" means that the respondent's social environment has the characteristic but not his/her sibling's environment, whereas a "5" indicates the opposite, that the characteristic belongs to the sibling's social environment but not to the respondent's.

On the SIDE, a correlation of sibling A's and B's scores has a particular interpretation—it represents siblings' *agreement* about their differential experiences. As shown in Table 14.1 using data from the Arizona Sibling Study (described in Rowe and Gulley 1992), the correlations on the SIDE's three differential peer exposure scales—peer achievement, peer popularity, and peer delinquency—fall into the -0.20 to -0.40 range. The negative sign results from the relative judgment requested of the siblings: if one sibling reported a *high* level of involvement in popular, delinquent, or achievement-oriented peer groups, then the other sibling should have reported a *low* level of involvement. In general, siblings show moderate agreement about their differential experiences (Baker and Daniels 1990).

As with measures of trait phenotypes, the causal structure of the SIDE measure can be analyzed into genetic, shared environment, and nonshared environment sources of variation. One would hope to find little evidence of genetic influence on a measure

Table 14.1 Sibling agreement for differential experience in the peer group.

SCALE	GROUP		
	Brothers	Sisters	Mixed-Sex
Peer popularity	−0.45	−0.26	−0.44
Peer deliquency	−0.35	−0.29	−0.38
Peer achievement	−0.27	−0.31	−0.49
N pairs	135	142	141

purporting to assess nonshared environment! For differential peer influence, this expectation has been disappointed. Baker and Daniels (1990) compared MZ twins', DZ twins', nontwin siblings', and adoptees' SIDE reports of the mean degree of difference in their peer groups. Adoptees, who have the least genetic similarity, were least alike in their choice of friends while MZ twins, with the most genetic similarity, were most alike in their choice of friends. These outcomes imply genetic influence on the degree to which siblings' peer groups share similar social orientations. Although Baker and Daniels' results did not give as clear a signal of genetic influence for differential parental treatment or sibling mutual treatment, these other dimensions of differential sibling experience may be partly "heritable" as well.

The presence of genetic variation in the nonshared environment is a serious concern for causal interpretations. If a nonshared environment correlates with some difference in siblings' behavior, then does it cause the behavioral difference, or do unshared genes elicit different environments?

In measures of nonshared environment, genetic influence can be understood in terms of the concepts of reactive or active genotype × environment (GE) correlations, i.e., siblings with particular genetic dispositions may receive responses from the social environment that reinforce these dispositions; or they may actively seek environments that reinforce heritable dispositions (Plomin et al. 1977). For instance, genetically brighter children reading more books at the library than their duller brothers or sisters actively reinforces siblings' differences in intellectual abilities. In this example, initial genetic differences between siblings have led to eventual IQ differences. The mixed direction of causality in GE correlations stands in contrast to the hypothesis that differential experiences are strong causal agents that are independent of heritable traits.

TESTS OF DIFFERENTIAL EXPERIENCE

Method 1: Identical Twin Differences

A natural research design for testing differential experience is the correlation of differences of experience and behavior *within* MZ twin pairs. This test has the positive feature that genetic confounds would be removed completely—as the twins match genetically, any difference between them must be due to environmental influences that differed from one twin to the other. Still, the direction of causality remains an obstacle for understanding results. Suppose, for example, that the MZ twin who is spanked more than the other also fights more with peers. The more spanked twin may have been made more aggressive than the co-twin; however, it is also possible that the more aggressive twin *elicits* spanking from a parent unable to control him. In the latter interpretation, some third variable—such as embryological development—may account for *within-family* variation in both spanking and aggression.

The MZ twin difference research design has been used rarely. Baker and Daniels (1990) report a number of correlations between experience and behavior within MZ twin pairs. They found, for example, that the identical twin in the more popular peer group was also more extroverted. The within-family correlations ranged from about 0.20–0.40. Because the correlations were based on twin differences, they can explain only that part of variation in which twins differ. If about half the personality variation is *within twin families*, then these within-family effects explain only about half as much of the total variation—from 2–8% of the *total* variation instead of from 4–16% of the *within* family variation. As noted earlier, the direction-of-effects issue is unresolved in cross-sectional data: Did personality cause the treatment differences, or vice versa?

In comparison to Baker and Daniels (1990), Loehlin and Nichols (1976) were less successful at identifying correlates of MZ twin pair differences. In their analyses, the number of correlations between MZ twin personality and treatment differences was at a chance level. In summary, the MZ twin difference method has much to offer. At the present, however, too few results have been replicated to be certain about nonshared environmental influences.

Method 2: Total Variances in One- and Two-Child Families

Differential treatment effects can be approached in another way using a research design based on different family sizes. If, as Dunn and Plomin argued, differential treatment effects result from social comparisons within the family—the sense by one sibling that his brother or sister gets more than a fair share of parental love—then the *total trait variance* in two-child families for a trait affected by differential treatment should be greater than its total variation in one-child families. This statement can be proved using mathematical derivations, which we do not give here. The advantage of this research design is that causality is unambiguous: The directionality must be from differential treatment $\rightarrow$ children's traits for trait variance to increase in larger family sizes. Its disadvantage is low statistical power: it takes large sample sizes to detect a 5–15% increase in total trait variance over the baseline established in only-child families.

We used this approach to study problem behavior (mothers' ratings) in 5- to 9-year-old children. The data source (the U.S. National Longitudinal Study of Youth) is a representative sample of American families stratified to contain more poor families than in the general population. We had information on about 250 single-born children, 900 children in two-child families (450 pairs), and 400 children in three-child families (200 pairs). According to our hypothesis, the variance of problem behavior should increase with family size, from only-children to children with one or more siblings. Note that this analysis treats all the children as *individuals*. For the convenience of more equal sample sizes, we compared the first-born children in two- and three-child families with the only-children; however, a second child or a random child would have yielded the same outcome.

To see whether total variances changed with family size, we took advantage of a method of comparing *variances* using the analysis of variance (ANOVA; O'Brien 1981). The problem behavior scores were transformed using a simple formula. After this transformation, an ANOVA analogous to a mean effects ANOVA now tests for group differences in variances among conditions of the research design. To equate families for differences in family environment better, we divided families into those above and below average on a measure of the quality of their home environments. The final analysis had two independent variables: family size and home environment quality.

Table 14.2 shows the cell means for each condition in the research design, where each mean is equal to the variance of the problem behavior scores prior to their transformation. The variances (cell means) of problem behavior did not differ significantly among conditions. Whereas the variances in low home quality condition fell in the predicted direction ($s^2 = 181$ to 202), those in the high home quality condition did not ($s^2 = 184, 176, 166$).

Admittedly, this comparison required a number of assumptions. It was assumed that other sources of trait variation—measurement error, genes, and other environmental influences—were about equal for different family sizes. If these assumptions hold true, then differential parental treatments must increase variance in problem behavior. We controlled for rearing influences by including home environmental quality in the research design. No reason exists to suspect differences in measurement error when problem behavior is measured for an only-child versus for a sibling. Selection on genotypes among family types would probably influence mean rates of problem behavior. However, we found that only-children and siblings had about the same mean rates of problem behavior. Thus, for these children, we concluded that differential parental treatment does not causally influence rates of problem behavior.

Table 14.2 ANOVA cell sample sizes and means (variances of untransformed scores).

	Mean	N	Mean	N
Only children	181	273	184	266
First-born in two-child families	186	436	176	536
First-born in three-child families	202	234	166	208

Method 3: Explaining Residuals from Genetic/Shared Environment Models

DeFries and Fulker (1985) developed a general multiple regression model for behavioral genetic studies. Our idea was to add to it nonshared environmental variables. Our approach is best understood by considering first the DeFries and Fulker regression equation:

$$Y = B_1 X_1 + B_2 G + B_3 G X_1 + B_o , \qquad (14.1)$$

where Y is sibling A's trait score, X_1 is sibling B's trait score, G is the coefficient of genetic relatedness (1.0 for MZ twins, 0.50 for DZ twins and nontwins siblings, 0.25 for half-siblings, 0.125 for cousins), and B_o is the intercept. The unstandardized regression coefficient on X_1 represents the percentage of variation explained by shared environment. It is statistically independent of B_3, the coefficient on the interaction term. This latter number represents a trait's heritability: the percentage of variation that owes to genetic influences. In other words, heritability is found when the "coefficient of genetic relatedness" *conditions* siblings' behavioral resemblance; siblings who are more alike genetically would be more alike in behavior. In unselected samples, B_2 is not particularly important to understanding genetic or environmental influences. It captures whether trait means are unequal across the different sibling types.

Nonshared influences can be sought in the "residuals"—the variation unaccounted for—in the basic DeFries and Fulker equation. In MZ twins, such residuals represent only nonshared environment, while in other sibling types, they may also include effects of nonshared genes.

To look at nonshared environmental influences using the DeFries and Fulker equation, we added two new terms to it on the right-hand side:

$$B_4 T + B_5 G T , \qquad (14.2)$$

where T is the signed difference score, sibling A minus sibling B, for some measure of differential treatment; and GT is the interaction of genetic relatedness with differential treatment. In our interpretation, the B_4 coefficient tests whether nonshared environment influences the trait. B_5 tests whether nonshared genes contribute to the strength of the influence of nonshared environment (for example, when differential treatments are in *response* to genetic differences among siblings).

This approach can be demonstrated using NLSY data on childhood antisocial behavior (the rate of minor delinquent acts). Our sample consists of twins, full siblings, half-siblings, and cousins. Because the former were of unknown twin type, we used 0.75 as their "coefficient of genetic relatedness." The remaining relatedness coefficients were 0.5, 0.25, and 0.125, respectively. Child A's behavior was predicted from child B's using Equation 14.1. The data were "double-entered" so that each child was entered once as A and once as B. The sample size was then the number of individuals rather than the number of pairs (No. individuals = 2 × No. of pairs). The regression equation (Eq. 14.1) explained 14.4% of the variation in antisocial behavior. The majority of variance, however, remained unexplained. Explained variance was limited to less than 100% for several reasons: (a) because of measurement errors and unmeasured nonshared environmental influences and (b) because of nonshared genetic influences. In this analysis, the heritability of antisocial behavior was about 60%, whereas the shared environment effect was about 12% of total variation (N = 764 pairs, $p < 0.05$).

Our nonshared environment measure was the HOME ENVIRONMENT total score. A nonshared score was calculated by computing the signed difference, HOME of sibling A –HOME of sibling B, for each sibling pair in the data set. With this new variable, we then added two terms (Eq. 14.2) to the basic (Eq. 14.1) DeFries and Fulker regression equation: one was the variable "HOME difference" (HE) and the other was the interaction, "HOME difference by coefficient of genetic relatedness" (HE × G).

Table 14.3 shows the unstandardized regression coefficients for these terms in the augmented equation. The estimates of heritability (54%) and shared environment (15%) changed when the home environment terms were added to the question. The equation also explained more variance than before, but this increase was only 0.5% (from 14.4% to 14.9%). Although a small increase, it was statistically significant. Based on the regression coefficient for the home environment term (Beta = – 0.02) in Table 14.3, children who received better home environmental quality appeared to engage in fewer antisocial behaviors than their siblings. But even here, some ambiguity was apparent, because the nonshared environmental term interacted with the coefficient of genetic relatedness (regression coefficient = – 0.04, $p < 0.025$). A purer environmental situation would be one in which the nonshared interaction term failed to produce a significant interaction effect.

Table 14.3 Nonshared environmental influences in the DeFries-Fulker regression model.

Variable	Regression Coefficient
Intercept	94.7*
Co-siblings antisocial (CA)	0.15*
Coefficient of relatedness (G)	– 64.36*
CA × G	0.54*
HOME ENVIRONMENT DIFFERENCE (HE)	– 0.02*
HE × G	– 0.04*
R^2	0.149

Note: 48 twin pairs, 902 full sibling pairs, 364 half sibling pairs, and 330 cousin pairs.
*$p < 0.05$.

SUMMARY AND CONCLUSION

We surveyed methods of deducing nonshared environmental influences on behavioral development. The main methods were (a) correlating behavioral differences *within* pairs of MZ twins with treatment differences, (b) comparing total behavioral variation for siblings versus only-children, and (c) adding nonshared environmental measures to a regression equation that first exhausted shared sources of behavioral variation. Certainly, this list of methods is incomplete, and other methods of evaluating nonshared influences may be proposed.

With regard to problem behavior in childhood, it is clear that nonshared environments and problem behaviors can be statistically associated. For instance, siblings with more behavior problems and delinquency belong to more delinquent peer groups relative to their better-behaved brothers and sisters. In addition, they also experience a less positive *within* family environment. Showing the differential treatment of children who are different in behavior is easy.

The more difficult issue is deciding on etiology: Are nonshared rates of problem behavior the result of, or cause of, differential treatments? Does some unmeasured, nonshared environmental variable account for variation between a measured nonshared variable and differences in siblings' outcomes? In our brief look at problem behaviors, we first tried method 2 above, comparing the total variance for only-children versus siblings. Because variation in problem behavior was not greater for the latter group, our analysis failed to support a causal influence of differential treatments on problem behavior. On the other hand, in the method 3 analysis, differential HOME environment did predict siblings' different outcomes for antisocial behavior (one dimension of total problem behavior). This second analysis suggests that differential parental warmth and encouragement of academic achievement may, indeed, affect problem behavior outcomes in childhood. However, even this nonshared environmental influence appeared to interact with genetic differences between siblings. Hence, our ambiguous results are less satisfying than would be a loud signal for the *causal* importance of nonshared influences. We suspect that active gene × environment correlations may be a dominant process involved in producing sibling differences in behavior—a process where genes cause environments, which cause behaviors, and so on, rather than differential treatments causing behaviors independently of genes. Nonetheless, we end this discussion with the routine call for more research—but a research effort that focuses more on etiology than on association.

ACKNOWLEDGEMENTS

This research was supported by U.S. grants DA06287 and HD21973.

REFERENCES

Baker, L.A., and D. Daniels. 1990. Nonshared environmental influences and personality differences in adult twins. *J. Pers. Soc. Psychol.* **58**:103–110.

Daniels, D., and R. Plomin. 1985. Differential experience of siblings in the same family. *Devel. Psychol.* **21**:747–760.

DeFries, J. C., and D.W. Fulker. 1985. Multiple regression analysis of twin data. *Behav. Genet.* **15**:467–473.

Dunn, J., and R. Plomin. 1990. Separate lives: Why siblings are so different. New York: Basic Books.

Goldsmith, H.H. 1983. Genetic influences on personality from infancy to adulthood. *Child Devel.* **54**:331–355.

Loehlin, J.C., and R.C. Nichols. 1976. Heredity, environment, and personality: A study of 850 sets of twins. Austin: Univ. of Texas Press.

O'Brien, R.G. 1981. A simple test for variance effects in experimental designs. *Psychol. Bull.* **89**:570–574.

Plomin, R., and D. Daniels. 1987. Why are children in the same family so different from one another? *Behav. Brain Sci.* **10**:1–60.

Plomin, R., J.C. DeFries, and J.C. Loehlin. 1977. Genotype-environment interaction and correlation in the analysis of human behavior. *Psychol. Bull.* **84**:309–322.

Rowe, D.C., and B. Gulley. 1992. Sibling effects on substance use and delinquency. *Criminology* **30**:217–233.

Rowe, D.C., and R. Plomin. 1981. The importance of nonshared (E_1) environmental influences in behavioral development. *Devel. Psychol.* **17**:517–531.

15

Prenatal and Perinatal Influences on Twin Children: Implications for Behavioral Studies

E.M. BRYAN

Multiple Births Foundation, Queen Charlotte's and Chelsea Hospital,
Goldhawk Road, London, W6 0XG, U.K.

ABSTRACT

Twins can only be used in childhood behavioral studies if account is taken of the many differences between twin and singleton children. This chapter describes some of the differences in the physical and emotional environment of twins and singletons from conception, through pregnancy, and during early childhood as well as reviews, where known, the effects these have on the development of twins. The significance of these differences in relation to behavioral studies is, however, still to be evaluated.

INTRODUCTION

Twins are clearly a rich source of information for childhood behavioral studies; however, research workers are understandably wary of the complicating factors twinship may introduce. To what extent can twin data be generalized to the singleton population?

This question can only be answered when the differences in the physical and emotional environment of twins and singletons, both before and after birth, have been carefully evaluated.

In this chapter, I describe the various prenatal and perinatal influences that are known to affect twins in a different way or to a different degree than they affect singletons, and explore other areas needing further study.

From the time of conception, the experiences of twins are likely to differ from those of singletons. First, twins tend to be born into larger families. Twins are also more likely to have older parents (MacGillivray et al. 1988).

Twins as a Tool of Behavioral Genetics
Edited by T.J. Bouchard, Jr. and P. Propping © 1993 John Wiley & Sons Ltd.

The intrauterine experiences of a twin may also differ and are likely to be much more hazardous than that of a singleton. The twin fetus is affected not only by the innumerable intrauterine environmental influences of a single fetus but also by its interaction with the second fetus. At best, it must compete for nutrition. At worst, it may be severely, even lethally, damaged by the co-twin.

GROWTH

In the U.K., between 1982–1984, just over 50% of multiple births weighed less than 2500 g, compared to 6% of singletons (Botting et al. 1987). As well as being of low birthweight, many of these infants—particularly the more mature—are light for their gestational age when compared with singletons. Despite sharing the maternal food supply and possibly having an additional handicap of an energy-wasting "third circulation," most twin fetuses grow as well as singletons during the first two trimesters (Fliegmer and Eggers 1984). From about the 28th week, the rate of growth decreases in comparison to that of a singleton.

The optimal weight for twins, as indicated by perinatal mortality rate, appears to be between 2.5 and 3.5 kg whereas for singletons it is between 3.5 and 4 kg (Butler and Alberman 1969).

Intrapair variation in weight is greater in monochorionic twins. In addition, these twins are generally lighter than dichorionic. This difference, however, cannot be entirely explained by the effect of placentation since dichorionic monozygotic (MZ) twins also tend to be lighter than dizygotic (DZ) twins, who must be dichorionic (Corney et al. 1972). These differences in birthweight between MZ and DZ dichorionic twins may have one or more possible explanations. These include the effect on early embryonic development of a reduction in cell mass through the division into two embryos. Second, the antigenic difference between DZ twins could beneficially affect intrauterine growth. Finally, the mothers of DZ twins are, on average, taller and heavier than those of MZ twins (MacGillivray et al. 1988), and so they are likely to have heavier babies.

One in four pairs of twins, that is those with a monochorionic placenta, have an important additional influence on their intrauterine development: the effect of a shared (often unequally) circulation through placental vascular anastomoses.

The majority of monochorionic twins harmoniously share this "third circulation" without apparent ill-effect. Occasionally, as in the case of acardia, the fetus is actually dependent on this intrapair circulation for survival. In other instances, such as the fetofetal transfusion syndrome, the results may be disastrous.

The Chronic Fetofetal Transfusion Syndrome

The chronic fetofetal transfusion syndrome is the result of an intrauterine blood transfusion between monochorionic twins. It is a condition that is more common than generally realized and occurs in about 15% of monochorionic pregnancies (Dudley and D'Alton 1986). In addition, it is the cause of some of the greatest disturbances in twin fetal development. Nutrients as well as hemoglobin are, of course, transfused. Thus the recipient may become much heavier than the donor.

Hydrops fetalis may occur in either baby due to anemia and hypoproteinemia in the donor and to circulatory overload in the recipient. Both babies are equally at risk of death, neonatal complications, and long-term morbidity. The donor suffers complications of intrauterine growth retardation and anemia. The recipient, on the other hand, may have cardiac failure, hyperbilirubinemia, and intravascular thromboses, which may damage any organ, particularly the brain.

Severe growth retardation, frequently seen in the donor twin, shows that the transfusion is a chronic process often lasting many weeks. Indeed, the syndrome has been recognized as early as the tenth week of pregnancy. Polyhydranmios develops in the recipient's amniotic sac, allowing free fetal movements, whereas oligohydramnios may severely restrict donor fetal movements and cause the so-called "stuck twin syndrome" (Mahony et al. 1985).

All organs of the body may be affected. The increased blood flow may cause hyperplasia of myocardial fibers and an increase in muscle mass in both systemic and pulmonary arteries. These could have a lasting effect on the cardiovascular system of the recipient.

Similarly, the renal glomeruli of the recipient are more mature than those not only of the donors' but also of singletons of the same gestational age (Naeye 1965).

Long-term Growth

To return to twins in general, there is disagreement in the literature as to how the smaller one fares in the long term, physically and mentally. Compared to the larger born, several workers have found, in MZ twins, that when there were large intrapair differences in birthweight, some discrepancy persists into adult life (Henrichsen et al. 1986). There are, however, examples of remarkable and unexplained catch-ups and even overtakes by the smaller born MZ twin (Bryan 1992a).

Why some smaller MZ twins should show this remarkable ability to catch up while others remain permanently smaller is uncertain. It may well be that there are two distinct groups of intrauterine growth-retarded twins. One, the "catchers-up" would be those who probably suffered from malnutrition only in the latter part of pregnancy and thus had already acquired their full cellular complement and therefore potential for growth. The others, the "laggards," would be those whose differential growth was established early in pregnancy, thereby reducing their long-term growth potential. The degree of intrauterine growth retardation in the smaller twin is probably more important in relation to later growth velocity than the size of the intrapair weight discordance (Keet et al. 1986), particularly, as with singletons, for those with a normal ponderal index. However, as yet, no longitudinal studies of growth in birthweight discordant twins have included data on their intrauterine growth patterns.

As with light-for-date singletons, the lighter born co-twin is likely to be less intelligent than his co-twin (Henrichsen et al. 1986), although the deficit may vary in degree according to the ability studied. The neurological outcome for the smaller twin may also be less good. There may be problems in fine motor performance, coordination, and visuomotor perception (Ylitalo et al. 1988). Birthweight discordant MZ twins are also—as described later—more likely to show differences in behavior (Hay and O'Brien 1987).

Postnatal growth in MZ pairs can depend, in part, on types of placentation. MZ monochorionic pairs become progressively more similar in weight between birth and four years, whereas dichorionic pairs maintain a remarkably similar intrapair discrepancy throughout (Falkner 1978).

DELIVERY

Prematurity

The incidence of preterm delivery (less than 37 weeks gestation) in a twin pregnancy is approximately 30% (MacGillivray and Campbell 1988) compared to 5–10% for singletons. The mean length of gestation is approximately 260 days compared with 280 days.

Prematurity is more common among monochorionic twins than dichorionic, possibly due to less efficient placental function. The "third circulation" may well reduce the efficiency of maternofetal transfer still further.

MZ boy twins are more likely to be delivered prematurely than MZ girls. This sex-related difference, however, has not been shown among DZ twins.

Presentation

Up to one-third of twin infants present by the breech in contrast to 3.4% of singletons. Kauppila et al. (1975) found the perinatal mortality to be twice as high in breech as opposed to vertex-delivered twins. In only 40–50% do both babies present in the most favorable position.

Birth Order

Birth order used to have a strong influence on perinatal mortality and neurological morbidity, yet in good obstetric units this difference is now minimal. There may be more subtle long-term psychological differences in being a first- or second-born twin (see below).

Other Hazards

Exsanguination of the second twin in a monochorionic placenta, where the first cord has been unclamped, is a recognized hazard for twins. Other complications may arise from placental abnormalities, such as vasa praevia, velamentous insertion of the cord, and single umbilical artery, which are all more common in twin pregnancies.

Monoamniotic twins are at particular risk because of the danger of entanglement of the umbilical cords, which may cause umbilical vessel occlusion in one or both fetuses. Finally the perinatal mortality rate is four to six times higher in twins than in singletons (Botting et al. 1987).

CONGENITAL MALFORMATIONS

Despite the difficulties of studying congenital malformations and therefore of undertaking accurate comparisons, there is no doubt that congenital malformations are more common in twins than singletons (Little and Bryan 1988). This higher incidence is probably limited to MZ twins. Perhaps surprisingly, discordancy for an anomaly is not uncommon in MZ and DZ twins, particularly for congenital heart disease. Discordancy in MZ twins may also occur as a result of asymmetry in the X inactivation process in the cells of the inner cell mass at the blastocyst stage. This was demonstrated by Burn and his colleagues (1986) in a pair of MZ girls discordant for Duchenne muscular dystrophy.

DISABILITY

Disability in twins is more common than in singletons, although the exact prevalence is difficult to obtain (Bryan 1992b). Congenital abnormalities, mental retardation, and cerebral palsy are all more common in twins. The latter two may result from intrauterine insults or from the intrauterine twin relationship itself. Some, although fewer than was previously thought, may result from lasting damage obtained through perinatal hazards.

Cerebral palsy is more common among MZ twins. Second-born and lighter-born twins, whether MZ or DZ, are at greater risk than the first-born or heavier twin.

An etiological factor peculiar to twinning is the damage resulting from the intrauterine death of a monochorionic co-twin (Durkin et al. 1976). In the past, it was suggested that the cerebral damage was caused by intrauterine-disseminated intravascular coagulation in the surviving twin as a result of the monochorionic fetus sharing a placental circulation with a macerated dead fetus. More recently it has been suggested that the damage could be due to a hypotensive crisis in the survivor at the time of the demise of the twin, due to a partial exsanguination of the survivor's blood into the low-resistant circulation of the dead twin.

RELATIONSHIPS

Intrauterine

Little is known about the long-term effects of the physical and emotional relationship of twins during intrauterine life. The profound loss felt by some adults whose twin died before birth, however, suggests that the influences may be very real (Woodward 1988).

There is certainly evidence of definite behavior patterns *in utero*. Piontelli (1989) found that the behavioral tendencies of twins, such as aggression or affection, as shown on ultrasound filming during fetal life, persisted at least during the first year of life. Another sonographic study of triplets (Arabin et al. 1991) showed two fetuses interacting regularly during fetal life, whereas the third baby was isolated and ignored beneath them. The third baby remained the odd one out after birth.

Mother–Infant

Mothers of twins often notice marked differences in the personalities of their babies within a few days of birth, and many mothers talk and behave differently to each infant. Mothers may be attracted to one baby more than the other and relative sizes or different temperaments may particularly influence this. It has been observed, however, that the amount of time a mother spends with each baby may not reflect the degree of attraction. The mother may compensate for her negative feelings by spending more time with the less attractive baby.

In their study of very low birthweight twins, Minde and colleagues (1990) found that the majority of mothers develop a preference for one twin within two weeks after birth and maintain this preference for at least four years. Mothers tended to prefer the stronger, healthier baby. This agrees with Spillman's (1984) finding of a maternal preference for the heavier twin but is in contradiction to the findings in the small study by Allen and colleagues (1971), who found that eight out of ten mothers were closer to the secondborn, or the smaller or neurologically inferior twin.

Both birthweight and birth order are factors that may influence the self-esteem of a child later on; the strongest influence appears to be coming home first from the hospital (Hay and O'Brien 1987). If one baby goes home with their mother while the other remains, the latter's relationship with his mother may suffer as also may the development of his self-esteem.

The attitude of parents towards a twin child may differ from that towards a singleton right from the start. Most parents find it difficult to avoid both making and voicing comparisons between the children—partly perhaps to foster the development of individuality—but this polarization can lead to stereotyping and caricaturing of only slight differences between the children. This, in turn, can lead to the children's behavioral patterns becoming a kind of self-fulfilling prophecy.

The postnatal environment is also different in that parents of twins cannot help spending less time with each twin than they would with a singleton. The twins therefore have much less one-to-one communication with their parents and when this does occur it is likely to be given in shorter bursts as the mother's attention flits from one child to the other.

In practice, twins must constantly share their mother and thus usually communicate in a triad. Language delay and reading difficulties are common in twins, and triadic communication is one of the many factors that may contribute to this delay. Despite the fact that it is over fifty years since the language development of twins was first noticed to be delayed, there is still much to be discovered about the relative importance of the causative factors.

Infant–Infant

It has been argued that a measure of solitude is probably crucial to the full development of the individual personality (Storr 1988). Few people have less opportunity for solitude than a twin. A twin is rarely alone and is forced to live almost constantly in the presence of another. Self-awareness cannot easily grow if the person concerned is constantly absorbed by relationships or bombarded by sensations from a constant companion. Twins tend either to copy each other or to react against each other. Either way, they respond to each other rather than gradually and quietly discover their own needs and wants. Nor do they learn sometimes to be contentedly on their own, hence laying the groundwork for later self-reliance and self-discovery.

MATERNAL DEPRESSION

It is well recognized that a child's development may suffer if his mother is depressed. Recent studies have shown a higher incidence of depression and anxiety in mothers of twins (Thorpe et al. 1991). This depression may well continue long after the time of acute fatigue. Thorpe and her colleagues (1991) found that mothers of five-year-old twins had a higher malaise score, indicating depression, than mothers of singletons, even when the singleton mothers had two closely spaced children.

CONCLUSIONS

Twins are a potentially rich source of data for child behavioral studies. In using this data, however, it is essential to consider the important ways in which the perinatal lives of twins differ from those of singletons and the lasting effect that this may have on their development. The extent to which any particular factor must be taken into account will depend on both the measure itself and the age at which it is being studied. For example, in some studies, it may be necessary to exclude birthweight discordant MZ pairs.

There are many physical factors in the perinatal period that are different for twins than for singletons. It is not so widely recognized that there are various different psychological factors which are also likely to play a significant part in a twin's development. These include the complex nature of the intrapair relationships, both *in utero* and after birth, and the triadic setting within which so much communication with the mother, and father, is conducted.

REFERENCES

Allen, M., W. Pollin, and A. Hoffer. 1971. Parental birth and infancy factors in infant twin development. *Am. J. Psychiat.* **127**:1597–1604.

Arabin, B., J.v. Eyck, J. Wisser, H. Versmold, and H.K. Weitzel. 1991. Fetales Verhalten bei Mehrlingsgravidität: methodische, klinische und wissenschaftliche Aspekte. *Geburtsh. Frauenheilk.* **51**:869–875.

Botting, B., I. Macdonald-Davies, and A. Macfarlane. 1987. Recent trends in the incidence of multiple births and associated mortality. *Arch. Dis. Child.* **62**:941–950.

Bryan, E. 1992a. Newborn twins. In: Twins and Higher Multiple Births: A Guide to Their Nature and Nurture, pp. 97–111. Sevenoaks: Edward Arnold.

Bryan, E. 1992b. The Disabled Twin. In: Twins and Higher Multiple Births: A Guide to Their Nature and Nurture, pp. 195–170. Sevenoaks: Edward Arnold.

Burn, J., S. Povey, Y. Boyd, E.A. Munro, L. West, K. Harper, and D. Thomas. 1986. Duchenne muscular dystrophy in one of monozygotic twin girls. *J. Med. Genet.* **23**:494–500.

Butler, N., and E. Alberman. 1969. Perinatal Problems— The Second Report of the 1958 British Perinatal Mortality Survey. London: E. and S. Livingstone Ltd.

Corney, G., E. Robson, and S. Strong. 1972. The effect of zygosity on the birth weight of twins. *Ann. Hum. Genet.* **36**:45–59.

Dudley, D., and M. D'Alton. 1986. Single fetal death in twin gestation. *Sem. Perinatol.* **10**:65–72.

Durkin, M., E. Kaveggia, E. Pendleton, G. Neuhauser, and J. Opitz. 1976. Analysis of etiologic factors in cerebral palsy with severe mental retardation. Analysis of gestational, parturitional and neonatal data. *Eur. J. Pediatr.* **123**:67–81.

Falkner, F. 1978. Implications for growth in human twins. In: Human Growth: I. Principles and Prenatal Growth, ed. F. Falkner and J.M. Tanner, pp. 397–413. London: Bailliere Tindall.

Fliegner, J., and T. Eggers. 1984. The relationship between gestational age and birthweight in twin pregnancy. *Aust. N.Z. J. Obstet. Gyn.* **24**:192–197.

Hay, D., and P. O'Brien. 1987. Early influences on the school social adjustments of twins. *Acta Genet. Med. Gem.* **36**:239–248.

Henrichsen, L., K. Skinhoj, and G. Andersen. 1986. Delayed growth and reduced intelligence in 9–17 year old intrauterine growth retarded children compared with their monozygous co-twins. *Acta Paediatr. Scand.* **75**:31–35.

Kauppila, A., P. Joupila, M. Koivisto, I. Moilanen, and O. Ylikorkala. 1975. Twin pregnancy: A clinical study of 335 cases. *Acta Obstet. Gyn. Scand.* **44**:5–12.

Keet, M., A. Jaroszewicz, and C. Lombard. 1986. Follow-up study of physical growth of monozygous twins with discordant within-pair birthweights. *Pediatrics* **77**:336–343.

Little, J., and E. Bryan. 1988. Congenital anomalies in twins. In: Twinning and Twins, ed. I. MacGillivray, D.M. Campbell, and B. Thompson, pp. 207–240. Chichester: Wiley.

MacGillivray, I., and D.M. Campbell. 1988. Factors affecting twinning. In: Twinning and Twins, ed. I. MacGillivray, D.M. Campbell, and B. Thompson, pp. 111–142. Chichester: Wiley.

MacGillivray, I., M. Samphier, and J. Little. 1988. Factors affecting twinning. In: Twinning and Twins, ed. I. MacGillivray, D.M. Campbell, and B. Thompson, pp. 67–98. Chichester: Wiley.

Mahony, B.S., C.N. Petty, D.A. Nyberg, D.A. Luthy, D.E. Hickok, and J.H. Hirsch. 1990. The "stuck twin" phenomenon: Ultrasonographic findings, pregnancy outcome, and management with serial amniocenteses. *Am. J. Obstet. Gyn.* **163**:1513–1522.

Minde, K., C. Corter, S. Goldberg, and D. Jeffers. 1990. Maternal preference between premature twins up to age four. *J. Am. Acad. Child Adol. Psych.* **29**:367–374.

Naeye, R. 1965. Organ abnormalities in a human parabiotic syndrome. *Am. J. Path.* **46**:829–842.

Piontelli, A. 1989. A study on twins before and after birth. *Int. Rev. Psycho-Anal.* **16**:413–426.

Spillman, J. 1984. The role of birthweight in maternal-twin relationships. MSc Thesis, Cranfield Institute of Technology.

Storr, A. 1988. Solitude. London: Fontana.

Thorpe, K., J. Golding, I. MacGillivray, and R. Greenwood. 1991. Comparison of prevalence of depression in mothers of twins and mothers of singletons. *Br. Med. J.* **302**:875–878.

Woodward, J. 1988. The bereaved twin. *Acta Genet. Med.* **37**:173–180.

Ylitalo, V., P. Kero, and R. Erkkola. 1988. Neurological outcome of twins dissimilar in size at birth. *Early Hum. Devel.* **17**:245–255.

Standing, left to right:
C. Kirschbaum, J. Hebebrand, M. Rutter, D.A. Hay, D.L. Pauls, D.C. Rowe, R. Derom
Seated, left to right:
B. Kracke, A.M. Macdonald, E.M. Bryan

16

Group Report: What Can Twin Studies Contribute to the Understanding of Childhood Behavioral Disorders?

A.M. MACDONALD, Rapporteur

E.M. BRYAN, R. DEROM, D.A. HAY,
J. HEBEBRAND, C. KIRSCHBAUM,
B. KRACKE, D.L. PAULS, D.C. ROWE,
M. RUTTER

INTRODUCTION

While twins have been a focus of interest in mythology, art and obstetric medicine for centuries, their usefulness as a research tool for advancing the understanding of the interplay of genetic and environmental factors in both normal variation and disease was recognized only a century ago. The development of analytical methods (Falconer 1965; Jinks and Fulker 1970; Smith 1974) dependent on the comparison of data from monozygotic (MZ) and dizygotic (DZ) twins has led to a huge scientific literature on twin studies of vastly variable quality. There has been a polarization of twin researchers: those interested in the value of twins as a "tool" of genetic epidemiology intent on refining the analytical methods and parameterizing sophisticated models (Eaves et al. 1989) and others interested in twins themselves, the unique circumstances of their pre- and postnatal development, and the problems and advantages of twinning. In some ways this has become like the nature-nurture debate of earlier decades with the development of separate languages; to advance understanding of the etiology of childhood disorders in particular, the findings of either group of researchers must influence the methods of the other. The background papers and discussion at this

Twins as a Tool of Behavioral Genetics
Edited by T.J. Bouchard, Jr. and P. Propping © 1993 John Wiley & Sons Ltd.

meeting provided a field for defining the areas of interest and topics for future empirical research.

Many of the issues covered in discussions have general relevance for twin studies. Thus, for example, changing rates of twin births have implications for all studies because they affect our expectations about the relative numbers of MZ and DZ twins. Other issues are much more specific to childhood: events in the pre- and perinatal periods are more recent in the life span and are perhaps more likely to influence the etiology and severity of childhood disorders than is the case with adult conditions. The development of language, which does seem to differ in twins, is an integral part of normal child development, as is the acquisition of cognitive and social skills.

In setting the agenda for our discussion, the topics raised by the background papers were many; three main areas were delineated. First, the differences between twins and singletons have begun to be documented, but there have been few studies that systematically examine either the implications of these differences for classic twin studies or the ways in which they might affect the psychological development of twins and psychopathology. The second area covered the way in which twin methodology may be used to approach issues of current interest in childhood genetic epidemiology; these included the definition of phenotypes, the study of co-morbidity, the use of categorical versus continuous measures, the lessons from adult psychiatric genetics, and innovative approaches to analysis. Finally, the group considered how future study designs might expand knowledge of childhood behavioral development and psychopathology, including the use of modifications of the classical twin study, the relevance of molecular genetics, and improved assessment of nonshared environmental effects.

TWIN/SINGLETON DIFFERENCES

The Pre- and Perinatal Periods

As discussed by Bryan (this volume), there are three types of variables that appear to differ in twins and singletons. These are biological (lower birthweight, discordant birthweight, younger gestational age, feto-fetal transfusion syndrome, placentation), parenting (maternal attachment to two babies, triadic communication, less individual attention) and the twin relationship (sharing a womb, competition/cooperation, lack of solitude, attention from others).

While these differences are now well documented, there is little or no systematic research on the longer-term implications for twin development and later psychopathology, nor for genetic analysis. Such issues are of considerable importance if the results of twin studies of psychopathology are to be generalized to the nontwin population. Considerable effort is needed to test whether such environmental effects are of importance to the particular disorder being studied. There is no a priori reason to suppose that, for example, differences in parenting important in the development of one type of disorder are necessarily so for another.

Many of the differences are measurable and hence easily incorporated into twin designs and analyses. In developing studies of young twins, it was considered to be extremely important to obtain basic obstetric information; birthweight, length of gestation, presentation, and mode of delivery are all variables which should be recorded. In addition, infant twins should be classified according to placentation as well as zygosity. Zygosity should be determined routinely in all obstetric units (R. Derom et al. 1991). Where studies do exclude the more extreme cases of prematurity and birth complications, this should be clearly stated.

In MZ twins, placentation profoundly influences the growth and feto-fetal blood circulation. Whereas dichorionic pairs are in all antenatal respects comparable to DZ twins (R. Derom 1993), this is not the case for monochorionic pairs. One should possibly also separate the rare monoamnionic from the diamnionic variety. In pairs of equal gestational age, monochorionic twins tend to have lower birthweights than dichorionic twins. Probably of more importance is the often large intrapair difference in birthweight. It is known that many of the lighter twins catch up in growth after birth. Yet subtle differences in central nervous system functions have been demonstrated, and it is likely that with the increased chances of survival of the very low birthweight babies, more pairs discordant in birthweight will also be discordant in some aspects of brain function. The majority of monoamnionic twins are female, as are conjoined twins (Milham 1966), and some 75% now survive (C. Derom et al. 1988). Analyses of twins classified according to placentation have shown that some 40% of variation in birthweight is due to maternal factors and 40% to genetic factors, so that birthweight cannot be considered as a purely environmental risk factor (Vlietinck et al. 1989).

This brings us to the conclusion that, in the future, more attention should be paid to placentation where the behavioral effects are largely unexplored. As an example of how specific such effects can be, O'Brien and Hay (1987) have shown that the effects of the feto-fetal transfusion syndrome are much more on nonverbal than on verbal ability. While most of the biological factors are likely to have lasting effects only at the extremes, this implies, at the very least, that twins experiencing such extreme factors might be considered as a separate group. This may be particularly important when the recent changes in twinning patterns and differential survival rates are taken into account.

Changes in Twinning Rates and Differential Survival

Three potentially important changes have characterized the last decade: (a) treatment of infertility with ovulation-inducing agents, (b) the increasing rate of cesarean sections, and (c) the substantial improvement in neonatal care since 1975.

Ovulation-induction drugs are used on a large scale in the industrial world, resulting in a dramatic increase in iatrogenic multiple pregnancies. For instance, in Belgium in 1991, one third of the twin deliveries resulted from infertility treatment (C. Derom et al. 1993). Consequently, the prevalence of twin births is going up (+ 60% in Belgium; C. Derom et al. 1993). Additionally, the distribution of MZ versus DZ twinning is changing (almost all iatrogenic twin pregnancies are DZ). From an earlier ratio of approximately one-third MZ twins to two-thirds DZ, the proportions changed to 45% MZ and 55% DZ with the secular decrease in DZ twins in the 1970s, and are now reversing. Another effect has been an inflation of the difference in age of mothers of MZ and DZ twins since the average age of mothers of iatrogenic DZ twins is higher (Salat-Baroux et al. 1988; Medical Research Intl. 1991).

1975 was the turning point in the care of very premature babies. Since then the survival of very low birthweight babies (below 1500 g) has improved dramatically (Botting et al. 1987). In the birthweight category of 1000–1500 g, approximately 90% of babies survive; in the extremely low birthweight category (less than 1000 g) that percentage is 50% to 60%. Even with the best of care, 6–8% of the surviving babies develop a severe, life-long disability (Alberman et al. 1985; Veen et al. 1991; Lipper et al. 1990).

Knowing that 10–15% of twins (versus 0.8% of singletons) fall into the very low birthweight category, one expects that, in the future, groups of twins subjected to studies of behavioral development or psychopathology will include an excess of individuals whose brain development will differ from that of singletons.

The number of twins being delivered by cesarean section has greatly increased, particularly in the U.S. Thus, the effect of birth order on physical outcome is reduced. However, the psychological advantage of being the firstborn may remain.

Parents conceiving twins usually suffer from shock and often concern over the practical and financial implications. Those conceiving as a result of treatment for infertility are likely to be better prepared and to be pleased at the prospect of a ready-made family. However, they are also often much more anxious about the outcome of their pregnancy, which may be of relevance where data on child psycho-pathology are based partly or wholly on parental reports. The high cesarean section rate among this group (Macfarlane et al. 1985) may reflect an increased level of anxiety in the obstetricians as well.

The Postnatal Period

Parental stereotypes or comparisons may well exaggerate differences in temperament within twin pairs, but are unlikely to affect intrapair differences in intelligence in the same way. It is important, therefore, to demonstrate that such influences do not distort the twin resemblance on the particular measures under investigation and not to rely on results from studies of other measures.

Comparisons between twins by parents and others may reinforce longitudinal stability of behavior (Hay and O'Brien 1987). If one twin is regarded as, say, "the more troublesome," it may be difficult for this child to break out of this stereotype and expectation of his/her behavior. This may affect twin development in a way that comparisons of singleton siblings of different ages do not.

Over the last fifty years, language delays have emerged as the most consistently reported difference in development between twins and singletons, especially in the preschool years. Possible explanations lie both in pre- and perinatal complications more often experienced in multiples and in the unique "twin situation." At present, data are contradictory as to which of these is more important or whether they are cumulative; however, such questions are open to empirical investigation. Since the action of these twin-specific effects is largely environmental (birthweight may have a more complex and partially genetically mediated role), twin studies of language delays are likely to underestimate seriously the influence of genetic factors as applied to singleton populations.

Summary

It is now recognized that there are many differences between the physical and emotional experiences of twins and singletons during the intrauterine, perinatal, and childhood periods. However, very little is known about the significance of these differences for the later behavior of the twins.

Two points must be stressed. First, it is unlikely that any of these influences are general and apply across the whole spectrum of behavior. Second, twin-singleton differences or events unique to twins, such as the twin-transfusion syndrome, do not necessarily lead to an overestimation of heritability as some of the critics of twin method would have us believe.

We considered three possible scenarios as examples:

1. If both MZ and DZ twin similarity are increased, this will be apparent from differences in the DZ and sibling correlation necessitating inclusion of a twin effect in biometrical genetic models, as has been demonstrated for schizophrenia by McGue et al. (1986).
2. Effects that exaggerate MZ differences but do not affect DZs, such as comparisons of the twins in attempts to differentiate them (Hay and O'Brien 1984), will lower heritability estimates and increase apparent environmental effects.
3. It is possible that differences in DZ twins may be exaggerated in some circumstances and the DZ intrapair correlation reduced, perhaps through maternal ratings (Wilson and Matheny 1983). In this case, heritability will be inflated.

The changing rates of twinning and differential survival will have important cohort effects on twin designs. If one is using a cohort-sequential design following samples of different ages, then the cohort effects on obstetric management of multiples becomes a serious issue. In replicating a twin study of infants from, say, 1970, the pattern of twinning in 1995 will be quite different, with many more iatrogenically induced twins, a higher rate of DZs, but also more MZs surviving who might not have done so at the earlier date. For longitudinal studies, the results from twins sampled years ago may not reflect what would happen today.

A further caution concerns the use of single-surviving twins. They cannot be considered as normal singletons. Not only may they have been exposed to the same prenatal and perinatal hazards that caused the death of their twin, but they have been brought up by bereaved parents. A recent study (Thorpe et al. 1991) found that mothers who have lost a newborn twin have a high rate of depression five years later, which may affect the child's development. Thus, in using such a sample, care will need to be taken to determine whether such factors may be confounded with the variables of interest.

TWIN STUDY METHODOLOGY AND CHILDHOOD PSYCHOPATHOLOGY

The Phenotypes

Discussions about the important issues for childhood behavioral disorders fell into two main concerns. First, there is a great need to define phenotypes that achieve some balance between, on the one hand, a cluster of clinical symptoms and observations and, on the other, categories or dimensions derived from theory-driven research. The twin method could be used to assess the definition of different phenotypes, co-morbidity, and the use of continuous and discontinuous variables in assessment.

Second, there is a considerable problem for genetic research of childhood disorders in that the phenotype for a disorder at age five may differ in the same individual at age ten. In addition, the question of whether an adult from another generation of the same family has a related phenotype is frequently impossible to answer. There are a few disorders with onset in childhood (e.g., autism and Tourette's syndrome) where the adult phenotype is quite similar to that seen in childhood. However, for the majority of other disorders, the adult outcome of childhood conditions is not yet clearly established. For example, children with Attention Deficit-Hyperactivity Disorder (AD-HD) may develop an antisocial personality disorder, anxiety disorders, affective disorders, or no disorder at a later stage.

Studies are needed to define valid childhood phenotypes and to determine the phenotype that represents the manifestation of the disorder in adulthood. The longitudinal study of twins and singletons would be useful in this endeavor, allowing inspection of phenotypic changes within individuals, within pairs of different degrees of genetic relationship, and between pairs. Discordant MZ twins followed longitudinally provide a chance to examine behavioral variability in the co-twin and to assess whether the observed behaviors reflect variable expressivity of the disorder.

In the absence of robust twin data on many childhood disorders, can we learn anything by the successes (and failures) of the adult twin studies? There are a number of approaches to adult psychopathology that so far have not been tried with children. The use of twin concordance rates to examine different approaches to diagnosis is illustrated in McGuffin et al. (this volume). This approach, though attractive, should be applied with caution; the criterion of finding a symptom cluster which maximizes heritability as a way of establishing the genetic validity of a clinically defined phenotype is a deceptively simple one (Kendler 1989). The possibility of capitalizing on chance effects is a potentially serious issue that may complicate rather than add to the understanding of the phenotype.

It is a truism that birth has been a more recent event in the lives of children than adults and we have to consider carefully the possibility that obstetric complications may create phenocopies of the behavior of interest in one or both members of the pair of twin children. If just one member of an MZ pair is affected, this may be the differential expression of a genetic effect or it may be a purely environmental phenocopy. The familial-sporadic distinction is probably worth examination in MZ twin families and particularly for more common disorders (Kendler 1988), with the caveat that, as described above, the assessment of older family members is hampered by changing phenotypes. Such studies must thus rely on recall of childhood disorder by adults. While the concerns in adult twin research about this distinction (Kendler 1988; Farmer et al. 1990) are likely to apply even more strongly in children, the possibility of this as one line of approach to twin data should not be dismissed.

Twin studies should not restrict assessment to one disorder. For example, assessment of AD-HD, reading disorders, and clumsiness in a large sample of twins will allow initial univariate analyses of the genetic and environmental contributions to the different syndromes; then multivariate approaches may be used to assess the extent of genetic overlap underlying associations. The use of opposite sex twins in such studies will allow male and female differences in expression to be examined. This is important, for example, in conduct disorder where longitudinal studies have shown that conduct disturbance in girls leads to different patterns of adult behavior than in boys.

The Use of Categories and Dimensions

In distinguishing normality and disorder, the continuity/discontinuity distinction is a crucial one. In medicine, disorder cannot be considered as the end of a trait continuum because most disorders are defined in terms of a constellation of factors not on a single dimension. The fact that a variable functions as a dimension in one respect does not mean that it is not a category in another; for example, IQ is a dimension, but severe retardation has different biological implications. One must expect to find dimensionalization within categories; almost all conditions show a range within a category.

It was proposed that multiple measures should be used to assess phenotypes and these subjected to latent class analysis to derive theory-driven constructs. It was considered that the strategy should always be to start with a dimensional approach, albeit multidimensional, and only relinquish it when specific syndromes are delineated, providing categories that are both clinically meaningful and validated by substantial research.

The implications for twin studies of a categorical versus continuous approach lie in the differing analyses. Quantitative genetics has developed models analyzing the covariances of twin scores on dimensional assessments or, alternatively, regression approaches developed by DeFries and Fulker (1985). A categorical approach relies on the derivation of tetrachoric correlations assuming a multifactorial threshold model and requiring accurate knowledge of prevalence rates if twins come from clinical samples. The latter approach has considerable limitations, and it is easy to see why behavioral genetics has focused on dimensional measures of normal traits amenable to model-fitting and the large sample sizes required. Twin studies of childhood disorder would be strengthened by the use of multidimensional approaches, amenable to latent class analysis, and the application of biometrical genetic approaches.

The Concept of Liability: Can We Operationalize It?

Much genetic research in the field of psychopathology is grounded on the reasonable assumption that categories of disorder are based on a continuously distributed underlying liability and that there is a threshold beyond which this liability is translated into a manifest disorder (Falconer 1965). It is important to remember that, however plausible, liability is a latent construct derived from the central limit theorem, and it does not just represent genetic vulnerability, but the liability due to all sources, genetic and environmental, acting additively. Nevertheless, if the goal is an understanding of genetic mechanisms (and that must be the objective if practical use is to be made of genetic findings), we must seek to move from "black box" assumptions of latent liability to a direct study of the postulated processes. It is likely that quite often there will be an observable index of the underlying liability—if only we knew how to conceptualize and measure it. So far the search for biological trait markers for psychopathological liability has been frustratingly unrewarding, but the endeavor remains an important one. Moreover, we should not assume that the marker must be "biological" in the narrow sense. For example, if autism should prove to be a multifactorial disorder (one of several possibilities), evidence of the underlying liability might be detectable in some behaviorally manifest social deficit, a particular cognitive pattern, or a biochemical feature (such as serotonin level). The implication would be that if the broader phenotype observed in "discordant" MZ co-twins were indeed an expression of the same liability, then these co-twins should also show marked differences from unaffected individuals on the measure of liability. Similarly, the liability to emotional disorder might be based on a trait of behavioral inhibition or that to conduct disorder on reduced physiological reactivity and/or hyperactivity/impulsivity. Possible underlying dimensions should not be thought of only in terms of levels of "symptoms."

FUTURE RESEARCH

The value of a systematic approach to genetic epidemiology lies in the use of a combination of different approaches, each with its own advantages and drawbacks. Integration of data from, say, classical twin studies, twin/family studies, sibling designs, and surviving twins can provide far more information than any one approach alone. The choice of which studies are needed will depend on the current knowledge about the disorder, the hypotheses to be tested, and the extent to which twin-specific factors relevant to the disorder are found to be operating. So, for example, in studies of language disorders, it would be difficult to extrapolate genetic findings from twin studies alone because of the higher rates of disorder in twins. However, in combination with studies of surviving twins who are brought up as singletons, and with familial data from other siblings, we may be able to use the twin method to provide extra information about such disorders. For autism, there is no apparent excess in twins over single-born children, so the classic twin study has been successful in establishing that this is a highly heritable disorder insofar as any twin study can when dealing with a very rare condition. In this case, the classic twin method provides the information that a family study cannot and suggests that the next step may be further examination of the broader phenotypes found in "discordant" MZ co-twins and a search for indications of the "liability."

Extensions of the Twin Design

In considering how twin studies may be extended, there are three general cautionary principles which apply. First, some studies that plan to learn more about twins may have different needs to those that focus on genetics. In light of the information discussed by Bryan (this volume), the latter may choose to exclude very premature twins or those very discrepant in birthweight, if these events turn out to be associated with the phenotypes of interest. On the other hand, these children are essential for studies of twins and twinning. It is important that the sampling frame be clearly defined because of its implications for comparisons between studies.

Second, large twin studies are expensive, and therefore replication—usually a prerequisite for scientific results to be accepted—may be difficult. While total replication is therefore unlikely, there are good reasons to include some common questions and instruments to enable comparison across studies (Hay et al. 1990).

Third, mention has already been made of the special advantage for childhood psychopathology studies in that twins are of the same age, obviating the problem that the same underlying pathology may have different manifestations at different ages. Extensions to siblings of different ages and even to relatives of different generations reintroduce all these age-related issues.

A number of extensions to the twin design may be considered.

1. *Twin/family studies.* Since studies of twin children may be especially susceptible to twin-specific influences, it is important to include other relatives, such as siblings and cousins with the previously discussed caveats. A further advantage of this is that recruitment of control families without twins is often more difficult than finding multiple birth families, and thus it is much easier to utilize siblings of twins. The correlation of an MZ co-twin and his/her single-born sibling should be the same as that within DZ twin pairs if all the assumptions of the classic twin study are met. As behavioral problems may be increased in

the older siblings of twins as a reaction to the twins (Hay et al. 1988), their use may have to be viewed with caution.

2. *Offspring of twins.* These studies may have only modest power for the analysis of childhood psychopathology, especially if the children are of different ages. However, as Foley and Hay (1993) have demonstrated, the study of offspring of twins discordant for a disorder may be particularly suitable for analyzing the relationship between parental and child psychopathology. The question is, then, whether abnormal behavior in the child relates more to the environment than to the genes. Such studies may also be especially useful, for example in genetic designs for examining temperament, where there is a danger of reporting biases in assessing twins, however careful the researchers are (Wilson and Matheny 1983).

3. *Surviving twins.* Given the earlier question of whether obstetric complications or the unique social environment in which twins develop contribute to the prevalence of language problems in this group, one design would appear to be the use of surviving twins (Rutter et al., this volume). As discussed in the section on differences, surviving twins may not be just like singletons. Thus, while finding that surviving twins are no different from singletons is strong evidence that pre- and perinatal events contribute mainly to particular twin-singleton differences, the finding of some difference from singletons is open to differing interpretations.

The Use of Registries

Problems of ascertainment bias in twin studies are overcome, to some extent, through the use of population-based twin registries; for childhood disorders, where the need for examination of continuity with adult disorders has already been stressed, the need for longitudinal genetic designs requiring a registry is paramount. Registries are expensive to maintain and strict ethical codes must be followed in their use. It was recognized that the use of registers for twin studies provided more representative samples than found in most of the social sciences, and there are strong arguments for continuing their use. The possibility that European privacy laws may alter to rule that personal records be destroyed after completion of the immediate use for which they were collected was felt to be shortsighted, not only on research grounds, where it would make longitudinal studies impossible, but also for clinical continuity.

The maintenance of registries is also the only way to organize the large sample twin studies necessary to generate the analytical power required to discriminate various genetic models. It is often assumed that the main problems with unduly small samples that lack statistical power are that their findings will be inconclusive or that true effects will be missed. Pocock (1983) has pointed out that the real problem is the reverse of that, namely, that they are likely to lead to an increased rate of false positive findings. This is because the rate of true positives increases with sample size, whereas the rate of false positives remains constant (i.e., regardless of sample size 5% of findings will be statistically significant at the 5% level by chance alone). Thus, the ratio of true positives to false positives will be much lower in small rather than in large samples. Of course, the tendency to publish positive but not negative findings serves to increase this bias still further. Without the maintenance of large twin registries, more small-scale twin studies with low discriminatory power would be conducted, and, on the whole, those with positive findings would be published. The implications are obvious.

On the other hand, the group recognized the need for detailed observational studies of behavior and especially of the environment and the dilemma in choosing between a small study with good measures and large twin studies where constraints may lead to less adequate measures. The use of a twin registry is still advisable in small studies to reduce ascertainment biases.

Molecular Genetics: The Possibilities and Problems

For most childhood disorders, the application of traditional molecular genetic techniques (i.e., family genetic linkage aimed at detecting the influence of a particular genetic locus) was considered to be premature. Apart from the unreliability of definition of some childhood conditions, the primary reason for this is that too little is known about the adult phenotype of childhood disorders. Furthermore, insufficient information is available about the extent to which genetic factors are involved, as well as the mode of their transmission, to allow correct specification of the models for linkage analysis. For traditional linkage studies to be successful, it is necessary to know the affected status of all individuals in the extended family, and hence we return again to the need to know the adult phenotype of a childhood disorder (see above). Other molecular genetic strategies (such as sib-pair studies and association studies) might have a better chance for success; however, the definition of the phenotype is still a critical issue for these.

On the other hand, by using molecular association type approaches, studies could be designed that might help delineate the inherited phenotype. By using multidimensional assessment strategies and genetic markers, it may be possible to identify a number of markers associated with specific dimensions of behaviors. Traditional linkage studies will not be successful if several loci acting simultaneously are important for the manifestation of a condition. Only if there are genes of major effect will current linkage methods be successful. Population association studies might be successful in identifying multiple loci. However, sample sizes necessarily will be very large and the control samples will need to be chosen with care where the frequency of an allele in a patient sample is compared with that in a control sample. Differences in allele frequencies may spuriously result from stratification if, e.g., the two samples are ethnically different. Stratification effects can be avoided by the method proposed by Falk and Rubinstein (1987). Here, in addition to a series of independent cases, blood samples are collected from their parents. The alleles of the index patients are determined and those parental alleles that were not transmitted are used as controls. Thus, the parental alleles transmitted and not transmitted to the affected child are compared to one another; a simple chi-squared test can be applied. This approach can be used even when one of the parents is homozygous.

The Nonshared Environment

Several means exist for improving environmental assessment in twin studies. The first improvement recognizes the important finding of behavioral genetic studies that shared rearing influences are relatively weak, whereas nonshared influences are relatively strong (Rowe and Plomin 1981). However, there are exceptions to this generalization, for example, for conduct disorders and for social class and IQ when extremes are considered, as shown by adoption studies. These effects may to some extent be mediated by assortative mating which will reduce heritability estimates from twin studies (see Rutter 1991 and this volume). Because of the apportionment of environmental influence, generalized measures of family emotional climate will not do. Rather, the environment must be assessed separately for each twin: what is parent-child interaction like for that twin, and how do patterns differ from that for the other twin? Intellectual stimulation, parental warmth, and other aspects of "environment" may well vary from one twin to another, and assessment procedures must find a way to make the assessment of each twin as independent as possible. Thus, it is good that the possibility of sibling contrast effects has been raised in relation to sibling studies (Dunn and Plomin 1990). The suggestion is that children may be affected more by the fact that they are, say, consistently less favored than their sibs by their parents than by the overall level of parental approval in the home. It will be important for this possibility to be considered in future twin research into family influence. In addition, particular nonshared environmental influences should be assessed, such as the different birth conditions within MZ twin pairs.

A second means of improvement is to consider the possible influence of twin mutual interaction on behavioral development. Direct measures of the qualities of the twins' mutual interaction are absolutely necessary. In older twin pairs, the degree of contact with mutual versus separate friends should be assessed. Minor delinquency and the acquisition of substance use occur in a context of peer support, opportunity, and encouragement. Siblings close in age can function as agents of social influence in much the same way as other adolescent peers. If so, the degree of mutual sibling contact and the degree to which siblings may belong to the same peer group may condition sibling resemblance for substance use and delinquency. Using siblings' reports of mutual friends as a moderator variable, Rowe and Gulley (1992) found that same-sex sibling pairs with mutual friends were considerably more alike in number of delinquent acts and levels of substance use than those pairs with separate friendship networks. Moreover, the direction of influence was asymmetrical, with the younger siblings in the mutual friends groups more advanced in delinquency and substance use than other younger siblings. Because only about 10%–20% of siblings report mutual friends, the moderating influence of sibling contact and mutual friends would probably go unnoticed in standard models fitting research designs. Here, a measured variable approach is clearly more powerful than one based upon statistical inference alone.

A third improvement is to include measurements of environment as moderators of trait expressivity, i.e., measures of the general environment may condition twins' behavioral resemblance. These environmental measures may include qualities of the community or schools rather than only the immediate family context, as has traditionally been the case.

Nonshared effects will also include effects that vary because their consequences differ according to the children's age at the time and effects that stem from within-individual biological features that operate through some environmental mechanism. The effects on norm-breaking behavior in girls in very early puberty would seem to be an example of the latter, because the effects seem to be mediated by peer group influences and much influenced by whether the girls attend all-girl or coeducational schools (Magnusson 1988; Caspi et al. 1993). The effect is brought about by a variable (timing of puberty) that is strongly genetic, but the impact is nevertheless mediated by an environment outside the immediate family. It is important to recognize that there is no necessary connection between the genetic or environmental origins of a factor and the mechanism of its effect. A particular twin case is that of MZ girls who reach puberty at different times; both the biological event and the psychological impact of the realization that one is less mature than one's co-twin must be taken into account.

Finally, a fourth improvement is the careful assessment of environmental differences within MZ twin pairs for what they may unambiguously reveal about nonshared environmental influences.

Analytically, it is important for behavioral genetic studies to recognize that genetic research methods may be applied to environmental measures as well as to children's behavioral phenotypes. In a multivariate behavioral genetic design, the covariation of an environmental measure and a behavioral phenotype may be apportioned to genetic, shared environmental, and nonshared environmental variation. The expectation is that environment-behavior covariation is environmental; however, this expectation should be addressed through the twin research design, together with experimental research into mechanisms.

A number of points should be made in relation to nonshared environmental effects. First, it is important to appreciate that the proportion of population variance explained provides no guide to the strength of an effect at an individual level. Second, the very nature of nonshared environmental effects involves an inferred construct and not anything directly measurable. Third, it cannot necessarily be assumed that nonshared environmental effects will be the same for twins as for nontwin siblings. Finally, up to now much environmental research has focused on the impact of events and experiences on the onset of disorder (e.g., this applies to the large literature on stressful life events and depressive disorder in adult life). However, many such disorders (particularly depression) are by their nature recurrent or chronic, and it is also necessary (and arguably more important) to consider their role in relation to the overall picture of individual variations in propensity or liability to the disorder in question. This means both attention to the assessment of environments, as they impinge over the long term, and methods of analysis that focus on liability as well as timing of onset.

CONCLUSIONS

To reiterate the point made in the introduction, this conference provided a setting for a dialogue between twin researchers from two very different perspectives: those interested in the use of twins as a tool for genetic epidemiology and those whose primary interest is the study of twins in their own right.

Many issues were covered, but it became clear that the differences raised by the study of pre- and perinatal factors in twinning are amenable to empirical research if the appropriate assessments are made. Variables of importance can be included in the analyses of twin data collected for genetic studies, and those which are shown to be important (which must be expected to vary from one childhood disorder to another) should be taken into account. It is also clear that the differences between twins and singletons should be used in their own right to learn more about the mechanisms leading to disorder.

The use of twin designs, including longitudinal ones, to refine phenotypes for childhood disorders was recommended, with the use of broad multidimensional assessments and attempts to measure liability more directly. Multiple strategies were considered to be necessary, using different combinations of relatives in conjunction with twins, to overcome the different problems of each design and to address specific hypotheses. The application of molecular genetic methods was considered premature on the whole, because of the lack of necessary information on well-defined phenotypes and modes of transmission. However, in some cases, such as autism, molecular approaches were thought to be possible.

The recognition that nonshared environment need not be another "black box" was important, and recommendations for directly assessing environment were made.

REFERENCES

Alberman, E., J. Benson, and W. Kani. 1985. Disabilities in survivors of low birthweight. *Arch. Dis. Child.* **60**:913–919.

Botting, B., I. Macdonald-Davies, and A. Macfarlane. 1987. Recent trends in the incidence of multiple births and associated mortality. *Arch. Dis. Child.* **62**:941–950.

Caspi, A., D. Moffit, T.E. Moffitt, and P.A. Silva. 1993. Unraveling girls delinquency: Biological, dispositional, and contextual contributions to adolescent misbehavior. *Devel. Psychol.,* in press.

DeFries, J.C., and D.W. Fulker. 1985. Multiple regression analysis of twin data. *Behav. Genet.* **15**:467–473.

Derom, C., R. Vlietinck, R. Derom, H. Van Den Berghe, and M. Thiery. 1988. Population-based study on sex proportion in monoamniotic twins (letter). *New Eng. J. Med.* **319**:119–120.

Derom, C., R. Derom, R. Vlietinck, H. Maes, and H. Van Den Berghe. 1993. Iatrogenic multiple pregnancies in East Flanders, Belgium. CITY: PUBLISHER, in press.

Derom, R. 1993. Risks to the fetus in multiple pregnancies. In: A Critical Appraisal of Fetal Surveillance, ed. H. van Geijn. CITY: PUBLISHER, in press.

Derom, R., R. Vlietinck, C. Derom, L.G. Keith, and H. Van Den Berghe. 1991. Zygosity determination at birth: A plea to the obstetrician. *J. Perinat. Med.* **19(1)**:234–240.

Dunn, J., and R. Plomin. 1990. Separate Lives: Why Siblings Are So Different. New York: Basic Books.

Eaves, L.J., H.J. Eysenck, and N.G. Martin. 1989. Genes, Culture and Personality: An Empirical Approach. New York: Academic.

Falconer, D.S. 1965. The inheritance of liability to certain diseases estimated from the incidence among relatives. *Ann. Hum. Genet.* **29**: 51–76.

Falk, C.T., and P. Rubinstein. 1987. Haplotype relative risks: an easy way to construct a proper control sample for risk calculations. *Ann. Hum. Genet.* **51**:227–233.

Farmer, A., P. McGuffin, and I.I. Gottesman. 1990. Problems and pitfalls of the family history positive and negative dichotomy: Response to Dalen. *Schizo. Bull.* **16**:367–370.

Foley, D., and D.A. Hay. 1993. Genetics and the nature of the anxiety disorders. In: Handbook of Anxiety, ed. G. Burrows et al., vol. 5. Elsevier, in press.

Hay, D.A., C. Clifford, P. Derrick, J. Hopper, B. Renard, and T.M. Theobald. 1990. Twin children in volunteer registries: Biases in parental participation and reporting. *Acta Genet. Med. Gem.* **39**:71–86.

Hay, D.A., R. MacIndoe, and P.J. O'Brien. 1988. The older sibling of twins. *Aust. J. Early Child.* **13**:25–28.

Hay, D.A., and P.J. O'Brien. 1984. The role of parental attitudes in the development of temperament in twins at home, school and in test situations. *Acta Genet. Med. Gem.* **33**:191–204.

Hay, D.A., and P.J. O'Brien. 1987. Early influences on the school social adjustment of twins. *Acta Genet. Med. Gem.* **36**:239–248.

Jinks, J.L., and D.W. Fulker. 1970. Comparison of the biometrical, genetical, MAVA and classical approaches to the analysis of human behavior. *Psychol. Bull.* **73**:311–349.

Kendler, K.S. 1988. The sporadic vs. familial classification given aetiological heterogeneity: II. Power analyses. *Psychol. Med.* **18**:991–999.

Kendler, K.S. 1989. Limitations of the ratio of concordance rates in monozygotic and dizygotic twins (letter). *Arch. Gen. Psychiat.* **46**:477–478.

Lipper, E.G., G.S. Ross, P.A.M. Auld, and M.B. Glassman. 1990. Survival and outcome of infants weighing <800 g at birth. *JAMA* **163**:146–150.

Macfarlane, A.J., F.V. Price, and E.G. Daw. 1991. The delivery. In: Three, Four and More: A Study of Triplet and Higher Order Births, ed. B.J. Botting, A.J. Macfarlane, and F.V. Price. London: HMSO.

Magnusson, D. 1988. Individual Development from an Interactional Perspective. Hillsdale, NJ: Erlbaum.

McGue, M., I.I. Gottesman, and D.C. Rao. 1986. The analysis of schizophrenia family data. *Behav. Genet.* **16**:75.

Medical Research Intl. and Soc. of Assisted Reproductive Technology. 1991. *In vitro* fertilization embryo transfer in the U.S.: 1989 results from the Natl. IVT-FT Registry. *Fert. Steril.* **55**:14–21.

Milham, S. 1966. Symmetrical conjoined twins: An analysis of the birth records of 22 sets. *J. Pediatr.* **69**:643–647.

O'Brien, P.J., and D.A. Hay. 1987. Birthweight differences, the transfusion syndrome and the cognitive development of monozygotic twins. *Acta Genet. Med. Gem.* **36**:181–196.

Pocock, S.J. 1983. Clinical Trials: A Practical Approach. Chichester: Wiley.

Rowe, D.C., and B.L. Gulley. 1992. Sibling effects on substance use and delinquency. *Criminology* **30**:217–233.

Rowe, D.C., and R. Plomin. 1981. The importance of nonshared (E_1) environmental influences in behavioral development. *Devel. Psychol.* **17**:517–531.

Rutter, M. 1991. Nature, nurture and psychopathology: A new look at an old topic. *Devel. Psychopathol.* **3**:125–136.

Salat-Baroux, J., J. Aknin, J.M. Antoine, and R. Alamovitch. 1988. The management of multiple pregnancies after induction of superovulation. *Hum. Reprod.* **3**:399–401.

Smith, C. 1974. Concordance in twins: Methods and interpretations. *Am. J. Hum. Genet.* **26**:454–466.

Thorpe, K., J. Golding, I. MacGillivray, and R. Greenwood. 1991. Comparison of prevalence of depression in mothers of twins and mothers of singletons. *Br. Med. J.* **302**:875–878.

Veen, S., H. Ens-Dokkum, A.M. Schreuder, S.P. Verloove-Vanhorick, R. Brand, and J.H. Ruys. 1991. Impairments, disabilities, and handicaps of very preterm and very low birthweight infants at 5 years of age in the Netherlands. *Lancet* **ii**:33–36.

Vlietinck, R., R. Derom, M.C. Neale, H. Maes. H. Van Loon, C. Derom, and M. Thiery. 1989. Genetic and environmental variation in the birth weight of twins. *Behav. Genet.* **19**:151–161.

Wilson, R.S., and A.P. Matheny. 1983.. Assessment of temperament in infant twins. *Devel. Psychol.* **19**:172–183.

17

Twin Studies as Vital Indicators of Phenotypes in Molecular Genetic Research

P. McGUFFIN[1], R. KATZ[2], J. RUTHERFORD[3], S. WATKINS[1], A.E. FARMER[1], and I.I. GOTTESMAN[4]

[1]Department of Psychological Medicine, University of Wales, College of Medicine, Heath Park, Cardiff CF4 4XN, U.K.
[2]Department of Psychology, Toronto General Hospital, Toronto, M5G 2C, Canada
[3]Institute of Psychiatry, De Crespigny Park, London SE5 8AF, U.K.
[4]Department of Psychology, University of Virginia, Charlottesville, VA 22903, U.S.A.

ABSTRACT

Modern diagnostic criteria afford good interrater reliability in studies of the major psychiatric disorders. However reliability does not necessarily ensure validity and studies of twins provide a useful method of refining the criteria for heritable disorders and of exploring new definitions. Such approaches are important forerunners of molecular studies that aim to uncover the etiology of psychiatric disorders at the genotypic level.

THE PROBLEM OF DIAGNOSIS IN PSYCHIATRIC RESEARCH

It is frequently suggested that the main problem in the diagnosis of psychiatric disorders is unreliability. This is a misconception. The introduction of operational criteria for use in psychiatric research (Feighner et al. 1972) and, more recently, in clinical practice (American Psychiatric Association 1987) has made the goal of reliability, as reflected in level of agreement between pairs of clinicians, readily achievable. However, operational definitions, like traditional descriptive accounts of psychiatric disorder, rely always on symptoms, sometimes on signs, but almost never on objective laboratory tests or the demonstration of specific lesions. The central problem concerning diagnosis of psychiatric disorders with a genetic component is

Twins as a Tool of Behavioral Genetics
Edited by T.J. Bouchard, Jr. and P. Propping © 1993 John Wiley & Sons Ltd.

therefore not one of reliability but of validity; i.e., do our modern definitions of clinical entities (phenotypes) accurately reflect some underlying pathophysiological substrates (genotypes)?

An answer to this question becomes particularly pressing at a time when the main stream of psychiatric genetic research is moving rapidly into exploring molecular genetic methods for the investigation of abnormal behaviors. The most commonly applied of such methods are linkage studies. Unfortunately, linkage studies are most efficient in identifying the genes for simple traits, where there is a clear and regular correspondence between phenotype and genotype, but they become inefficient and may fail altogether when applied to more complex traits where the expression of the phenotype is variable and the pattern of segregation irregular (Clerget-Darpoux and Bonaiti-Pellie 1992). In this chapter, we will primarily discuss the problem of phenotypic definition and what can be learned from the study of twins. We must also, at least in passing, touch on some of the problems in specifying modes of transmission and how these relate to studies using genetic markers. However, before we go further we must spend a little time considering the usefulness of current diagnostic methods in research.

THE MERITS AND DEMERITS OF OPERATIONAL CRITERIA

The recent history of the development of explicit criteria for such disorders as schizophrenia and depression has been described in detail elsewhere, and the advantages and disadvantages of "operationalism" have been discussed (Farmer et al. 1991; Farmer and McGuffin 1989). Essentially, the approach involves defining a concept in terms of the set of operations that are performed in order to determine its applicability. Despite a degree of circularity, the notion of operational definitions, originally devised for research in physics, provides a means of capturing the apparently intangible within a clear experimental framework. Thus, if we can accept the clinician's eliciting a specific set of signs and symptoms as the approximate equivalent of a physicist performing a series of experiments, such a set of operations can be grouped together in a prescribed way to provide an explicit definition of the disorder, even where the underlying pathophysiology is obscure.

In practice, an operational definition of a psychiatric disorder usually turns out to be a sort of clinical algorithm, which may even lend itself to rewriting as a computer algorithm (McGuffin et al. 1991). Definitions of disorders, such as those embodied in the 3rd edition (revised) of the Diagnostic and Statistical Manual of the American Psychiatric Association (DSM IIIR; American Psychiatric Association 1987), have been welcomed by researchers because of their (apparent) clarity and their facilitation of inter-clinician agreement. By contrast, for the doctor in the consulting room, the definitions of disorder in DSM III may appear cumbersome and two-dimensional, ignoring much of the patient's biographical details and avoiding difficult-to-define but clinically important symptoms (Farmer and McGuffin 1989).

However, it is principally the advantages and disadvantages of operational definitions *in research* that concern us here. DSM IIIR is so far the most comprehensive set of explicit definitions of psychiatric disorders, but there have been many other attempts in recent years to write down operational definitions of schizophrenia and affective disorders (Berner et al. 1983). The work of Brockington and colleagues (1978) on schizophrenia was the first attempt to compare a variety of different operational definitions of the disorder. All ten of the definitions investigated allowed significant and satisfactory interrater reliability. However, the agreement between definitions concerning which patients were or were not classed as schizophrenic was alarmingly low. Some definitions, such as the PSE–CATEGO definition, were comparatively broad, while others such as the St. Louis criteria (Feighner et al. 1972), proved to be very restrictive. Subsequent research has produced similar results concerning the great variability in breadth of criteria. Farmer et al. (1991) point out, however, that the degree of overlap between differing definitions of schizophrenia is not always predictable, i.e., all restrictive definitions are not necessarily "nested" within all broad ones. Furthermore, a definition of disorder that appears to be restrictive in one series may be fulfilled by the majority of patients in another. Presumably this can reflect differences in the quality and type of information available in different studies as well as "real" differences in the mix of patient types.

The pioneering work by Brockington and colleagues attempted to use clinical outcome as a means of validating different diagnostic criteria. However they recognized that predictive or outcome validity was not necessarily the same as biological or "genotypic" validity and suggested that studies of twins could provide one useful means of comparing competing sets of diagnostic criteria. Here they echoed the sentiments of twin researchers such as Gottesman and Shields (1972) who had used their Maudsley Hospital twin series ascertained via schizophrenic probands to explore the utility of a range of clinical and descriptive definitions of schizophrenia. In this chapter we focus on twin studies of both schizophrenia and unipolar depression and review recent results which shed light on the usefulness of modern explicit approaches to the classification of these disorders.

SCHIZOPHRENIA

Twin studies of schizophrenia have provided a highly consistent pattern of greater concordance in monozygotic (MZ) than in dizygotic (DZ) pairs favoring an important genetic contribution to the disorder (Gottesman 1991). Although early studies suffered from various methodological flaws, four systematic, register-based studies reported in the 1960s and early 1970s overcame most of the major shortcomings. The introduction of systematic ascertainment, objective assessments of zygosity, inclusion of nonchronically hospitalized cases, and an attempt in some studies to introduce blind diagnosis resulted in lowered MZ concordance as compared to early studies; however, the weighted average MZ concordance on combining the studies was 48% as compared to the DZ concordance of 17% (Gottesman 1991). If a population lifetime risk of around 1% is assumed, then these concordances translate to a heritability of about 66%.

Among these studies, the one that attempted the most comprehensive inspection of the effects of differing diagnostic criteria was that of Gottesman and Shields (1972). In addition to using one of the first standardized measures in a twin study of schizophrenia, the Minnesota Multiphasic Personality Inventory (MMPI), Gottesman and Shields prepared detailed case abstracts based on tape-recorded personal interviews, hospital case notes, and a compilation of other sources of clinical information. A panel of clinical experts was then asked to supply diagnoses blind to the identity of the probands, the co-twins, and their respective zygosities. Although the overall finding of higher MZ than DZ concordance was stable across all raters, the size of the difference in concordances for MZ and DZ twins varied considerably. The highest MZ:DZ concordance ratio was achieved for a "middle-of-the-road" definition of schizophrenia, whereas those clinical raters who had either a very inclusive or a very exclusive view of the disorder produced results showing smaller MZ/DZ differences. Consequently, Gottesman and Shields argued that a concept of schizophrenia that was neither too liberal nor too conservative provided a phenotypic definition that was in affect the "most genetic."

The Maudsley Hospital twin study of Gottesman and Shields was carried out before the era of operational definitions of psychiatric disorder; however, the material was collected in sufficient detail for a subsequent reexamination to be possible (McGuffin et al. 1984; Farmer et al. 1987). Therefore, two clinical researchers (A.E. Farmer and P. McGuffin) independently rated the case abstracts, again with personal identifiers and zygosity information removed, but now with the aid of a standardized operational criteria checklist. This approach afforded good interrater agreement and indeed for one set of diagnostic criteria—those based on Schneider's (1959) first rank symptoms—the Kappa coefficient was 1.0, indicating perfect agreement. The MZ and DZ concordance rates for 8 different definitions of schizophrenia are summarized in Table 17. 1.

Whereas both versions of the research diagnostic criteria (RDC), St. Louis criteria and DSM III, resulted in similarly high MZ compared with DZ concordances, two definitions—those of Carpenter and colleagues (1973) and the criteria of Taylor et al. (1975)—proved more restrictive, giving DZ concordances of zero. The definition based on Schneider's first rank symptoms also turned out to be highly restrictive and actually resulted in a DZ concordance higher than the MZ rate. Although the sample size was very small and the standard errors large, this translated to a heritability best estimate of zero. Clearly therefore the most reliable definition in this study turned out to be the one with the lowest utility and defined the "least-genetic" phenotype. Reassuringly, the DSM III definition, like its near relatives the St. Louis and RDC definitions, gave heritabilities of close to, or a little under, 80% (albeit again with high standard errors because of small sample size).

Despite the comfort afforded by modern definitions of schizophrenia, such as that contained in DSM III, to describe a highly heritable syndrome, the exercise of applying

Table 17.1 Probandwise concordance for a variety of definitions of schizophrenia (data from McGuffin et al. 1984; Farmer et al. 1987).

definition*	MZ		DZ	
	n probands	concordance%	n probands	concordance%
RDC				
broad	22	46	23	9
narrow	19	53	21	10
St. Louis				
probable	21	48	22	9
definite	19	47	18	11
DSM III	21	48	21	10
Schneider				
First rank symptoms	9	22	4	50
Carpenter et al.	14	29	15	0
Taylor et al.	9	33	4	0

* see McGuffin et al. (1984) and Farmer et al. (1987) for references.
RDC = research diagnostic criteria.

diagnostic criteria to a body of data not specifically designed for that purpose is open to criticism. However, a new twin series from Norway, its purpose being the application of operational criteria, has now been reported. This has produced similar results. Onstad and colleagues (1991) conducted standardized interviews with schizophrenic probands and their co-twins which enabled the application of DSM IIIR criteria. The probandwise concordance for MZ twins ($n = 31$) was 48% and for DZ twins ($n = 28$) was 4%, so that the findings show a fair resemblance to the results reported by Farmer et al. on applying DSM III criteria to the Maudsley twin series.

Further similarities between the two studies emerged when the effect of including disorders other than schizophrenia in the definition of concordance was explored. Thus, Farmer and co-workers (1987) calculated MZ:DZ concordance ratios for phenotypes of differing breadth. Nonetheless, they are admittedly a crude indicator of the degree of genetic determination of a trait, and the approach has been subsequently criticized on statistical grounds (Kendler 1989). The MZ:DZ concordance ratios could, in these data, be shown to have a simple linear relationship with broad heritability, with which it was highly correlated (Farmer et al. 1991). The main results of the exploratory analysis are summarized in Figure 17.1. Here we can see that the highest MZ:DZ concordance ratio is achieved when schizotypal personality, affective disorder with mood incongruent psychosis, and atypical psychosis are included with schizophrenia in defining the phenotype. Greater broadening to include any other form of affective disorder resulted in a marked fall in the ratio, which reached its lowest value when any axis-1 DSM III category was included.

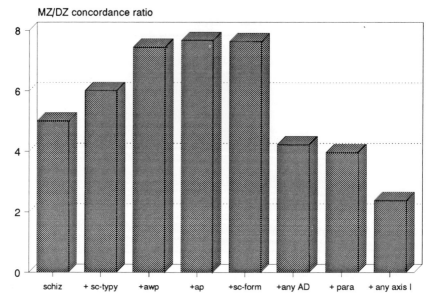

Figure 17.1 DSM III categories and MZ:DZ concordance ratio in the Maudsley twin series reanalysis (Farmer et al. 1987): schiz = schizophrenia; sc-typy = schizotypal personality; awp = affective disorder with (mood-congruent) psychosis; ap = atypical psychosis; sc-form = schizophreniform disorder; any AD = any form of affective disorder; para = paranoid disorder; any axis I = any axis-I DSM III category.

Although not precisely comparable, the data of Onstad and colleagues (1991) can also be interpreted as showing that a DSM IIIR definition slightly broader than schizophrenia alone, but not so broad as to include any axis-1 diagnosis, resulted in the highest degree of genetic determination, as reflected in MZ:DZ concordance ratios. Adding schizoaffective disorder, delusional disorders, and atypical psychosis increased the MZ concordance to 58%, while leaving the DZ concordance unchanged to 4%. Adding any axis-1 diagnosis to the definition of the phenotype boosted the MZ concordance further to 77%, but also greatly increased the DZ concordance to 32%. Later we will discuss these findings and their implications for future research, particularly molecular genetic studies. However we will now turn to a different phenotype, that of unipolar depression.

UNIPOLAR DEPRESSION

Although (inexplicably) the proposal that there may be a genetic basis for affective disorders has never proven as controversial as has been the case for schizophrenia, the twin data, at least until recently, have been less comprehensive and thorough. This is particularly true of unipolar depression. Earlier studies on affective disorders were reviewed by Gershon and colleagues (1976). Although the results were fairly consistent in showing markedly higher MZ than DZ concordances, little attempt was made to differentiate unipolar from bipolar illness, and little description was given of diagnostic procedures. The first study to include systematic ascertainment via a register and other methodological safeguards already incorporated in schizophrenia studies, as well as to differentiate unipolar from bipolar disorder, was that of Bertelsen and colleagues (1977) in Denmark. With an overall probandwise MZ concordance of 67% and a DZ concordance of 20%, the findings were close to the compilation of earlier studies produced by Gershon and co-workers (1976). There was some tendency towards homotypia in the study of Bertelsen et al. (1977), with 15 out of 19 affected identical co-twins of unipolar probands having unipolar disorder and 21 out of 27 identical co-twins of bipolar probands having bipolar disorder. However, Bertelsen et al. clearly demonstrated for the first time that bipolar and unipolar disorders are not completely genetically distinct, since 10 out of 46 co-twins had a different form of affective illness from their genetically identical probands. There was a suggestion, however, that a bipolar disorder has a higher MZ:DZ concordance ratio and may be more genetically influenced than unipolar disorder. This receives support from most family studies, which have tended to find more affected relatives in the families of bipolar than unipolar probands (McGuffin and Katz 1989).

The first study to apply operational criteria and to focus explicitly on moderately severe depression was carried out in Norway by Torgersen (1986). He applied DSM III criteria for major depression and reported probandwise concordance rates in MZ twins ($n = 37$) of 51% compared with 20% in DZ twins ($n = 65$). On the face of it, these results are markedly similar to preliminary findings from a recent study in the U.K., again using the Maudsley twin register. McGuffin and colleagues (1991) reported a concordance rate of DSM III major affective disorder of 53% in MZ twins ($n = 62$) as opposed to 28% in DZ twins ($n = 79$). However, we then went on to a rigorous blind diagnostic reassessment of the twins and a series of model-fitting analyses, which we will briefly describe. The results would seem to carry a more complicated message about how we should best define the phenotype in future genetic studies of unipolar depression.

Our sample was ascertained via the Maudsley Hospital Twin Register (the same register as had been used earlier by Gottesman and Shields in their study of schizophrenia). The register was founded by Eliot Slater in 1948 and contains information about all psychiatric patients who have ever attended the Bethlem Royal and Maudsley Hospital in London (either in- or outpatients) since the register's beginning and who upon questioning admit to having been born a twin. All available written material on proband with a possible diagnosis of depressive disorder was screened using the syndrome checklist of the PSE (Wing et al. 1974) and a DSM III checklist. Of 408 probands with a register-recorded primary diagnosis of depression, 214 fulfilled the criteria for major depression. On further investigation it was found that 34 of these had had one or more episodes of mania. The results of preliminary analyses on this entire sample have been reported previously elsewhere (McGuffin et al. 1991). We have recently gone on to assess the 180 unipolar probands and their co-twins blind to identifying information and zygosity much as we did earlier with the Gottesman and Shields schizophrenia series. Both DSM IIIR and ICD 9 diagnoses were made with the aid of a check-list for which two of us (S. Watkins and P. McGuffin) were able to achieve good interrater agreement. Three probands did not fulfill DSM IIIR criteria for major depression and have therefore been set aside. The analyses reported here are therefore based on 177 probands and their co-twins (68 MZ and 109 DZ) and will focus on two alternative definitions of concordance: one fairly broad and the other narrow. The broad criteria include any PSE–CATEGO classes of depression (Wing et al. 1974), provided that the subject has also received hospital treatment. The narrow criteria are those of strictly applied DSM IIIR, irrespective of whether hospital treatment had been received (in practice, however, all co-twins fulfilling these criteria also received hospital treatment).

For both definitions the MZ concordance was significantly higher than the DZ. The broad definition yielded an MZ concordance of 66% and a DZ concordance of 42% (chi square = 6.52, $p = 0.011$). For the narrow definition, the MZ concordance was 46% and the DZ concordance 20% (chi square = 12.88, $p = 0.0003$). We carried out model-fitting (for a detailed description, see McGuffin et al. 1991) using a Fortran computer program TWIN, which makes use of the GEMINI optimization routine to arrive at a maximum likelihood solution for each model fitted.

It was assumed that depression is a threshold trait such that liability to develop the disorder is normally distributed within the population (or can be readily transformed to normality), and only those individuals whose liability at some point exceeds the threshold actually manifest the disorder. Liability was assumed to have a mean of zero and variance, $Vp = 1$, and to be made up of a simple additive combination of genetic and environmental sources of variation such that:

$$Vp = h^2 + c^2 + s^2 , \tag{17.1}$$

where h^2 is the heritability or proportion of variance in liability contributed by additive genetic effects, c^2 is the proportion of variance due to environment shared by twins, and s^2 is the proportion of variance due to environmental factors that are specific to the individual and not shared with other family members.

The results are summarized in Tables 17.2 and 17.3. We compared a full model, in which there are both additive genetic and shared environmental affects, with reduced models, in which only additive genetic factors ($c^2 = 0$) or only shared environmental factors are allowed ($h^2 = 0$). Both reduced models can be compared with the full model using a likelihood ratio test (based on the property that twice the difference in log likelihoods is asymptotically distributed as a chi square). In turn, all models can be compared with a null or "no transmission" model where both h^2 and c^2 are constrained to be 0.

Table 17.2 Model-fitting using a narrow definition of unipolar depression (blind DSM IIIR).

Model	h^2	(se)	c^2	(se)	$-2LnL$	Pearson X^2	p
	Parameter Estimates						
Full model	0.79	(0.05)	0.0*		204.19	0.85	
Additive genetic effects only	0.79	(0.05)	[0]		204.19	0.85	0.36
Shared environment only	[0]		0.61	(0.06)	216.07	12.88	0.0003
No transmission	[0]		[0]		346.75	359.06	0.0000

* set at a bound during iteration.
[ ] indicate fixed parameters, i.e., [0] are fixed at zero.

Table 17.3 Model-fitting using a broad definition of unipolar depression (hospital-treated CATEGO).

Model	h^2	c^2	$-2LnL$ + cons	Pearson X^2	p
	Parameter Estimates				
Full model	0.45	0.43	235.46	0.0	
Additive genetic effects only	1.0*	[0]	261.61	33.65	0.000
Shared environment only	[0]	0.76	245.23	9.64	0.002
No transmission	[0]	[0]	456.38	424.44	0.000

* set at a bound during iteration.
[ ] indicate fixed parameters, i.e., [0] are fixed at zero.

From Tables 17.2 and 17.3. we can see that rather different patterns emerge as a result of the model-fitting using the two different definitions of unipolar depression. On applying the broad definition, there is a substantial reduction in the maximum log likelihood when either h^2 or c^2 are set at zero. Table 17.2 also shows the Pearson goodness-of-fit chi square. Whether we focus on this or compute the likelihood ratio chi square, comparing either reduced model with the full model, the results indicate that the reduced models are unsatisfactory. The least satisfactory of all is the null model, where both the likelihood ratio chi-square comparison with the full model and Pearson chi square are very large. We can conclude that this fairly broad definition of depression, where having hospital treatment forms part of the diagnostic criteria, is moderately heritable but that common environmental effects account for a substantial part of the resemblance between twins and, possibly (if we are permitted to extrapolate), other family members.

The situation is more straightforward for the narrow definition of depression. Here, when we attempt to fit the full model, c^2 becomes fixed at zero (and exactly the same result is achieved if we deliberately fix c^2 at zero and only search on the value of one parameter, h^2). Again, the no-transmission model can be resoundingly rejected. We can, therefore, conclude that there is strong evidence for the transmission of narrowly defined unipolar depression based on DSM IIIR criteria, and this can be explained entirely in terms of additive genetic effects with the heritability approaching 80%.

To date there has only been one other twin study of depression in which a formal genetic model-fitting exercise has been carried out and where a variety of definitions of the phenotype have been explored. Kendler and colleagues (1992) examined over a thousand female twin pairs from a nonpsychiatric population-based registry in Virginia, U.S.A., using structured psychiatric interviews and applying nine different definitions of major depression. For most of these definitions, including DSM IIIR, there was strong support for the presence of additive genetic factors; however, the heritabilities were only moderate, in the range of 33–45%. For two definitions of depression based on the St. Louis criteria (Feighner et al. 1972), the heritability was even more modest, at a little over 20%. There was no evidence of shared environmental effects for any of the definitions of depression.

The broadly similar pattern of results in the two studies from the U.K. and the U.S. (i.e., evidence of heritability, whatever the definition of the phenotype) are of interest given the marked differences in the way the two samples were ascertained. However, there were also some large contrasts. Kendler and co-workers found no evidence whatsoever of common environmental effects for any definition and found a much lower heritability overall. Although one of the definitions of depression, that of DSM IIIR, was common to the two studies, the estimates of population frequency were markedly different. In the U.K. study, the lifetime risk calculated as a simple proportion of the population risk of depression reported by Sturt et al. (1984) was 4.2% for men and women combined. In the U.S. study, based on a sample of women alone, the lifetime prevalence of DSM IIIR was 31%. This population frequency is actually considerably higher than the DZ concordance in the British study. Presumably, unless there are very marked real differences in the frequency of major unipolar depression in the U.S. and the U.K., the DSM IIIR criteria have been interpreted very differently in the two centers.

DISCUSSION AND CONCLUSIONS

The results just outlined for studies of schizophrenia and depression demonstrate that the appropriate selection of operational criteria is not a trivial matter rather, one that can greatly affect theories about the etiology of psychiatric disorders and the size of the estimated genetic component of the disorder. Thus apparently reliable criteria can define varieties of "schizophrenia" with best estimates of heritability as different as zero and 80%. None of the definitions of unipolar depression considered here have heritabilities quite as low as zero, but the range is again wide, from just over 20% to just under 80%. The most obvious implication is that biological studies of psychiatric disorder, including those using genetic markers in linkage or association strategies, need to focus not just on reliable clinical definitions but on definitions which maximize the chances of finding a biological substrate. Although the twin study results outlined above provide important pointers and suggest a common theme in the exploration of heritable phenotypes, there are differing and specific problems relating to the definitions of schizophrenia on the one hand and depression on the other. There is also the theoretical question, which may have practical importance, of whether results obtained under a simple multifactorial model of inheritance can truly inform studies employing linkage analysis, where typically the aim is to locate genes of major effect. We will briefly consider each of these in turn.

The twin data, in terms of operational definitions of schizophrenia, cannot be considered to have yielded definitive results. Operational definitions have been used only in two studies, and for both of these the sample size was small. In the earlier study based on Gottesman and Shields (1972) Maudsley Twin series, data not specifically collected for this purpose were opportunistically recycled (McGuffin et al. 1984). This procedure may not have been altogether fair as a means of comparing competing sets of criteria. For example, the definition based on Schneider's first rank symptoms may have appeared over-restrictive because we were re-rating material that was not originally collected within a "Schneider-oriented" framework. Nevertheless, the reassessment of the Maudsley schizophrenia twin series suggests that operational definitions such as DSM III which include longitudinal information (e.g., on duration and outcome) and not just cross-sectional accounts of psychopathology may be the most robust in delimiting a phenotype strongly influenced by genes (Farmer et al. 1987). Onstad and co-workers (1991) explored only one system of classification, that of DSM IIIR, but produced results which were similar to those of Farmer et al. (1987) when DSM III criteria were applied. Again both studies agreed that the definition of the phenotype might be "improved" by broadening it slightly to include other forms of psychosis. Furthermore, marked broadening of the definition in both studies resulted in an apparent weakening of the genetic effect. This is in keeping with the theoretical expectation that irrelevant items, not genetically correlated with liability to schizophrenia, are unlikely to be selected for inclusion in a definition of the disorder which is based upon maximizing heritability.

Ideally the next stage should be to extend such work using a larger sample, which would enable a more sophisticated model-fitting exploration of the precise phenotype that maximizes heritability. It might also be prudent to include family data as a safeguard against the possibility that twins, both MZ and DZ, will show concordance for schizophrenia greater than other pairs of relatives (McGue et al. 1985).

The two recent studies of unipolar depression have certain advantages over the studies of schizophrenia. Both were based on larger samples and in the case of the U.S. study (Kendler et al. 1992), the numbers involved were particularly impressive. There were, however, marked differences in the way the samples were ascertained in that the U.K. study is a clinical sample based on a hospital register while the U.S. study was population based. There was broad agreement between the two over the question of whether or not unipolar depression is heritable. Heritable it is, apparently irrespective of which definition is used (even though, as we have already noted, the range of values of heritability was large). What is more surprising and problematic is the wide spread of population frequencies. These range from a morbidity risk to age 65 of 4.2% in men and women combined in the U.K. study, to a lifetime prevalence estimate (not age corrected) of 31% for DSM III major depression in women in the U.S. study. Although, as already noted, this might mean that there are very marked differences in the actual frequency of depression between the U.S. and U.K., it is more likely that the two groups of workers were applying ostensibly the same definitions, but in different ways. Thus, in the U.S. study, nearly a third of the general population of women have had major depression while in the U.K., less than half of those probands who have already received a hospital diagnosis of depression fulfill the criteria. If we take the lifetime prevalence from the U.S. study and use it in an attempt at model-fitting on the U.K. twin data we obtain an estimate for h^2 of 0.1 (± 0.12), a c^2 of 0.0, and a poor overall fit (chi square = 11.43, $p = 0.0007$). This result is scarcely surprising since, as we noted, the U.S. lifetime prevalence actually exceeds the U.K. DZ concordance rate.

Presumably, therefore, although the operational criteria for unipolar depression in DSM IIIR appear reassuringly explicit, they are actually open to very different interpretations. This means that comparing the results of biological research, such as linkage studies, across different centers will require not only the use of the same criteria but also some method of "calibrating" researchers' applications of the criteria followed by subsequent checks on reliability. In the past, reliance on hospital-treated cases has only sometimes been seen as providing at least a partial safeguard. The assumption has been that receipt of specialist psychiatric treatment will usually mean that there has been obvious impairment of functioning and hence will denote a more severe type of disorder. Sometimes there has also been a less explicit assumption that more severe is equivalent to "more biological." However, the U.K. twin study strongly suggests that when receipt of hospital treatment is included as part of the diagnostic criteria, evidence of an important common environmental effect emerges, and the heritability is lower than when strict DSM IIIR criteria are applied. One plausible interpretation of this pattern of results is that the strictly defined syndrome of major depression is familial because of genetic effects alone, but more broadly defined depression, where an element of help-seeking is incorporated in the definition, is substantially influenced by family environmental factors. For example, this would mean that someone's probability of seeking hospital treatment or of being referred for treatment once affective symptoms have developed is influenced by whether or not their twin (or other relative) has already received hospital treatment. If this is correct, then referral to hospital becomes a poor criterion for deciding who is or is not "affected" when performing a linkage analysis.

The discussion throughout this chapter has been concerned with categorical approaches to diagnoses which assume that current classifications, although imperfect, are broadly speaking "correct." The twin methods outlined here could, of course, be applied in a much more radical way to explore dimensional approaches to psychopathology or to search for brand-new categories. This can be facilitated by collecting data in such a way that the emphasis is on component signs and symptoms rather than on preset diagnostic hypotheses (McGuffin et al. 1991a).

Finally, we need to return to the problem of mode of transmission. To date the mode of transmission of schizophrenia and affective disorders remains obscure. The models that we and others have applied in comparing definitions of these two groups of disorders have tended to make the assumption that genetic effects are additive and involve several, perhaps many, loci. By contrast, linkage studies of disease are performed under the assumption that there is a gene of large effect in at least some of the families under investigation. Such a gene need not necessarily segregate in a regular Mendelian fashion but, as implemented in most current linkage analysis programs, the assumption is of a general single locus model where a major gene is the only source of resemblance between relatives. Providing that the segregation parameters for the gene to be detected are correctly specified, the existence of other loci affecting the disease trait (either other components of oligogenic transmission or a polygenic "background") should not prevent detection of linkage (Ott 1991). In practice, for dichotomous traits, such as being affected or unaffected by a disorder, it often proves very difficult or impossible to discriminate between single locus, polygenic or "mixed" model inheritance purely on the basis of segregation analysis. It is also unclear whether the results of analyzing twin data under a threshold model will vary greatly if it is assumed that the genetic contribution of liability is oligogenic, due to a major locus plus polygenic background, or due to many loci of small effect. We propose, therefore, that although simple additive models as applied to twin data are not necessarily "correct" and indeed actually ignore the possible existence of major genes, such analyses remain useful. Indeed until major loci are reliably identified, twin studies are perhaps still the *most* useful way of defining and refining phenotypes for molecular genetic research into psychiatric disorders.

REFERENCES

American Psychiatric Association (APA). 1987. DSM-III-R: Diagnostic and Statistical Manual of Mental Disorders, 3rd ed. Washington, D.C.: APA.

Berner, P., E. Gabriel, H. Katschnig, et al. 1983. Diagnostic Criteria for Schizophrenia and Affective Psychoses. World Psychiatric Association. New York: American Psychiatric Press, Inc.

Bertelsen, A., B. Harvard, and M. Hauge. 1977. A Danish twin study of manic-depressive disorders. *Br. J. Psychiat.* **130**:330–351.

Brockington, I.F., R.E. Kendell, and J.P. Leff. 1978. Definitions of schizophrenia: Concordance and prediction of outcome. *Psychol. Med.* **8**:387–398.

Carpenter, W.T., J.S. Strauss, and J.J. Bartko. 1973. Flexible system for the diagnosis of schizophrenia: Report from the WHO pilot study of schizophrenia. *Science* **182**:1275–1278.

Clerget-Darpoux, F., and C. Bonaiti-Pellie. 1992. Strategies based on marker information for the study of human diseases. *Ann. Hum. Genet.* **56(2):**145–153.

Farmer, A.E., and P. McGuffin. 1989. The classification of the depressions: Contemporary confusion revisited. *Br. J. Psychiat.* **155**:437–443.

Farmer, A.E., P. McGuffin, and I.I. Gottesman. 1987. Twin concordance for DSM-III schizophrenia: Scrutinising the validity of the definition. *Arch. G. Psychiat.* **44**:634–641.

Farmer, A.E., P. McGuffin, I. Harvey, and M. Williams. 1991. Schizophrenia: How far can we go in defining the phenotype? In: The New Genetics of Mental Illness, ed. P. McGuffin and R. Murray, pp. 71–84. Oxford: Heinemann.

Feighner, J.P., E. Robins, S.B. Guze, R.A. woodruff, G. Winokur, and R. Munoz. 1972. Diagnostic criteria for use in psychiatric research. *Arch. G. Psychiat.* **26**:57–63.

Gershon, E., W.E. Bunney, J.F. Leckman, M. Van Eerdweegh, and B.A. De Bauche. 1976. The inheritance of affective disorders: A review of data of hypotheses. *Behav. Genet.* **6**:227–261.

Gottesman, I.I. 1991. Schizophrenia genesis. Origins of Madness. San Francisco: Freeman.

Gottesman, I.I., and J. Shields. 1972. Schizophrenia and Genetics: A Twin Study Vantage Point. London: Academic.

Kendler, K.S. 1989. Limitations of the ratio of concordance rates in monozygotic and dizygotic twins (letter). *Arch. G. Psychiat.* **44**:477–478.

Kendler, K.S., M.C. Neale, R.C. Kessler, A.C. Heath, and L.J. Eaves. 1992. A population based twin study of major depression in women: The impact of varying definitions of illness. *Arch. G. Psychiat.* **49**:273–281.

McGue, M., I.I. Gottesman, and D.C. Rao. 1985. Resolving genetic models for the transmission of schizophrenia. *Genet. Epidem.* **2**:99–110.

McGuffin, P., A.E. Farmer, and I.I. Gottesman. 1984. Twin concordance for operationally defined schizophrenia: Confirmation of familiality and heritability. *Arch. G. Psychiat.* **41**:541–545.

McGuffin, P., and R. Katz. 1989. The genetics of depression and manic-depressive illness. *Br. J. Psychiat.* **155**:294–304.

McGuffin, P., R. Katz, and J. Rutherford. 1991. Nature, nurture and depression: A twin study. *Psychol. Med.* **20(2)**:329–335.

Onstad, S., I. Skre, S. Torgersen, and E. Kringlen. 1991. Subtypes of schizophrenia: Evidence from a twin-family study. *Acta Psychiat. Scand.* **84**:203–206.

Ott, J. 1991. Analysis of Human Genetic Linkage, 2nd ed. Baltimore: The John Hopkins Univ. Press.

Schneider, K. 1959. Clinical Psychopathology. Translated by M. W. Hamilton. London: Grune.

Sturt, E., N. Kumakura, and G. Der. 1984. Life-long morbidity risk for depressive disorder in the general population. *J. Affec. Dis.* **7**:109–122.

Taylor, M.A., R. Abrams, and P. Gazbanga. 1975. Manic depressive illness and schizophrenia: A partial validation of recent criteria utilizing neuropsychological testing. *Comp. Psychiat.* **16**:91–96.

Torgersen, S. 1986. Genetic factors in moderately severe and mild affective disorders. *Arch. G. Psychiat.* **43**:222–226.

Wing, J.K., J.E. Cooper, and N. Sartorius. 1974. The Measurement and Classification of Psychiatric Symptoms. Cambridge: Cambridge Univ. Press.

18

Critical Review of Psychopathology in Twins: Structural and Functional Imaging of the Brain

M.S. BUCHSBAUM

Department of Psychiatry, Mount Sinai School of Medicine,
New York, NY 10029, U.S.A.

ABSTRACT

Brain imaging techniques can provide specific brain anatomical morphometry and regional functional information essential for understanding the mediating phenomena between genes and behavior. Family and twin studies in psychiatry have lacked emphasis on mechanisms, depending largely on examining familial patterns of symptom-based diagnostic groupings. Anatomical studies with magnetic resonance imaging have suggested important genetic influences. Functional studies with the electroencephalogram have long confirmed genetic factors; however, these studies have not exploited the additional specificity of multichannel topography. Positron emission tomography studies have reported twin and quadruplet case studies, but significant samples have only recently been studied.

INTRODUCTION

Since the elegant methodology of the Danish adoption studies confirmed the importance of genetic factors in psychopathology a quarter of a century ago, the problem of understanding what biological diathesis is inherited has been left to solve. A dominant approach has been to examine family pedigrees in the hope of determining whether the gene for the disorder under study is dominant or recessive and, more recently, whether it is linked to a marker gene and thus could identify the chromosome. The success of this approach has not been definitive; the unreliability of diagnosis has limited this approach, because we lack an external validating standard for symptom

Twins as a Tool of Behavioral Genetics
Edited by T.J. Bouchard, Jr. and P. Propping © 1993 John Wiley & Sons Ltd.

cluster-based diagnostic paradigms and/or genetic heterogeneity and phenocopy incidence . Even if molecular genetics establishes linkage markers, detailed knowledge of the pathophysiology is still necessary for physiological interventions. Regional brain function, together with its structural framework, remains one of the leading approaches to understanding the action of genes on the central nervous system. The study of twins also provides a way of choosing among biological parameters to identify specific measures, and brain regions that can be the focus of increasingly detailed study in psychiatric populations. In this chapter, I review the existing studies of twins and present new, unpublished data analyses from the study of the schizophrenia-concordant Genain quadruplets as a case study.

THE GENAIN QUADRUPLETS: A CASE STUDY

Clinical and Psychosocial Assessment

The quadruplets (Nora, Iris, Myra, and Hester, in birth order) were reared together by their parents (Figure 18.1). By age 24, all four had received the diagnosis of schizophrenia. Nora, Iris, and Myra graduated from high school, but Hester had withdrawn from high school in the 11th grade because of emotional problems. Nora and Iris became ill within a few months of each other, about two years after graduation, and both were hospitalized for "schizophrenic reaction, catatonic type" in 1951–1952. Myra and Hester had their first hospital admission at the National Institute of Mental Health (NIMH) in 1955, but had had major psychiatric symptoms between 1951 and 1955. The quadruplets' birth, childhood, and initial NIMH studies in 1955 are described in a fascinating book by David Rosenthal (1963) and presented in tabular form in the reports from their 1981–1982 NIMH studies (DeLisi et al. 1984). All 14 red blood cell antigens and HLA antigen titers were identical (DeLisi et al. 1984). Lifetime psychiatric diagnoses using the research diagnostic criteria (RDC) were made on each patient in 1982 (Nora: schizophrenia, chronic undifferentiated; Iris: schizophrenia, chronic; Myra: schizoaffective disorder, schizophrenia residual type; Hester: schizophrenia, chronic undifferentiated). Neurochemical, attentional, social, and historical data are presented elsewhere (DeLisi et al. 1984; Mirsky et al. 1984).

Brain imaging with positron emission tomography (PET) and computerized electroencephalographic (EEG) topography was used during the second period of intensive study of the quadruplets when they returned to NIMH in 1981. Our major scientific interest at that time was the reduction of metabolic rate in the frontal lobe of patients with schizophrenia, consistent with the blood flow findings of Ingvar and Franzen (1974). Since this was expressed as a hypofrontality ratio, the frontal lobe/occipital lobe quotient, we were also interested in the contribution of the occipital lobe. There had been earlier reports of diminished alpha activity in patients with schizophrenia. Since alpha activity increases in normal subjects at rest with their eyes closed and diminishes with visual activation when the eyes are open, we expected patients with low frontal/occipital ratios for metabolic rate to demonstrate the reduced alpha

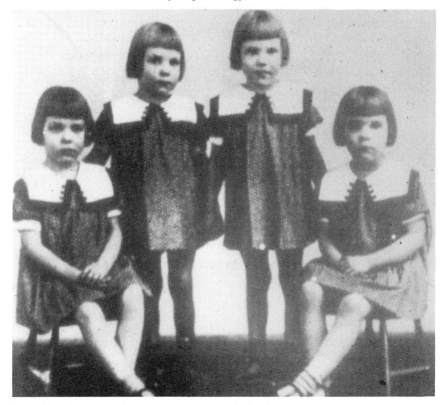

Figure 18.1 Genain quadruplets in preschool period.

consistent with occipital activation. Since a marked genetic component to the alpha predominance had been established in twins, we also expected to see alpha activity consistent across the four quadruplets.

Positron Emission Tomography

Cerebral glucography was obtained with PET scanning exactly as described elsewhere (Buchsbaum, Ingvar et al. 1982). The subjects sat in a darkened room with their eyes closed. 18F-deoxyglucose (3–5 mCi) was injected intravenously. The subjects remained still during the 35-minute radiotracer uptake period to match our six concurrent normal controls and eight patients with schizophrenia. After the uptake period, the subjects were moved to our ECAT scanner, where six to eight horizontal planes were obtained. Scans were reprocessed for this report using our current anatomical approach, which identifies four specific gyral regions within each lobe of the brain for each hemisphere (Buchsbaum et al. 1989; see also Figure 18.2). We also explored 63 subcortical regions obtained stereotaxically (Gottschalk et al. 1991).

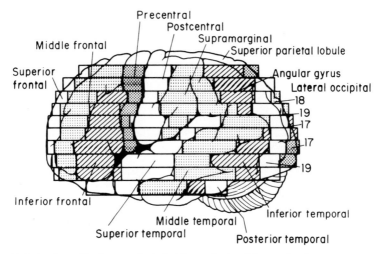

Figure 18.2 Outline of brain areas included for analysis of the quadruplets.

Resting EEG

We recorded EEG from 16 leads placed on the left hemisphere for topographic analysis (Buchsbaum, Cappelletti et al. 1982). EEG was recorded during uptake of the 18F-deoxyglucose as described above. Ten-second epochs of EEG were digitized at 102.4 cps, and each epoch was screened for artifacts by visual inspection and by a 100 v excursion automatic deletion rule. The data were filtered with autoregressive methods, power spectra computed, and topographic maps generated (Buchsbaum, Rigal et al. 1982). Results for 1 cps resolution are reported here for the first time (Figure 18.3).

EEG Topography

The significant reduction in occipital alpha power that had been found in patients with schizophrenia (Figure 18.3) also characterized the quadruplets (Figures 18.4 and 18.5). This effect seems maximal in the fast alpha range at 11 cps (Figure 18.6). The variance was much lower within the quads; however, this effect was most marked at the occipital region and the peak of the normal/schizophrenic differences, 11 Hz (Figure 18.6).

Figures 18.4 and 18.5 present EEG maps in normals and the quadruplets concordant for schizophrenia. Note that while the maps for 7 and 15 cps are similar, alpha activity in the 8–12 cps range is markedly diminished in the quadruplets. This difference is similar to the difference seen between the normal controls and singleton schizophrenics in Figure 18.3.

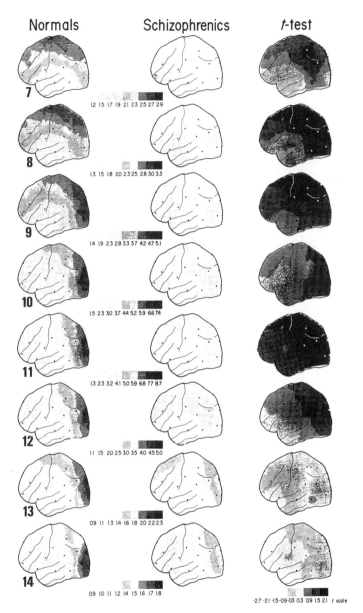

Figure 18.3 Computerized topographic EEG maps for normal subjects and eight unmedicated patients with schizophrenia. The maps portray an approximately equal-area projection of the side view of the brain (Buchsbaum, Rigal et al. 1982). Each map shows activity for a single frequency from 7 to 14 cps. Right: Differences between normal subjects and schizophrenics showing greater differences in the alpha frequency from 8 to 14 cps.

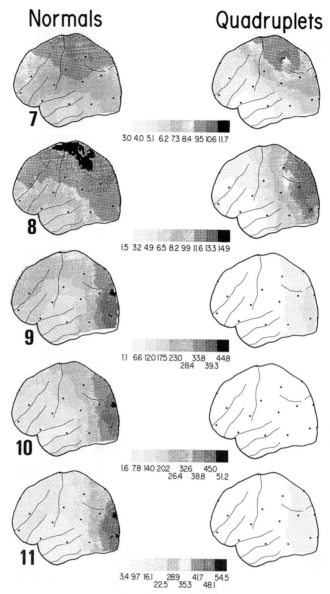

Figure 18.4 EEG maps in normals and the quadruplets concordant for schizophrenia. Note that while the maps for 7 and 15 cps are similar, alpha activity in the 8 to 12 cps range is markedly diminished in the quadruplets. This difference is similar to the difference seen between the normal controls and singleton schizophrenics in Figure 18.3.

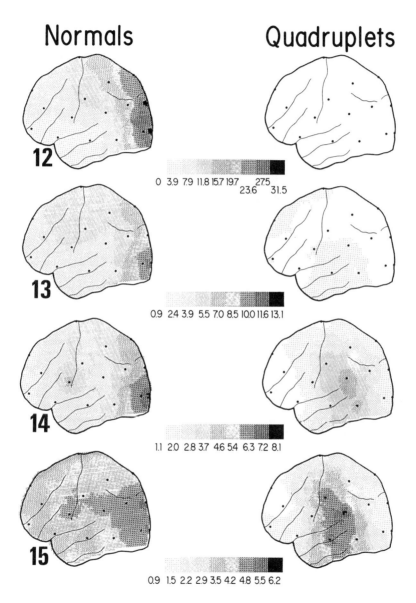

Figure 18.4 *continued.*

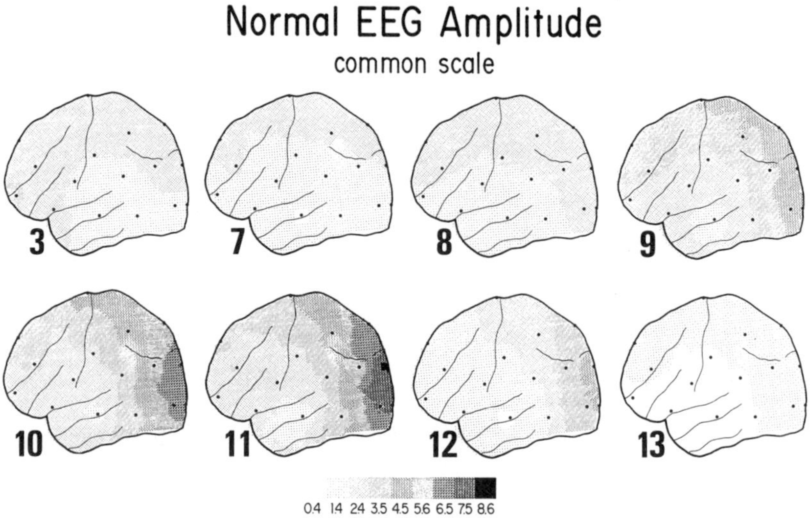

Normal EEG Amplitude
common scale

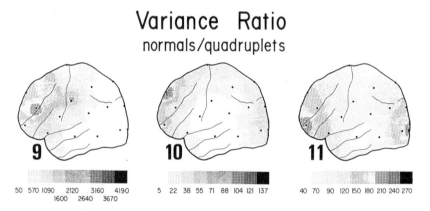

04 14 24 35 45 56 65 75 86

Figure 18.5 EEG maps of the alpha rhythm in normal controls with all maps expressed on a common scale. Note that alpha reaches maximum amplitude for 9–11 cps and that other EEG frequencies are of very much smaller amplitude.

Variance Ratio
normals/quadruplets

50 570 1090 2120   3160   4190   5 22 38 55 71 88 104 121 137   40 70 90 120 150 180 210 240 270
      1600   2640 3670

Figure 18.6 Topographic maps of ratio of variance in normal subjects/variance in quadruplets. Scale bar is F-ratio × 1000 for 9 cps and × 10 for 10 and 11 cps. A ratio greater than 4.35 indicates that normals had significantly ($p < 0.05$, $df = 3.7$) greater variability than the quadruplets, indicating a familial effect. Thus no area at 9 cps reached statistical significance but much of the map for 10 and 11 cps was highly significant with a peak over the occipital part of the brain at 11 cps. Compare the variance map for 11 cps with the alpha peak distribution seen in Figure 18.6 and the t-test comparison maps of normal subjects and patients with schizophrenia in Figure 18.3.

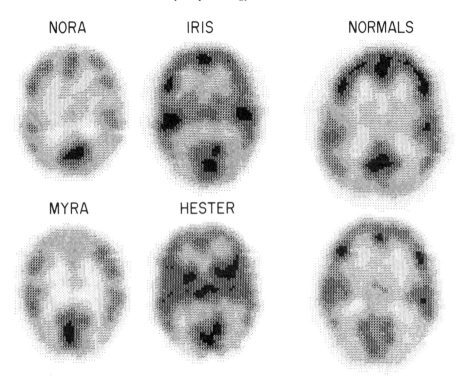

Figure 18.7 Positron emission tomography scans in the quadruplets and two normal subjects for comparison. Note the lower metabolism in the frontal lobe (top of slice) compared with the occipital lobe (bottom of slice, especially midline calcarine region) for Nora, Myra, and Hester, and the equal metabolism in Iris.

PET Scans

The quadruplets were markedly hypofrontal and differed significantly from our control group in the prefrontal region but not the motor strip (Table 18.1). The variance of frontal lobe relative metabolic rates was significantly lower within the quads for the left superior frontal gyrus. On an exploratory basis, the left anterior and posterior cingulate gyrus, left frontal white matter (slice #8) and the left hippocampus (#11 slice; Matsui and Hirano 1979) showed significantly smaller variances in the quadruplets than in normal subjects.

The PET scans show reduction in frontal metabolic rate, prominent in the cingulate and entire frontal cap (Figure 18.7). Employing the same edge-finding algorithm used for outlining the brain, we increased the glucose metabolic threshold until the algorithm outlined the frontal lobe (Figure 18.8) in normal subjects. However, the same series of steps could not produce a contiguous frontal area in the quadruplets except

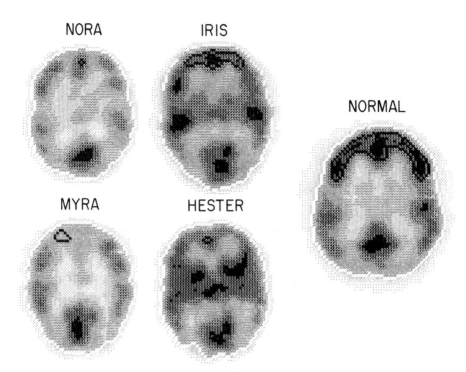

Figure 18.8 Application of edge-finding algorithm to assessment of hypofrontality. Using the brain edge-finding method (Buchsbaum, Ingvar et al. 1982), we first outlined the brain slice (white contour line.) Next, the threshold was gradually increased until the first time the edge broke through the temporal region and outlined the interior edge of the frontal cortex and cingulate gyrus (normal subject). The same method failed to produce any symmetrical outline of the frontal lobe in any of the quadruplets because the frontal lobe did not have a distinct region of high metabolic rate. Thus, all four quadruplets showed qualitative hypofrontality.

for the partial outline seen in Iris. This provides qualitative evidence for the differential pattern of organization of the frontal lobes.

Taken together, these data provide evidence from two imaging modalities—EEG and PET—for similarity among the quadruplets in dysfunction of a neural system linking the frontal and occipital lobes. This system has been shown to have similar deficits in studies of clinical populations of schizophrenics.

VENTRICULAR SIZE

Reveley et al. (1982) reported that ventricular size measurements from x-ray computed tomography (CT) on healthy monozygotic (MZ) and dizygotic (DZ) twin pairs revealed much higher correlations between MZ than DZ pairs, consistent with a very

Table 18.1 Relative metabolic rate in the frontal lobe for the Genain quadruplets and normal controls.

Gyrus	Hemi.	Quads		Normals		t	p	F	p
		mean	SD	mean	SD				
Superior	Left	1.065	0.0058	1.123	0.683	2.08	0.09	33.5	0.0004
	Right	1.015	0.0412	1.086	0.668	2.10	0.07	3.18	0.11
Middle	Left	1.0475	0.031	1.14	0.704	2.83	0.02	2.13	0.18
	Right	1.0375	0.0236	1.1317	0.0826	2.64	0.03	2.40	0.16
Inferior	Left	1.0925	0.025	1.1583	0.0445	2.99	0.02	1.96	0.19
	Right	1.0575	0.027	1.1217	0.611	2.23	0.05	2.64	0.14
Precentral	Left	1.0975	0.0908	1.12	0.632	0.41	0.70	0.64	0.44
Combined	Both	1.0605	0.0299	1.12	0.431	2.59	0.03	0.23	0.64

t-test compares normals and quadruplets; F-ratio is Levine's test for differences in variance between quadruplets and normals.

high degree of genetic control. In seven MZ pairs discordant for schizophrenia, the schizophrenic twins had larger ventricles than their well co-twins; however, there was a tendency for the co-twins to have larger ventricles than normal subjects. Reveley et al. suggest that the study of discordant pairs can identify individuals for whom environmental insults, such as perinatal damage, may have been especially important in contributing to the emergence of schizophrenia. In a CT family study, the healthy siblings of patients with schizophrenia had larger ventricles than controls but smaller ventricles than their ill siblings (Weinberger et al. 1981). Again, these data are consistent with the importance of environmental factors in the expression of clinical schizophrenia of the type that is associated with enlarged ventricles. Also consistent with this interpretation are the data obtained from magnetic resonance imaging studies of 15 discordant MZ pairs by Suddath et al. (1990). They found smaller ventricles and hippocampal volume in the affected MZ co-twins than in their unaffected co-twins but did not provide statistical contrasts with concordant pairs or normal pairs. There are three case studies of identical multiple births. The Genain quadruplets had identical but small ventricles (Buchsbaum et al. 1984). McGuffin et al. (1982) did not find CT abnormalities in identical triplets, of which two had been diagnosed schizophrenic and one manic-depressive. However, in an MZ twin pair in which one was diagnosed with schizophrenia and one with bipolar affective disorder, both had significantly elevated ventricle-brain ratios (VBR) (Lohr and Bracha 1992).

CORPUS CALLOSUM

Some studies have shown enlargement of the size or thickness of the corpus callosum in schizophrenia (for a review, see Casanova et al. 1990). In 12 pairs of discordant twins, no differences in the area, length, or vertical thickness were observed between affected and unaffected pairs, but an upward bowing of the shape in affected individuals was confirmed (Casanova et al. 1990) through the ingenious use of harmonic analysis on the length of lines joining the centroid and the structure edge. This is interpreted as possibly related to the ventricular enlargement in these individuals rather than as a developmental abnormality, but the authors did not support their speculation by presenting correlation coefficients between their measure of callosal shape and the lateral ventricular measurements made on these twins and reported elsewhere (Suddath et al. 1990).

HYPOFRONTALITY AND FUNCTIONAL IMAGING

Beginning with the studies of Ingvar and Franzen (1974), a series of studies has found that patients with schizophrenia had lower cerebral blood flow and metabolic rates in their frontal lobes than normal controls (see review by Buchsbaum 1990). Four studies have examined hypofrontality in twins or multiple births. The first was the study of quadruplets described above. In a second study of three sets of identical twins with psychotic symptoms, reduced metabolism in the orbitofrontal cortex was observed (Clark et al. 1989). In a third study, cerebral blood flow data showed significant hypofrontality in MZ twins discordant for schizophrenia who were only slightly less hypofrontal than concordant MZ twin pairs (Berman et al. 1992). Weinberger et al. (1992) found frontal flow in this sample correlated with hippocampal volume.

INSULT-DIATHESIS SCHIZOPHRENIA VS. HEREDITARY FRONTO-STRIATAL DYSFUNCTION SCHIZOPHRENIA (regional frontal, basal ganglia, or lateralized schizophrenia)

Do the discordant twin data suggest that an insult-diathesis (brain shrinkage) subtype of schizophrenia is associated with some environmental insult to an individual with a genetic diathesis? And that hypofrontal/striatal schizophrenia is less dependent on environmental triggers? Anatomical studies of the brains of MZ twins discordant for schizophrenia have shown a large difference between healthy and ill twins in ventricular and hippocampal size. These differences are larger than those between healthy MZ co-twins and normal controls (Reveley et al. 1982). Family studies showed little familial effect on ventricular size in families of schizophrenic probands (Weinberger et al. 1981). Thus, cortical shrinkage, perhaps greatest in the temporal lobe, might characterize the subtype of schizophrenia that is associated with birth damage, viral infection, or other environmental factors. Earlier studies of the neurological findings in discordant MZ twins by

Mosher et al. (1971) also seemed to implicate environmental factors, associating low birthweight and neurological dysfunction with the ill twin. Other studies, however, have suggested that a relatively low proportion of schizophrenics are affected by organic causes (see Kaprio et al., this volume).

In contrast to the anatomical studies, PET studies have tended to implicate frontal heritability. Regional cerebral metabolic rate in normal MZ pairs showed significant prefrontal, orbital frontal, and striatal similarity; no temporal area reached statistical significance (Clark et al. 1988). The Genain quadruplets showed significant similarity in the prefrontal cortex but not the lateral temporal lobe or hippocampus. Both the schizophrenia-concordant Genain quadruplets and the psychosis-concordant McGuffin triplets showed no ventricular enlargement, again consistent with brain shrinkage being associated with insult-diathesis schizophrenia rather than fronto-striatal dysfunction schizophrenia. However, Berman et al. (1992) reported cerebral blood flow data showing significant hypofrontality in MZ twins discordant for schizophrenia who were only slightly less hypofrontal than concordant MZ twin pairs. The fact that both twins in the eight concordant pairs were medicated may have diminished differences between discordant and concordant pairs if medication had the effect of ameliorating hypofrontality. In this study, pairs who were receiving high neuroleptic doses were significantly less hypofrontal than pairs who were receiving low doses. However, most PET studies of neuroleptic-treated schizophrenia have revealed little effect of medication on hypofrontality (Buchsbaum et al. 1992). While the differentiation of insult-diathesis and fronto-striatal dysfunction as two forms of schizophrenia is consistent with most data, we need far more data than currently available. First, functional studies contrasting significant numbers of unmedicated concordant and discordant MZ twin pairs are crucial. Current sample sizes (3–8) are so small that the failure to discriminate concordant and discordant members is compromised by unacceptably high Type II error rates. Second, systematic measurements of VBR and the size of the hippocampus, frontal lobe, and basal ganglia within the same subjects are not yet available. Third, comparisons of frontal and basal ganglia metabolic rates within groups of schizophrenics with positive and negative family histories of schizophrenia would usefully address the issue. Similar problems are present in the discordant twin pair studies of organic brain syndrome (Jarvik et al. 1980) and Alzheimer's disease (Kumar et al. 1991).

Three unsolved methodological problems remain with diagnosis. First, if a diagnostic threshold of lesser severity were used or if the pairs are early in the age of risk, some discordant pairs might become concordant, changing the analyses of discordant pairs. Interviewer bias in assessing MZ twins is a second hazard of uncertain directionality from study to study. Only the McGuffin triplets received blinded diagnostic interviews with unrelated patients included in the series. Third, there is no external validating criterion for diagnosis; only half of the patients of Berman et al. (1992) were hypofrontal. Analyses of concordant MZ pairs that contrast the hypofrontality and VBR enlargement of co-twins with extreme values on these measures may be especially informative.

REFERENCES

Berman, K.F., E.F. Torrey, D.G. Daniel, and D.R. Weinberger. 1992. Regional cerebral blood flow in monozygotic twins discordant and concordant for schizophrenia. *Arch. G. Psychiat.* **49:**927–934.

Buchsbaum, M.S. 1990. The frontal lobes, basal ganglia, and temporal lobes as sites for schizophrenia. *Schizo. Bull.* **16:**377–387.

Buchsbaum, M.S., J. Cappelletti, R. Coppola, F. Rigal, A.C. King, and D.P. van Kammen. 1982. New methods to determine the CNS effects of antigeriatric compounds: EEG topography and glucose use. *Drug Devel. Res.* **2:**489–496.

Buchsbaum, M.S., J.C. Gillin, J. Wu, E. Hazlett, N. Sicotte, and R. DuPont. 1989. Regional cerebral glucose metabolic rate in human sleep assessed by positron emission tomography. *Life Sci.* **45:**1349–1356.

Buchsbaum, M.S., D.H. Ingvar, R. Kessler, R.N. Waters, J. Cappelletti, D.P. van Kammen, A.C. King, J.L. Johnson, R.G. Manning, R.W. Flynn, L.S. Mann, W.E. Bunney, Jr., and L. Sokoloff. 1982. Cerebral glucography with positron emission tomography. *Arch. G. Psychiat.* **39:**251–259.

Buchsbaum, M.S., A.F. Mirsky, L.E. DeLisi, J. Morihisa, C.N. Karson, W.B. Mendelson, A.C. King, J. Johnson, and R. Kessler. 1984. The Genain Quadruplets: Electrophysiological, positron emission, and X-ray tomographic studies. *Psychiat. Res.* **13:**95–108.

Buchsbaum, M.S., S. Potkin, J. Marshall, S. Lottenberg, C.W. Heh, R. Tafalla, C. Reynolds, and W.E. Bunney. 1992. Effects of clozapine and thiothixene on glucose metabolic rate in schizophrenia. *Neuropsychopharm.* **6(3):**155–163.

Buchsbaum, M.S., F. Rigal, R. Coppola, J. Cappelletti, C. King, and J. Johnson. 1982. A new system for gray-level surface distribution maps of electrical activity. *EEG J.* **53:**237–242.

Casanova, M.F., R.D. Sanders, T.E. Goldberg, L.B. Bigelow, G. Christison, E.F. Torrey, and D.R. Weinberger. 1990. Morphometry of the corpus callosum in monozygotic twins discordant for schizophrenia: A magnetic resonance imaging study. *J. Neur. Psychiat.* **53:**416–421.

Clark, C.M., H. Klonoff, J.S. Tyhurst, D. Li, W. Martin, and B. Pate. 1989. Regional cerebral glucose metabolism in three sets of identical twins with psychotic symptoms. *Can. J. Psychiat.* **34:**263–270.

Clark, C.M., H. Klonoff, J.S. Tyhurst, T. Ruth, M. Adam, J. Rogers, R. Harrop, W. Martin, and B. Pate. 1988. Regional cerebral glucose metabolism in identical twins. *Neuropsychol.* **26:**615–621.

DeLisi, L.E., A.F. Mirsky, M.S. Buchsbaum, D.P. van Kammen, K.F. Berman, C. Caton, M.S. Kafka, P.T. Ninan, B.H. Phelps, F. Karoum, G.N. Ko, E.R. Korpi, M. Linnoila, M. Scheinin, and R.J. Wyatt. 1984. The Genain quadruplets 25 years latter: A diagnostic and biochemical followup. *Psychiat. Res.* **13:**59–76.

Gottschalk, L.A., M.S. Buchsbaum, J.C. Gillin, J.C. Wu, C.A. Reynolds, and D.B. Herrera. 1991. Anxiety levels in dreams: Relation to localized cerebral glucose metabolic rate. *Brain Res.* **538:**107–110.

Ingvar, D.H., and G. Franzen. 1974. Abnormalities of cerebral blood flow distribution in patients with chronic schizophrenia. *Acta Psychiat. Scand.* **50:**425–462.

Jarvik, L.F., V. Ruth, and S.S. Matsuyama. 1980. Organic brain syndrome and aging: A six-year follow-up of surviving twins. *Arch. G. Psychiat.* **37:**280–286.

Kumar, A., M.B. Schapiro, C.L. Grady, M.F. Matocha, J.V. Harby, A.M. Moore, J.S. Luxenberg, P.H. St. George-Hyslop, C.D. Robinette, M.J. Ball, and S.I. Rapoport. 1991. Anatomic, metabolic, neuropsychological and molecular genetic studies of three pairs of identical twins discordant for dementia of the Alzheimer's type. *Arch. Neur.* **48**:160–168.

Lohr, J.B., and H.S. Bracha. 1992. A monozygotic mirror-image twin pair with discordant psychiatric illnesses: A neuropsychiatric and neurodevelopmental evaluation. *Am. J. Psychiat.* **149**:1091–1095.

Matsui, T., and A. Hirano. 1979. An Atlas of the Human Brain for Computerized Tomography. Tokyo: Ikakn-Shoin.

McGuffin, P., A. Reveley, and A. Holland. 1982. Identical triplets: Non-identical psychosis? *Br. J. Psychiat.* **140**:1–6.

Mirsky, A.F., L.E. DeLisi, M.S. Buchsbaum, O.W. Quinn, P. Schwerdt, L.J. Siever, L. Mann, H. Weingartner, R. Zec, A. Sostek, I. Alterman, V. Revere, S.D. Dawson, and T.P. Zahn. 1984. The Genain Quadruplets: Psychological studies. *Psychiat. Res.* **13**:77–93.

Mosher, L.R., W. Pollin, and J.R. Stabenau. 1971. Identical twins discordant for schizophrenia: Neurologic findings. *Arch. G. Psychiat.* **24**:422–430.

Reveley, A.M., M.A. Reveley, C.A. Clifford, and R.M. Murray. 1982. Cerebral ventricular size in twins discordant for schizophrenia. *Lancet* **I**:540–541.

Rosenthal, D. 1963. The Genain Quadruplets. New York: Basic Books.

Suddath, R.L., G.W. Christison, E.F. Torrey, M.F. Casanova, and D.R. Weinberger. 1990. Anatomical abnormalities in the brains of monozygotic twins discordant for schizophrenia. *New Eng. J. Med.* **322**:789–794.

Weinberger, D.R., K.F. Berman, R. Suddath, and E.F. Torrey. 1992. Evidence for dysfunction of a prefrontal-limbic network in schizophrenia: A magnetic resonance imaging and regional cerebral blood flow study of discordant monozygotic twins. *Am. J. Psychiat.* **149**:890–897.

Weinberger, D.R., L.E. DeLisi, A.N. Neophytides, and R.J. Wyatt. 1981. Familial aspects of CT scan abnormalities in chronic schizophrenic patients. *Psychiat. Res.* **4**:65–71.

19

What Can We Learn about the Determinants of Psychopathology and Substance Abuse from Studies of Normal Twins?

A.C. HEATH

Department of Psychiatry, Washington University School of Medicine,
4940 Children's Place, St. Louis, MO 63130, U.S.A.

ABSTRACT

Areas in which behavioral genetic approaches to the inheritance of psychopathology may be particularly fruitful are reviewed, using illustrations from the field of alcoholism research. These include testing of multivariate and multidimensional models for gene action, investigation of genotype-environment correlation and genotype × environment interaction, identification of mediating and moderator variables, and modeling of the role of genetic and environmental factors not merely in the onset of a disorder, but also in the timing of onset and course of the disorder.

INTRODUCTION

Can twin, twin-family, adoption, and other kinship studies play a role in understanding the inheritance of psychopathology? In considering this question, it is helpful to consider two alternative research strategies, which may be characterized as the "medical genetic" and the "behavioral genetic" approaches. The *medical genetic* approach is motivated by studies of the inheritance of relatively rare disorders having a simple Mendelian pattern of inheritance. For such disorders molecular genetic studies are advancing our knowledge at a rapid pace. The *behavioral genetic* approach typically focuses on characters that exhibit a multifactorial mode of inheritance, where genetic, epidemiologic, and behavioral risk factors may all play an important role. The behavioral genetic approach is typically multivariate, since it is concerned

Twins as a Tool of Behavioral Genetics
Edited by T.J. Bouchard, Jr. and P. Propping © 1993 John Wiley & Sons Ltd.

not merely with demonstrating genetic influence on an outcome variable (e.g., occurrence of a particular psychiatric disorder) but also with identifying variables, be they behavioral or biochemical, which may account for that genetic influence. Here, issues about genotype-environment (GE) correlation (how genetic and environmental risk factors covary) and GE interaction (how environmental factors may moderate the effects of inherited risk, or how genetic predisposition may increase vulnerability to epidemiologic risk factors) become important. Here, also, the search for intervening or "mediating" variables in the causal chain from genotype and epidemiologic risk factos to outcome—perhaps heritable temperamental or personality traits or other constructs much loved by psychologists, which may account for some of the genetic influence on risk of disorder—becomes important. In the behavioral genetic approach, developmental questions, about how genetic and environmental effects unfold through time and contribute to consistency and change in behavior, or to the course of an illness, become important issues.

For many medical geneticists, given the powerful molecular tools that are now available for "finding the genes" for simple Mendelian disorders and the singular lack of success of linkage studies of psychiatric disorders, the key question is why we should "divert" funds from studies of simple Mendelian disorders. Fortunately for behavioral scientists, the economic costs of the psychiatric disorders to society are tremendous. In the United States alone, the direct health costs of alcohol misuse, and indirect costs through lost productivity, are estimated to be close to $140 billion per year (Sullivan 1990). In this chapter, I will explore some of the questions about the inheritance of psychopathology that can be profitably addressed by behavioral genetic methods. Our focus will be primarily on using an epidemiologic sampling strategy, i.e., surveying clinically unselected samples of twins or other relatives ("normal twins"), rather than on clinic-based studies, which screen a systematic series of patients to identify twins and study their relatives. Consequently, our discussions will be most pertinent to relatively common disorders (e.g., alcoholism, depression, conduct disorder) and disorders exhibiting rather strong familial aggregation (e.g., hyperactivity), as well as to behavioral traits that are continuously distributed in the general population.

AN EXAMPLE: ALCOHOL USE AND ABUSE

To explore the utility of twin studies for investigating the inheritance of psychopathology, we shall consider work on the inheritance of alcoholism and patterns of alcohol use. Most issues that arise apply equally to other forms of psychiatric disorders. Evidence for a genetic contribution to alcoholism risk from twin, adoption, and half-sibling studies has been the subject of a recent review by McGue (1993; see also contributions in Goedde and Agarwal 1989; Begleiter and Kissin 1993). In brief, alcoholism shows a strong tendency to run in families, and the data are consistent with a significant genetic contribution to alcoholism risk, at least in males. (Findings for female alcoholism are more ambiguous, though the recent study of Kendler and colleagues (Kendler et al. 1992) suggests that genetic factors may be equally important for females.) Identical twin pairs, who share identical genotypes, show increased concordance for alcoholism compared to fraternal twin pairs, who on average share only 50% of their genes in common. Adopted-away children of alcoholics still show an increased risk of alcoholism, compared to adopted-away children of nonalcoholics. Alcoholism in adoptive parents, in contrast, is not predictive of alcoholism in their adopted children. This evidence for a genetic involvement in alcoholism risk has led to the initiation of a multi-center collaborative linkage study in the U.S., in an attempt to identify individual genes that may contribute to alcoholism risk. It has also stimulated an active tradition of high-risk research, comparing offspring of alcoholics and controls in an attempt to identify mechanisms that may explain the differences between individuals who are genetically at high versus low risk (e.g., Sher 1991). Once the importance of the role of genetic factors has been demonstrated, is now the time for behavioral geneticists to leave the field to others, who may find the genes, or identify the mechanisms, associated with increased alcoholism risk? When we consider the questions raised in the introductory section above, it appears that retreat is probably premature!

GENETICS OF ALCOHOL EXPOSURE

Under the conventional medical genetic model, the genetics of self-exposure to alcohol (i.e., the initiation and progression of alcohol use, including quantity and frequency measures of consumption pattern) is viewed as irrelevant to the inheritance of the alcoholism disease. Clearly, we cannot use measures of consumption as a proxy for measures of alcohol abuse or dependence. For alcoholism and other substance use disorders, however, the genetics of self-exposure may be an important component of the developmental pathway leading to alcoholism, with many individuals being genetically "protected," i.e., maintaining very low levels of consumption (in part because of the influence of genetic factors), levels which never put them at risk for alcohol-related problems. Findings from large sample twin studies in Scandinavia, the U.S.A., and Australia all support a substantial genetic contribution to variation in levels of alcohol consumption in general population samples (e.g., Heath, Eaves, and Martin, unpublished; Kaprio et al. 1987; Heath et al. 1991). Alcohol consumption patterns beyond age 25 show substantial long-term stability at follow-up intervals as long as 8–10 years, with genetic factors accounting for as much as 70% of the longitudinally stable variation in consumption levels in both sexes (Heath and Martin, unpublished; Carmelli et al., unpublished; Kaprio et al. 1992). Evidence for an important genetic influence remains whether we rely only on self-report data or combine this with ratings by informants (Heath and Martin, unpublished).

Modeling the inheritance of patterns of substance use raises explicit questions that are often ignored in analyses of other behavioral traits. Can we assume that the same genetic and environmental factors are affecting consumption levels at all points on the continuum (or "scale") of consumption, or do different factors determine (a) abstinence versus use of alcohol, (b) light versus moderate drinking, and (c) heavy drinking? Incorrect assumptions can lead to nonsensical inferences about the inheritance of drinking behavior. Nonmetric multidimensional scaling applied to twin pair contingency tables (adjusted for marginal differences in category endorsement frequency) provides a useful tool for exploring the dimensionality of alcohol consumption patterns and in an analysis of data from the 1981 Australian twin survey, suggested the existence of separate abstinence, frequency, and quantity dimensions with no evidence for a dimension separating very heavy drinkers from more moderate drinkers (Heath et al. 1991). Parametric model-fitting, using a hierarchical extension of the conventional threshold model (Heath et al. 1991), confirmed a relatively independent determination of initial self-exposure to alcohol and level of consumption in those who became drinkers, with abstinence versus use strongly determined by familial environmental factors and the level of consumption by drinkers showing a strong genetic influence. Demonstrating the important role of genetic factors, however, is only the beginning, not the final goal, of a behavioral genetic research strategy.

GENOTYPE–ENVIRONMENT CORRELATION

Behavioral geneticists have written extensively about the importance of GE correlation and the "genetic environment" (e.g., Eaves et al. 1977; Plomin et al. 1977; Plomin and Bergeman 1991). A case could be made that most behavioral genetic research has more to say to those who study environmental influences on behavior than to those interested in the biological underpinnings of behavior. Many supposed "environmental" measures or "epidemiologic" risk factors (e.g., perceptions of parental rearing style) show patterns of aggregation in the families of adoptees and of twins, which suggest that they are partly under genetic control. There are many instances where we suspect that GE correlation is probably important, a few instances where we know for certain that it does occur, and there may even be exceptional examples where we understand the mechanism by which it comes about. It may be helpful, following Eaves et al. (1977) (with modifications), to distinguish between three mechanisms by which genetic risk may come to be correlated with environmental risk: (a) GE autocorrelation, whereby an individual at high genetic risk also exposes himself to high-risk environments (e.g., individuals at high risk for alcoholism who frequent bars rather than churches), or perceives his environment as being a high-risk environment, or simply elicits a high-risk environment from the world at large (e.g., an individual who has inherited physically attractive features may attract many potential sexual partners); (b) primary GE correlation, whereby biologic relatives influence the environment to which an individual is exposed (e.g., sibling effects, parent-to-offspring environmental transmission, or even offspring-to-parent transmission); and also (c) secondary GE correlation, whereby an assortment process (e.g., assortative mating or assortative friendship) leads to the selection of individuals who will have an important environmental influence on a subject (spousal interaction, peer effects). For each of these forms of GE correlation, the twin study (e.g., Neale and Cardon 1992) or extended twin-family designs (studying twin pairs and their spouses or peers, twin pairs and their parents, or twin pairs and their offspring, e.g., Heath et al. 1985) is potentially very informative.

If we consider these issues with reference to the inheritance of alcoholism and patterns of alcohol use, then we are left with many unanswered questions. It would not be surprising to discover that individuals with high genetic liability to sensation-seeking or impulsive behavior are also more likely to expose themselves to high-risk environments. However, it has not yet been demonstrated conclusively that this is the case. It would be more surprising, and controversial, to discover that the association between risk of psychiatric disorder and history of rape or sexual assault arises because aspects of personality or behavioral style of individuals at increased genetic risk of psychiatric disorder make them less able to avoid high-risk situations. This hypothesis of GE autocorrelation has not yet been disproved (Eaves, pers. comm.). Although there is good evidence for parent-to-offspring environmental influences on abstinence versus use of alcohol, evidence that parental drinking levels have any *environmental* influence on consumption patterns of their nonabstaining adult offspring remains elusive (Heath 1992). Reciprocal sibling environmental effects have been often suspected but never convincingly confirmed and replicated. Spousal concordances for alcohol consumption patterns are higher than for most other behavioral or lifestyle variables; however, preliminary analyses have failed to find evidence for any reciprocal environmental influence of drinking patterns of one spouse on that of the other, and vice versa, once spousal concordance arising through assortative mating is controlled for. Although behavioral geneticists have powerful tools for dissecting such effects, evidence that living in the same household as a parent or partner has any lasting effects on behavior is astonishingly weak.

GENOTYPE × ENVIRONMENT INTERACTION

After GE correlation, perhaps the importance of genotype × environment interaction is the most preferred topic for discussion of behavioral geneticists (e.g., Eaves et al. 1977; Plomin et al. 1977). From the early research traditions of plant and animal breeders, biometrical geneticists (e.g., Mather and Jinks 1971; Eaves 1987) have long recognized that gene expression may differ under different environmental conditions, that there may be genetic differences in sensitivity to the environment, and that different genes may be involved in the control of environmental sensitivity versus average level of response across a range of environments. Convincing demonstrations of G × E interaction in behavioral genetic research have been rare. Cloninger et al. (1981) have reported significant G × E interaction for alcoholism risk, but it is not clear that explanations of their findings not involving G × E have been excluded. In the simplest case of a dichotomous environmental exposure variable, demonstration of differences in genetic influence under low-risk versus high-risk environmental conditions would involve showing significant heterogeneity of genetic and environmental influences as a function of exposure condition that could not be accounted for merely by overall differences in response variability between conditions (e.g., Heath et al. 1989; Neale and Cardon 1992). Reanalyzing data on alcohol consumption levels of female Australian twin pairs, we found an apparent $G \times E$ interaction associated with marital status (Heath et al. 1989), with increased genetic influence on consumption levels in those without a marital or other partner. In young adult twins, aged 18–30, genetic differences accounted for only 31% of the variance in consumption levels of married respondents, but for 60% of the variance of unmarried respondents; these figures increased to 46–59% and 76% of the variance, respectively, in older twins. Changes in environment apparently can be important moderators of the genetic influence on normal variation in drinking patterns.

It can be argued that the strongest evidence for $G \times E$ interaction will come from laboratory settings, where the environment can be manipulated experimentally. Research on genetic differences in alcohol reactivity, assessed using an alcohol challenge paradigm, provides one example of such research (e.g., Martin et al. 1985). In the Australian alcohol challenge twin study of Martin and colleagues, for example, using a standard body weight-adjusted dose of alcohol, even after controlling for genetic differences in baseline performance, a significant genetic contribution to post-alcohol performance on a variety of psychomotor tasks was observed (Martin et al. 1985): a clear example of genotype $\times$ alcohol interaction!

INSIGHTS FROM MULTIVARIATE GENETIC STUDIES

From the pioneering paper of Martin and Eaves (1977) onwards, methods of multivariate genetic analysis have become a standard tool of the behavioral geneticist for deciding whether univariate analyses of different variables are in fact assessing the same dimension of genetic variability. For example, since subjects were not alcohol-naive in the Australian alcohol challenge twin study, is it possible that the observed genetic differences in alcohol reactivity can be explained by genetically determined differences in history of alcohol use, with heavy drinkers showing a diminished deterioration in performance after alcohol (Heath and Martin 1991)? Whereas a traditional factor analysis uses only within-person correlations, and a univariate genetic analysis only within-trait correlations between relatives, multivariate genetic analysis exploits the additional information contained in the between-trait correlations between relatives to estimate separate genetic and environmental common and specific factors. Applied to the Australian twin alcohol challenge study, multivariate genetic analysis identified two orthogonal genetic factors: (a) a genetic factor that influenced both drinking history measures and post-alcohol challenge intoxication ratings, subjective willingness to drive, and objective body-sway measures; (b) a second genetic factor that influenced post-alcohol blood alcohol concentration and fine psychomotor coordination, but not alcohol consumption variables (Heath and Martin 1992). While individual differences in alcohol reactivity, as assessed by measures of body sway and by the subjective effects of alcohol, do seem to be influenced by the same genetic factors that influence consumption level, genetic differences in deterioration in psychomotor coordination cannot be thus explained. Multivariate genetic analysis allows us to progress beyond the question of whether genetic factors are influencing variation in drinking patterns (or risk of psychiatric disorder), to explore the question of how this influence may come about.

THE SEARCH FOR MEDIATING VARIABLES

Given the evidence for a genetic contribution to alcoholism risk, can we identify psychological constructs which might intervene in the pathway from genotype to alcoholism risk? There is clearly an important advantage, from the perspective of prevention research, in being able to identify individuals at high genetic risk who have not yet become abusers or dependent. Several theorists have hypothesized an important role of heritable temperamental or personality traits (e.g., Cloninger 1987). While personality differences between alcoholics and controls have been frequently reported, such differences might also be explained as a secondary consequence of the effects on behavior of chronic alcohol abuse or dependence. There is certainly extensive evidence for a genetic contribution to personality differences, from studies of twins, separated twins, and adoptees and their families (e.g., Eaves et al. 1989; Eaves et al., in review): insofar as the personality traits of family members are correlated, it appears that this is because of shared genes rather than shared environmental factors. However, it appears that the pattern of inheritance of alcoholism is quite different from that observed for personality traits such as *extraversion* or *neuroticism*. For personality traits, much of the genetic variance appears to be nonadditive, leading to relatively low parent-offspring correlations, even when both generations are assessed in adulthood. For alcoholism, high parent-offspring correlations in risk are observed. For most personality traits, spousal correlations that are effectively zero are observed. For alcoholism, strong marital concordance for alcoholism risk is observed, though it is not yet clear whether this reflects assortative mating or reciprocal spousal influence. Given these very contrasting modes of inheritance, it is unlikely that personality traits play more than a minor role as mediating variables in the inheritance of alcoholism. Of course, this does not preclude the possibility that personality traits play an important role as moderator variables that interact with inherited differences on other dimensions (e.g., alcohol reactivity—see above) to determine alcoholism risk.

Attempts to identify intervening or "mediating" variables in the causal chain from genotype to alcoholism risk have largely been conducted within the framework of high-risk research (e.g., see recent reviews by Sher 1991; Newlin and Thomson 1990). The rationale for such studies has been that in studies of alcoholics and controls, associations between alcoholism and a risk factor assessed retrospectively may be a consequence of alcohol misuse, or a reporting bias. By studying adult offspring of alcoholics versus controls, it is argued, a prospective study can be conducted that may identify risk factors which emerge prior to the onset of psychiatric disorder. In a typical study, a group of adult offspring of alcoholics and a control group with no family history of alcohol dependence are identified, most commonly from a population of white male students! The two groups may be compared with respect to such variables as personality traits, anticipated effects of alcohol ("alcohol expectancies"), neuropsychological deficits, neurophysiological characteristics such as event-related potentials, or objective performance measures and subjective responses in the laboratory to a challenge dose of alcohol. Typically, such studies are *not* family studies, in that most rely upon family history to assess parental alcoholism and only study a single individual per family. With rare exceptions (e.g., in the research of Schuckit [Schuckit and Gold 1988]), a long-term follow-up, to demonstrate that a variable which discriminates between offspring of alcoholics and controls is also predictive of future onset of alcoholism, has not been conducted.

High-risk studies have generated some interesting differences between offspring of alcoholics and controls. In alcohol challenge research, for example, many studies have found that offspring of alcoholics, given a standard body-weight-adjusted dose of alcohol, differ in their reactivity to alcohol—whether measured by subjective ratings of intoxication or more objective measures of body sway, deterioration in psychomotor performance, and hormonal changes—when compared to controls, although the precise circumstances under which heightened or dampened reactivity is observed remain a controversial issue (e.g., Newlin and Thomson 1990). Differences in baseline evoked potential responses in offspring of alcoholics and controls also appear to be replicable (Sher 1991). However, the conventional high-risk research strategy has important drawbacks, which make interpretation of such findings problematic. Although such differences could be secondary environmental consequences of growing up with an alcoholic parent, they certainly cannot be concluded to be measures of genetic risk, however "biological" the variables that are studied. They have not, in most cases, been shown to be predictive of future alcoholism risk. Even when a correlation with alcoholism risk is demonstrated, the experimental design employed does not allow us to determine whether this association is merely a reflection of cross-assortative mating, i.e., a difference between the nonalcoholic parent (usually the mother) in the high-risk group compared to the mothers of the control subjects, rather than being determined by the same genetic factors that determine alcoholism risk.

The primary goals of high-risk research—identifying variables that are determined by the same genetic factors which also determine alcoholism risk or risk of other psychiatric disorder (and which may therefore allow identification of individuals at high risk), and showing that these are not merely a secondary consequence of alcohol abuse—are goals which can be addressed, with considerable resolving power using the twin design. (In terms of statistical power, however, it may be necessary to use a two-stage sampling strategy, with an initial screening followed by oversampling of twin pairs with at least one alcoholic twin, rather than merely random sampling [cf. Neale and Cardon 1992]). It is not widely recognized that even cross-sectional twin data (or adoption or other family data) will, under certain circumstances, be informative about issues of direction of causation. To understand why this is the case, it is helpful to consider a hypothetical, and extremely simplified, example (e.g., Neale and Cardon 1992). We have noted above that the same dimension of genetic variability influences both the level of alcohol consumption by drinkers and certain measures of alcohol sensitivity in the alcohol challenge paradigm. It is possible that individuals who have inherited low sensitivity to alcohol are able to drink increased amounts (sensitivity→ consumption), or, alternatively, that individuals with increased experience of alcohol have diminished sensitivity (consumption→ sensitivity).

Let us consider the hypothetical case where, apart from any genetic influence mediated through the effects of alcohol sensitivity, level of habitual alcohol consumption is purely determined by shared and nonshared environmental factors; and, apart from any shared environmental influence mediated through the causal effects of drinking history, variation in alcohol sensitivity is purely determined by genetic factors plus within-family environmental factors. Under these strong assumptions, if we consider the cross-correlation between drinking history in one twin and alcohol sensitivity in the co-twin, then clearly the two alternative unidirectional causal hypotheses lead to quite different predictions. Under the hypothesis that differences in sensitivity cause differences in consumption, the magnitude of this cross-correlation will depend upon the magnitude of the twin pair correlation for sensitivity (as well as the effect of sensitivity on consumption) and hence will be higher in MZ than in DZ twin pairs. Under the hypothesis that differences in consumption are causing differences in sensitivity, the magnitude of the cross-correlation will be dependent upon the twin pair correlation for consumption and hence should be the same in MZ and in DZ pairs. Although this extreme example is likely to be unrealistic in practice, the same argument will apply even when the relative magnitudes of genetic and shared environmental influences, or additive and nonadditive genetic influences, on two traits differ between the two traits. It can be shown more formally, by developing a path model allowing for a reciprocal causal influence of two variables, that application of these methods depends critically upon two correlated variables having somewhat different modes of inheritance (e.g., strong additive plus nonadditive genetic effects on one character, additive genetic plus shared environmental effects on the second character) and that there is considerable loss of power if error variances for the two variables are very unequal, unless multiple indicator variables (i.e., multiple measures of consumption and multiple measures of alcohol reactivity) are used to assess the underlying "consumption" and "sensitivity" constructs (Heath et al. 1993).

We have used this approach to reanalyze the Australian twin alcohol challenge data on self-report average weekly alcohol consumption and subjective rating of intoxication after alcohol challenge (Heath and Martin 1991; Neale and Cardon 1992). In male twins, at least, model-fitting results gave better support to the hypothesis that differences in drinking history were leading to differences in alcohol sensitivity, rather than vice versa. In female twins, resolving power was too low to permit any conclusions to be drawn. Certainly these analyses suggest that the twin design has great potential for addressing the traditional goals of high-risk research.

NATURAL HISTORY OF ALCOHOL USE

Questions about the genetics of behavioral development have long played a central role in the field of behavioral genetics, leading to the elaboration of time-series, growth curve, and survival genetic models (e.g., Eaves et al. 1986; Meyer and Eaves 1988; Goldsmith and McArdle, unpublished). The "natural history" perspective is clearly an important one in genetic studies of substance use disorders, where we wish to understand how the progression occurs from initial exposure through casual use to dependence or abuse, recovery, and relapse. However, many of these same questions apply to the course of other psychiatric disorders. While an increasing amount is known about the genetic contribution to risk of first onset of a psychiatric disorder, almost nothing is known about the genetic contribution to speed of recovery or risk of relapse. Most genetic studies of alcoholism and of normal variation in drinking patterns (as well as of smoking or illicit substance use) have been performed on adult samples. We have very little data about the early course of substance use through adolescence into early adulthood. In this area, behavioral geneticists have powerful models to apply, but a paucity of data on the large samples that we need.

CONCLUSIONS

Behavioral geneticists have traditionally shown great enthusiasm for studies of the inheritance of cognitive abilities, personality traits, and similar psychological constructs. We have generally been slower to apply these methods to investigate the inheritance of psychiatric disorders. In this chapter, I have tried to show that for disorders such as alcoholism, many of the favorite research questions and research strategies of behavioral geneticists are highly pertinent. Most of the scientific questions which we have reviewed are not exclusively applicable to questions about polygenic inheritance: the same questions could be addressed with greater power if we were able to identify individual genes that contribute to genetic risk. In the frenzy to map rare Mendelian disorders, however, they are questions that are being unduly neglected.

ACKNOWLEDGEMENTS

Preparation of this manuscript was supported, in part, by ADAMHA grants AA03539, AA07535, AA07728, DA05588, MH31302 and MH40828.

REFERENCES

Begleiter, H., and B. Kissin, eds. 1993. Alcohol and Alcoholism: Genetic Factors and Alcoholism, vol. 1. Oxford: Oxford Univ. Press, in press.

Cloninger, C.R., M. Bohman, and S. Sigvardsson. 1981. Inheritance of alcohol abuse: Cross-fostering analysis of adopted men. *Arch. G. Psychiat.* **38**:861–868.

Cloninger, C.R. 1987. Neurogenetic adaptive mechanisms in alcoholism. *Science* **236**:410–416.

Eaves, L.J., K.A. Last, N.G. Martin, and J.L. Jinks. 1977. A progressive approach to nonadditivity and genotype-environmental covariance in the analysis of human differences. *Br. J. Math. Stat. Psychol.* **30**:1–42.

Eaves, L.J. 1987. Including the environment in models for genetic segregation. *J. Psychiat. Res.* **21**:639–647.

Eaves, L.J., H.J. Eysenck, and N.G. Martin. 1989. Genes, Culture and Personality: An Empirical Approach. New York: Academic.

Eaves, L.J., J. Long, and A.C. Heath. 1986. A theory of developmental change in quantitative phenotypes applied to cognitive development. *Behav. Genet.* **16**:143–162.

Goedde, H.W., and D.P. Agarwal, eds. 1989. Alcoholism: Biomedical and Genetic Aspects. New York: Pergamon.

Heath, A.C. 1993. Genetic influences on drinking behavior in humans. In: Alcohol and Alcoholism: Genetic Factors and Alcoholism, vol. 1, ed. H. Begleiter and B. Kissin. Oxford: Oxford Univ. Press, in press.

Heath, A.C., R. Jardine, and N.G. Martin. 1989. Interactive effects of genotype and social environment on alcohol consumption in female twins. *J. Stud. Alcohol* **50**:38–48.

Heath, A.C., K.S. Kendler, L.J. Eaves, and D. Markell. 1985. The resolution of cultural and biological inheritance: Informativeness of difference relationships. *Behav. Genet.* **15**:439–465.

Heath, A.C., R.C. Kessler, M.C. Neale, J.K. Hewitt, L.J. Eaves, and K.S. Kendler. 1993. Testing hypotheses about direction of causation using cross-sectional family data. *Behav. Genet.* **23**:29–50.

Heath, A.C., and N.G. Martin. 1991. Intoxication after an acute dose of alcohol: An assessment of its association with alcohol consumption patterns using twin data. *Alcohol Clin. Ex.* **15**:122–128.

Heath, A.C., and N.G. Martin. 1992. Genetic differences in psychomotor performance decrement after alcohol: A multivariate analysis. *J. Stud. Alcohol* **53**:262–271.

Heath, A.C., J. Meyer, R. Jardine, and N.G. Martin. 1991. The inheritance of alcohol consumption patterns in a general population twin sample: I, II. *J. Stud. Alcohol* **52**:345–352, 425–433.

Heath, A.C., M.C. Neale, J.K. Hewitt, L.J. Eaves, and D.W. Fulker. 1989. Testing structural WHERE?

Kaprio, J., M.D. Koskenvuo, H. Langinvainio, K. Romanov, S. Sarna, and R.J. Rose. 1987. Genetic influences on use and abuse of alcohol: A study of 5638 adult Finnish brothers. *Alcohol Clin. Ex.* **11**:349–356.

Kaprio J., R. Viken, M. Koskenvuo, K. Romanov, and R.J. Rose. 1992. Consistency and change in patterns of social drinking: a 6-year follow-up of the Finnish twin cohort. *Alcohol Clin. Ex.* **16**:234–240.

Kendler, K.S., A.C. Heath, M.C. Neale, R.C. Kessler, and L.J. Eaves. 1992. A population-based twin study of alcoholism in women. *JAMA* **268**:1877–1882.

Martin, N.G., and L.J. Eaves. 1977. The genetical analysis of covariance structure. *Heredity* **28**:79–95.

Martin, N.G., J.G. Oakeshott, J.B. Gibson, G.A. Starmer, J. Perl, and A.V. Wilks. 1985. A twin study of psychomotor and physiological responses to an acute dose of alcohol. *Behav. Genet.* **15**:305–347.

Mather, K., and J.L. Jinks. 1971. Biometrical Genetics: The Study of Continuous Variation. London: Chapman and Hall Ltd.

McGue, M. 1993. Genes, environment and the etiology of alcoholism. In: Development of Alcohol-Related Problems in High-Risk Youth: Establishing Linkages across Biogenetic and Psychosocial Domains, ed. R. Zucker, J. Howard, and G. Boyd. Rockville, MD: NIAAA Monograph, in press.

Meyer, J.M., and L.J. Eaves. 1988. Estimating genetic parameters of survival distributions: A multifactorial model. *Genet. Epidem.* **5**:265–276.

Neale, M.C., and L.R. Cardon. 1992. Methodology for Genetic Studies of Twins and Families, NATO ASI Series. Dordrecht: Kluwer.

Newlin, D.B., and J.B. Thomson. 1990. Alcohol challenge with sons of alcoholics: a critical review and analysis. *Psychol. Bull.* **108**:383–402.

Plomin, R., and C.S. Bergeman. 1991. The nature of nurture: Genetic influence on "environmental" measures. *Behav. Brain Sci.* **14**:373–385.

Plomin, R., J.C. DeFries, and J.L. Loehlin. 1977. Genotype-environment interaction and correlation in the analysis of human variation. *Psychol. Bull.* **84**:309–322.

Schuckit, M.A., and E.O. Gold. 1988. A simultaneous evaluation of multiple markers of ethanol/placebo challenges in sons of alcoholics and controls. *Arch. G. Psychiat.* **45**:211–216.

Sher, K.J. 1991. Children of Alcoholics: A Critical Appraisal of Theory and Research. Chicago: Univ. of Chicago Press.

Sullivan, L.W. 1990. Seventh Special Report to the U.S. Congress on Alcohol and Health. Rockville, MD: USDHHS, PHS.

APPENDIX: Current Controversies:

Which of the following statements are false?

1. The twin design is most useful for addressing nongenetic questions, e.g., resolving spousal interaction versus assortment effects, or assortative friend-ship versus peer effects, or testing hypotheses about direction of causation, or identifying mediating or moderator ("vulnerability") variables. It has been largely superceded as a means of studying genetic inheritance.

2. Within-family environmental effects ("sibling differential experience") are either measurement error effects, or transient effects, and contribute little to the longitudinal stability of behavior (including recurrence of psychopathology) in adulthood. Stability and recurrence are largely a function of genotype.

3. Once we allow for such complications as assortative mating (in twin data) or selective placement (in adoption data), there is NO evidence that the family environment shared by individuals growing up in the same family has any influence on risk of psychopathology in adulthood.

4. While there is strong evidence for a genetic contribution to personality differences, most of the genetic variance is epistatic, yielding very modest correlations between first degree relatives assessed as adults. Models for the familial transmission of psychopathology can safely ignore personality constructs.
5. The "genetic environment" is a behavioral geneticist's way of saying that we have found a correlation, but have not yet tried to model the underlying process.

Standing, left to right:
J. Kaprio, H.H. Stassen, A.C. Heath, P. McGuffin, P. Propping

Seated, left to right:
M.S. Buchsbaum, I.I. Gottesman, J. Körner, M. Rietzschel

20

Group Report: What Can Twin Studies Contribute to the Understanding of Adult Psychopathology?

J. KAPRIO, Rapporteur

M.S. BUCHSBAUM, I.I. GOTTESMAN, A.C. HEATH, J. KÖRNER,
E. KRINGLEN, P. McGUFFIN, P. PROPPING,
M. RIETZSCHEL, H.H. STASSEN

INTRODUCTION

The contribution of twin studies is perhaps best exemplified by research in schizophrenia where, despite long-standing evidence of familiality, the existence of an important genetic component has sometimes been viewed as controversial. Early results from twin studies were open to methodological criticisms. More recent register-based investigations, in particular from the Nordic countries, have addressed most of these and, although the concordance rates were reduced, the evidence of genetic influences has been remarkably consistent (Tienari 1963; Kringlen 1967; Fischer 1973). These results together with the findings in a series of well-known adoption studies in Denmark and the U.S.A., combine with family study data to provide a compelling body of evidence that the genetic effect in schizophrenia is relatively strong (Kringlen 1987; Gottesman 1991). Twin studies have also played an important role in demonstrating a genetic basis for the severe affective disorders. Again, the combination of family, twin, and adoption study results compels the conclusion that genetic factors contribute a prominent component in the etiology of schizophrenia (Bertelsen 1992). The existence of these findings will be taken as "givens" in this discussion paper and will not be examined further. Our brief was rather to explore the ways in which twin studies can be useful to further our understanding of psychopathology beyond simply demonstrating that a genetic component exists.

Twins as a Tool of Behavioral Genetics
Edited by T.J. Bouchard, Jr. and P. Propping © 1993 John Wiley & Sons Ltd.

Historical divergence in diagnoses among psychopathologists has been reduced; reliable, reproducible psychiatric diagnoses have been achieved, although their validity remains uncertain. McGuffin et al. (this volume) and Heath (this volume) document the value of twin studies in defining the phenotype in studies of psychopathology. Brain imaging of twins permits us to gain insight into the biological correlates of psychopathology (Buchsbaum, this volume), but findings in psychiatric patients overlap with findings in normal subjects. Heterogeneity of psychiatric disorders is assumed, but homogeneous entities cannot be disentangled because of a lack of validating criteria. Phenocopies are difficult to distinguish from "genetic" cases. These problems in taxonomy, together with the lack of knowledge of mode of inheritance and ascertainment bias, suggest that linkage analysis of psychiatric disorders will continue to be a highly challenging and difficult enterprise. Despite increasing numbers of genetic markers becoming available and much improved mapping and statistical technology, diagnostic uncertainty will remain a key issue in psychiatric genetics.

Studies of identical twin pairs concordant and discordant for psychopathology, in comparison to fraternal twins or full sibs similarly concordant or discordant, are useful for indicating genetically distinct categories of psychiatric disorders. Twin studies applying current diagnostic schemes for both schizophrenia and affective disorder indicate a tendency toward homotypia but show that there is also clear overlap, e.g., not only in subtype diagnoses but even between major diagnostic categories (McGuffin et al., this volume). The twin studies by the Richmond, Virginia group (Kendler et al. 1992a, b, c) address these same issues for such diagnoses as depression, generalized anxiety, and phobias.

The apparent existence of two diseases in the same individual or in both members of an identical twin pair has given rise to the concept of psychiatric co-morbidity, a term borrowed from internal medicine. Co-morbidity can arise for a number of reasons, and our group, in its discussion, considered that the usefulness of the idea of co-morbidity has sometimes been overstated. While the second disease may have a truly independent etiology from the first, many other explanations are possible. For example, a previously healthy twin might react to the co-twin's illness (e.g., schizophrenia) by becoming depressed. Alternatively, symptoms of one disease may appear as precursors to another disorder, but when followed up, the diagnosis becomes clear. While some co-morbidity may arise by chance occurrence of two independent diseases, diagnostic problems may arise because of methodologies. For example, operational criteria, as embodied in DSM IIIR, improve reliability by taking what has been termed a Chinese menu approach (Farmer et al. 1991). This, in turn, may produce a "Chinese menu artefact," i.e., some polysymptomatic patients may appear to conform to the bill of fare of two or more dinners (diagnoses as defined by DSM IIIR criteria) when actually they have simply been producing their psychopathology a la carte. In other words a single disorder manifested as multiple, perhaps atypical, symptoms is interpreted as having two apparent separate diagnoses.

Thus, depression, generalized anxiety, and alcoholism can be diagnosed as separate conditions but in a single individual might be more usefully regarded as different manifestations of the same disorder. Traditionally, European psychiatry has applied hierarchies of diagnosis in clinical practice. For current research purposes it might be prudent to retain the notion of clinical hierarchies alongside the idea of co-morbidity and use twin methods as a means of testing the merits and demerits of each approach.

PATTERN RECOGNITION APPROACHES TO SYMPTOM PROFILES: ALTERNATIVE CONSTRUCTIONS OF PSYCHOPATHOLOGY PHENOTYPES

Various twin studies point to the role of heredity in human brain waves. However, it turns out to be difficult to assess the genetic component reliably via traditional methods, thus motivating the application of pattern recognition techniques. EEG provides a very large number of potential variables (such as amplitude at 3, 5, 7, 9, ... cycles/second), analogous to the large number of variables from psychopathology assessment (such as MMPI profile scores, the 18 items of the Brief Psychiatric Rating scale, or neurological soft signs). To select the most useful variables for twin studies, Stassen (1985) chose the criterion of similarity of the profile over time. An iterative process of testing all possible permutations of subsets of variables was used. The sum of profile score differences between the first and second session was calculated for every possible pairing of every individual's first record with the second record of all other subjects and his own. The subset of variables that allowed the successful pairing of the largest number of the individual's first and second EEG recordings was selected. This method allowed 93% of individuals' records to be reliably identified after two weeks. A similar percentage was observed after five years, suggesting that EEG characteristics of the matured brain change slowly with age. The same methods could be applied to other behavioral data sets.

Once EEG spectral patterns were optimally constructed, the method of analysis was applied to samples of monozygotic (MZ) and dizygotic (DZ) twins brought up together and reared apart (Stassen et al. 1988a). The results yielded conclusive proof that the individual resting EEG pattern is predominantly determined by genetic factors: $h^2 = 0.825$ for adolescents ranging in age from 12 to 19 years (Minnesota twin/family study); $h^2 = 0.713$ for adults between 20 and 35 years old (Heidelberg twin study), and $h^2 = 0.655$ for adults aged 21 to 65 years (Minnesota study of twins reared apart). The corresponding quotients of within-pair DZ and within-pair MZ similarities were 0.499, 0.452, and 0.598, not too far from the theoretical value of 0.5 of the polygenic-additive model.

The success of adaptive strategies in EEG research as well as in various pattern recognition problems encourages the application of such a method to the question of reproducibly modeling the distinct individuality of psychopathology typically found in psychiatric patients. In particular, the consequent application of pattern recognition techniques allows one to decide upon (a) the optimum configuration of psychopathology features (including previous history, neurological, and somatic items); (b) the optimum measure of coincidence; (c) the existence of a "natural" partitioning of patients into homogeneous subgroups. After having calibrated the method with independent learn-and-test samples, it should be possible to assess reproducibly psychopathology coincidence in twins (Stassen et al. 1988b).

Conclusion

Twin studies are a powerful tool and can demonstrably be useful in refining psychiatric diagnoses where validity, not reliability, has been a key problem. Given the observation of subtypal variation in MZ twins (or even more puzzling in the case of the Genain quadruplets), it is difficult to accept qualitative-genotype distinctions. Spurious fractionation and proliferation of subtypes within a diagnostic category should be avoided until rigorous data are available. Application of this methodology to subtyping various psychiatric disorders (e.g., alcoholics) would be informative.

DO STUDIES OF MZ TWINS DISCORDANT FOR PSYCHOPATHOLOGY TELL US ABOUT ENVIRONMENTAL CAUSES OF DISEASE?

Discordant MZ Pairs as a Research Model

MZ twins discordant for a disease have been idealized as a model for studying environmental causes of a disease. The rationale has been that as MZ twins share all their genes, all differences between co-twins must arise from environmental causes. Thus, pairs discordant for a disease should permit the identification of environmental causes of the disease, in a research design comparable to a matched case-control design in epidemiology. Alternatively MZ pairs discordant for a risk factor or exposure of interest (e.g., smoking or alcohol use) can be studied to investigate the prevalence or incidence of disease. This is analogous to a matched cohort study in epidemiology. Studies of exposure discordant pairs on a large scale were initiated by Rune Cederlöf (1966) in the Swedish Twin Registry and have been utilized in the Finnish Twin Cohort study to demonstrate higher rates of mortality (Kaprio and Koskenvuo 1989), lung cancer (Kaprio and Koskenvuo 1990), atherosclerosis (Haapanen et al. 1989) and intervertebral disc changes (Battié et al. 1991) in smokers compared to their nonsmoking co-twins. The model can also be used in controlled clinical trials, such as that to test the hypothesized protective role of vitamin C against the common cold (Miller et al. 1977; Carr et al. 1981). In a longitudinal study of disease incidence involving the exposure discordant pairs, a strong inference on the causal relationship between exposure and disease may be made.

Causes of Discordance

In the disease discordant MZ pair model, other mechanisms may account for discordance, even when differences in environmental effects are found between affected and nonaffected co-twins. One concern is to assure that variability in age-of-onset of disease is not creating twin pairs apparently discordant for disease that upon sufficient follow-up become concordant. Knowledge of the natural history and distribution of age-of-onset of disease is necessary to form criteria for a sufficient time period of discordance.

Discordance may arise from other mechanisms. Studies of a large series of schizophrenics suggest that identifiable organic causes (e.g., cerebral injury, infections, drug use) account for less than 6% of all cases (Johnstone et al. 1987). Thus, is a healthy MZ co-twin of a schizophrenic index twin a control subject not carrying the liability to schizophrenia? This is not necessarily so, because (a) vulnerability might be expressed alternatively, e.g., as a neuropsychological or neurophysiological disturbance (smooth pursuit eye movement, SPEM) or as a personality trait (schizoidia), and (b) vulnerability might be present in the co-twin but strong protective factors are operating (consequently discordant MZ twins might oversample for protective factors in the unaffected twin). However, cases of complete discordance for schizophrenia in MZ twins without any schizoidia in the co-twin (Kringlen 1986, 1991) can be found and the study of such pairs is very important. Finally, twin pair discordance for exposure may in some cases be a consequence, not a cause, of twin pair discordance for psychopathology, e.g., due to differential recall of events by the affected and healthy twins. Interviews of parents, spouses, and siblings can be used to evaluate this type of bias.

Twin discordance has been observed in handedness and other measures of laterality (Buchsbaum 1974). A mirror image twin pair for a trait is defined by finding right lateralization in one twin and left lateralization in the other twin. Abnormal lateralization in schizophrenia has been observed with CT (Reveley et al. 1987), PET (Buchsbaum et al. 1991), and EEG (Flor-Henry 1979). While higher rates of left-handedness have not been shown in schizophrenia patients, left-handedness is different from brain asymmetry. Modern imaging techniques (Buchsbaum 1990) permit a more accurate, quantitative assessment of brain asymmetry and will resolve whether brain asymmetry is important in the etiology of psychopathology. The ongoing studies of Torrey, Gottesman and colleagues, on carefully defined MZ pairs concordant and discordant for schizophrenia, will be a very interesting sample for determining causes and consequences of disease discordance.

Finally, recent developments in medical genetics should remind us that "environmental" factors include random events, such as somatic mutation, that are not transmitted from one generation to the next but do involve changes in DNA. Such mutations are clearly relevant in disorders, such as cancer, that can result in discordance in identical twins, even in forms of malignancy where there is substantial inherited liability. Similarly, such phenomena as inactivation of gene expression by DNA methylation may occur on a random basis and thus result in phenotypic differences between individuals with an identical genotype.

Offspring of Discordant Pairs

The offspring of twin pairs discordant for psychopathology are uniquely informative for testing the hypothesis that the genotype for disease can be unexpressed and yet transmitted to the next generation. Comparisons of the risk of schizophrenia and affective disorder in offspring of healthy and diseased members of discordant identical twin pairs for schizophrenia (Fischer 1971; Gottesman and Bertelsen 1989; Kringlen and Cramer 1989) and affective disorder (Bertelsen 1992), respectively, indicate that the risk to offspring is not statistically different in the two groups. A strong role for genetic factors in schizophrenia and affective disorder is thus supported, while less evidence is found for the role of the rearing environment. It also suggests that other putative causes of discordance in MZ twins, such as new mutations, organic causes, or infections affecting only one member of a twin pair, are unlikely to account for a major part of schizophrenia (Gottesman and Bertelsen 1989). The study of the offspring of MZ pairs is a powerful design to disentangle the special effects of being an MZ twin from genetic effects. The offspring of MZ pairs are genetically half-sibs, while socially they are only cousins. They can be compared to the offspring of DZ pairs that are cousins both socially and genetically. Most of the power of the MZ half-sib derives from the intergenerational, not intragenerational, comparisons (Heath et al. 1985). However, offspring of MZ pairs are relatively uncommon, and the alternative strategy of obtaining half-sibs from second marriages of either parent has the advantage of being relatively common. However, such half-sib data may be biased due to the events (divorce, parental death) prior to the second marriage as well as the existence of a step-parent.

Conclusion

There is no single cause for discordance among MZ pairs discordant for psychopathology. Inferences about causality need to made carefully, even when differences in exposure experience are found between affected and nonaffected co-twins. Sufficient follow-up time is of course necessary to ensure that the unaffected co-twin does not become ill in the time period after the proband became affected. Careful studies of discordant twin pairs are nonetheless among the best methods available, because the genetic factor is under control.

Examination of the offspring of MZ discordant pairs are informative about unexpressed genotypes. Follow-up of the offspring of twins discordant for schizophrenia and affective disorder reveals rates of the same disorders in the children of normal co-twins as high as those in the children of the proband parent. It should be remembered, however, that these few studies are based on quite small numbers.

FROM GENOTYPE TO PHENOTYPE
Resolving the Genetic Architecture of Psychopathology

A growing number of monogenic diseases has been successfully analyzed down to the molecular level. Thus, monogenic diseases have taught us how a biochemical defect evolves from a single mutation, which parts of a gene are indispensable for normal function, and how phenotypes develop from different mutations. Based on these insights, molecular genetics can yield information on normal traits and common diseases. Normal traits and common diseases generally have a genetic contribution from more than one gene. The genetic architecture of a trait includes information on how many genes are involved, which genes are polymorphic, how many alleles exist, and what the allele frequencies are. Allele frequencies and average effects associated with the alleles determine the contribution of allelic variation (which can be further partitioned into additive genetic variance and variance due to dominance) to the overall genetic variation. Thus, the behavioral genetic approach should incorporate information on genetic variation in specific genes (the "measured genotype approach") (Sing and Moll 1990; Boerwinkle et al. 1986).

Biochemical and molecular genetics of coronary heart disease can serve as a useful example. The genetic architecture of traits involved in coronary artery disease is being studied in increasing detail (Sing and Moll 1990). Genes possibly involved in the etiology of disease are called candidate genes. Candidate genes may be used as targets with potential genetic variation leading to differences in the proteins encoded by these genes. These proteins are part of the physiological system that, when disturbed, gives rise to the disease being studied.

For example, genetic variation in only a few specific genes (apolipoprotein E, apolipoprotein B, and the LDL-receptor gene) accounts for about one-half of the total genetic variation (estimated by twin and family studies) in plasma cholesterol levels in the population. Further, for specific genes and some environments (diet, medications, and smoking), gene-gene interactions, gene-environment interactions and genotype dependent relationships of concomitants with serum lipid levels in lipid metabolism have been demonstrated (Reilly et al. 1992). Thus, simple linear models fail to grasp the complexity of the real world, indicating the difficulty of unravelling the contribution of genes and environment in the etiology of a complex disease.

A prerequisite for these "measured genotype" approaches is more knowledge of the intermediate physiological levels linking genes, gene products, and intermediate metabolites to biochemical and behavioral outcomes. Integration of information using data from specific genes with the behavioral genetic approach as it now stands, will be one of the main research strategies in the years to come.

In psychiatry, knowledge both of the responsible genes and of the intervening variables and relevant physiological systems is poor compared to knowledge of lipid metabolism. Knowledge of the base sequence of gene products essential for brain function (such as receptors, membrane proteins, enzymes, and gene control elements), however, enables the study of expressed genetic variation. This can be achieved by the examination of DNA extracted from blood samples. Variation in gene products, such as the series of dopamine or glutamate receptors that are expressed in defined brain areas (Anon 1992), will undoubtedly transform psychiatric genetics in the coming years. It remains to be shown, however, to what extent genetic variation in brain proteins is permitted, given the failure to uncover major differences in the enzymes of the Krebs cycle in the brain by protein electrophoresis (Cohen et al. 1973). In spite of the progress in our understanding of the molecular basis of brain function and its genetic variation, twin studies (e.g., examination of discordant MZ and DZ pairs) continue to possess their value in the field, since the relationship between genotype and phenotype promises to be the key issue of psychiatric genetics.

Causal Models and Twin Data

A twin model cannot distinguish between vertical and horizontal transmission of shared environmental factors unless data on parents of twins or offspring of twins are also obtained. Twin models can be used to test causal hypotheses (Heath, this volume), when the traits in question are reliably measured, preferably by multiple indicator variables. While cross-sectional analyses are not as good as longitudinal data, they are much cheaper and faster to carry out. Thus, they are useful to identify possible risk factors from cross-sectional data to test in future prospective studies, both of individuals and of families or twins.

Analysis of environmental events increasing the risk of disease has to consider whether there are genetic influences on the supposedly environmental agents. For example, in the Camberwell data on the relationship of recent life events to onset of depression, a positive association was found (McGuffin et al. 1988). However, life events were found to aggregate in families. Two mechanisms may account for this: either persons created life events or there is selective reporting of life events. The distribution of life events is not equal in the population, and there is twin data suggesting that there is a heritable component to controllable life events (Plomin et al. 1990; Kendler et al. 1991).

Conclusion

There are lessons to be learned in psychiatry from experiences with other medical diseases that have genetic components. The advent of molecular genetics has informed us of the genetic architecture of traits and diseases in much greater detail than earlier twin and family studies. The candidate gene approach is particularly promising for the analysis of common diseases, including psychiatric disorders. Although its relevance to disease is still unclear, molecular genetics has revealed complexities in the mechanisms of neurotransmission unknown to traditional biochemical and neuropharmacological approaches. For example, ligand-binding methods have revealed only two species of dopamine receptors, but recent molecular methods have revealed the existence of at least five subtypes on separate chromosomes. This may seem a long way from current twin studies but, in the near future, molecular methods will be integrated with twin studies. For example, DZ pairs can serve as age-matched sib pairs, while stratifying MZ pairs by genotype can illustrate the genotype-specific variability of the trait.

RESEARCH DESIGNS FOR FUTURE TWIN STUDIES OF PSYCHOPATHOLOGY

Until recently, complex designs in twin studies of psychopathology were not generally used. Thus, concordance studies provided us with first-hand impressions of the relative contribution of heredity and environment. Earlier twin studies investigated hospital-based series of schizophrenic twins, which led to an over-identification of concordant and severe cases. Population-based studies such as those from the Nordic countries are more valuable in establishing reliable concordance rates. In several studies, numbers of reported MZ twins are lower than expected, which means that some MZ twins most likely are incorrectly labelled DZ. This would tend to increase the difference between MZ and DZ twins with regard to concordance figures, since the majority of DZ twins are discordant. Also, sources of error due to unreliable psychiatric diagnoses were more common earlier. Since many studies have been carried out by only one investigator, knowledge about the clinical status of the index case might bias the evaluation of the second twin.

There are three principal sampling strategies for genetic studies of psychopathology using twins. Each of the three will be considered briefly.

Clinically Ascertained Twin Series

How representative are treatment samples of "cases" in the general population? There is, for example, overrepresentation of individuals with multiple diagnoses, or who are unresponsive to treatment in primary care settings. Reliance on "treated" samples may provide misleading results, e.g., the Virginia interview study of 1000 adult female twin pairs (Kendler et al., in review), which uses an epidemiologic sampling strategy, finds a heritability for narrowly defined DSM IIIR alcohol dependence on the order of 60%, while most studies of treatment samples find a moderately high heritability in males but little evidence for a genetic influence on alcoholism risk in females.

Clinical series may be the only feasible strategy for rare disorders, such as schizophrenia, attention deficit disorder, and bipolar disorder, unless we can screen very large populations of twin pairs. It is useful for resolving state vs. trait markers, e.g., in PET work comparing twins from discordant pairs.

Epidemiologic Twin Series

These are derived from data on birth certificates or population registries. Epidemiological series avoid confounding the genetics of treatment-seeking (or nonresponse to medication) with the genetics of incidence of the disorder. Treatment-seeking becomes a part of the course of the disorder. For a given sample size, the statistical power for resolving genetic and nongenetic hypotheses is a function of: (a) the prevalence of the disorder (e.g., 10% rather than 1%), and (b) the magnitude of twin pair correlations in liability under the threshold model (e.g., MZ correlations of 0.8 vs. 0.6 vs. 0.4). Therefore, they are impractical except for relatively common disorders (5–10% prevalence) or very highly familial disorders.

Epidemiological series are most useful for a "genetic epidemiologic" strategy, where we wish to explore how genetic factors covary and interact with other risk factors, and also for multivariate genetic analysis (e.g., the relationship between alcoholism and anxiety/affective disorders). Power is increased if we can achieve a continuous measure of liability in unaffecteds (cf. "blood-pressure") or can successfully identify "spectrum" cases which are intermediate in liability between cases and non-cases. Likewise, a "measured genotype" approach will have the same advantage.

"High-Risk" Strategies

"High-risk" strategies can be used to identify the offspring of twins from clinic series. It can also be a two-stage sampling strategy with an initial screening interview of an epidemiological sample followed by oversampling of pairs with (under different oversampling strategies) at least one affected parent, at least one affected twin, or at least one affected older sibling.

These high-risk strategies allow studies of the "precursors" of a disorder that are not so vulnerable to biases of retrospective data. Oversampling on twin affection status (e.g., conduct disorder) is a powerful strategy (but only if we are interested in later sequelae of conduct disorder, otherwise we would miss many cases of later onset alcoholism, etc.). Oversampling on parental affection status is only feasible for disorders with high parent-to-offspring transmission (e.g., alcoholism), otherwise the gain in efficiency over random sampling is slight. The sample size of our stage I population (prior to screening) is still the same as under epidemiological sampling. This procedure is only efficient if screening costs are low.

Additional Considerations in Research Design

In twin studies with multiple measurements and high factor loadings, genotypic and environmental factor scores can be estimated as part of a multivariate genetic analysis (Boomsma et al. 1990). This approach is useful for longitudinal studies in predicting which deviant scores arise from deviant genotypes and which from environmental effects. Complex segregation analysis (Perusse et al. 1991) can also be used on family data to identify major loci for continuously distributed traits, such as blood pressure.

In terms of *age-of-onset analysis*, the age-of-onset varies even in single gene disorders. The environmental and genetic determinants of age-of-onset may or may not be the same as for the disease itself, when a disease with a multifactorial etiology is analyzed.

Age-of-onset correlations for schizophrenia vary by degree of relatedness in twin and family samples (Kendler et al. 1987) and are very low for second-degree relatives.

Age-of-onset distribution should be taken into account in study design. The peak incidence of schizophrenia is in early adulthood, so feasible prospective studies of reasonable length could be designed to commence in mid-adolescence. Longer-term studies are of value, but researchers may find that all relevant data have not been collected at baseline, when the analysis of cases accumulated over time is undertaken.

For *rare disorders* that do not segregate in a Mendelian fashion, twin and family data have poor resolving power to distinguish traits with a few loci from traits with many loci. While discordance in MZ twins can reject single-locus traits with complete penetrance, the pattern of recurrence risk by degree of relatedness allows one to reject single locus or additive multilocus models (McGue and Gottesman 1989).

Assortative mating can bias estimates from twin analyses if it is not taken into account. Compared to a randomly mating population, assortative mating will increase additive genetic variance. However, if assortative mating exists in the parental generation and is ignored, a twin analysis will overestimate shared environmental effects, as assortative mating will increase the DZ correlation relative to the MZ correlation. Thus, the parents of twins or at least their spouses should be studied to assess the impact of assortative mating. By examining multiple variables from a data set, the degree and variation of the effect of assortative mating can be judged.

Conclusion

For rare disorders, or disorders which are only modestly familial, systematic ascertainment of twins from clinic series will usually be the only feasible strategy. The same approach is attractive when we seek to identify severe cases for biological studies, or to refine our diagnostic concepts for molecular genetic studies. The epidemiologic sampling strategies used by behavioral geneticists are very powerful if our primary focus is on more common disorders (e.g., alcoholism, depression), and particularly the role of genetic factors in co-morbidity, pharmacological response, or in the course of psychiatric disorders (e.g., relapse, treatment seeking). High-risk strategies, which screen an epidemiological sample in order to sample pairs with an affected family member, may improve the efficiency of such twin studies. Extensions of the twin design allow us to address such issues as assortative mating for psychopathology or nongenetic factors in parent-to-offspring influences.

REFERENCES

Anon. 1992. Glumate receptors: Genetic control of molecular and functional diversity. *Neurosci. F.* **3(10)**:1–2.

Battié, M.C., K. Gill, G.B. Moneta, R. Nyman, J. Kaprio, M. Koskenvuo, and T. Videman. 1991. Smoking and lumbar intervertebral disc degeneration: A study of identical twins using MRI. Winner of Volvo Award for research in basic science. *Spine* **16**:1015–1021.

Bertelsen, A. 1992. Twin research. In: Genetic Research in Psychiatry, ed. Mendlewiecz and Hippius, pp. 234–242. Berlin: Springer.

Boerwinkle, E., R. Chakraborty, and C.F. Sing. 1986. The use of measured genotype information in the analysis of quantitative phenotypes in man: I. Models and analytical methods. *Ann. Hum. Gen.* **50**:181–194.

Boomsma, D.I., P.C. Molenaar, and J.F. Orlebeke. 1990. Estimation of individual genetic and environmental factor scores. *Genet. Epidem.* **7**:83–91.

Buchsbaum, M.S. 1974. Average evoked response and stimulus intensity in identical and fraternal twins. *Physio. Psychol.* **2**:365–370.

Buchsbaum, M.S. 1990. The frontal lobes, basal ganglia, and temporal lobes as sites for schizophrenia. *Schizo. Bull.* **16**:377–387.

Buchsbaum, M.S., R.J. Tafalla, C. Reynolds, M. Trenary, L. Burgwald, S. Potkin, and W.E. Bunney, Jr. 1991. Drug effects on brain lateralization in the basal ganglia of schizophrenics. In: The Mesolimbic Dopamine System: From Motivation to Action, ed. P. Wilner and J. Scheel-Kruger. Chichester: Wiley.

Carr, A.B., R. Einstein, L.Y. Lai, N.G. Martin, and G.A. Starmer. 1981. Vitamin C and the common cold: Using identical twins as controls. *Med. J. Aust.* **2**:411–412.

Cederlöf, R. 1966. The twin method in epidemiologic studies on chronic disease. Ph.D. diss., Univ. of Stockholm.

Cohen, P.T., G.S. Omenn, A.G. Motulsky, S.H. Chen, and E.R. Giblett. 1973. Restricted variation in the glycolytic enzymes of human brain and erythrocytes. *Nature New Biol.* **241**:229–233.

Farmer, A.E., P. McGuffin, I. Harvey, and M. Williams. Schizophrenia: How far can we go in defining the phenotypes. In: The New Genetics of Mental Illness, ed. P. McGuffin and R. Murray, pp. 71–84. 1991. Oxford: Butterworth Heineman.

Fischer, M. 1971. Psychoses in the offspring of schizophrenic monozygotic twins and their normal co-twins. *Br. J. Psychiat.* **118**:43–52.

Fischer, M. 1973. Genetic and environmental factors in schizophrenia. *Acta Psychiat. Scand. Suppl. No.* 238.

Flor-Henry, P. 1979. On certain aspects of the localization of the cerebral systems regulating and determining emotion. *Bio. Psychiat.* **14**:677–698.

Gottesman, I.I. 1991. Schizophrenia genesis. The Origins of Madness. San Francisco: Freeman.

Gottesman, I.I., and A. Bertelsen. 1989. Confirming unexpressed genotypes for schizophrenia: Risks in the offspring of Fischer's Danish identical and fraternal discordant twins. *Arch. G. Psychiat.* **46**:867–872.

Haapanen, A., M. Koskenvuo, J. Kaprio, Y.A. Kesäniemi, and K. Heikkilä. 1989. Carotid arteriosclerosis in identical twins discordant for cigarette smoking. *Circulation* **80**:10–16.

Heath, A.C., K.S. Kendler, L.J. Eaves, and D. Markell. 1985. The resolution of cultural and biological inheritance: Informativeness of different relationships. *Behav. Genet.* **15**:439–465.

Johnstone, E.C., J.F. Macmillan, and T.J. Crow. 1987. The occurrence of organic disease of possible or probable aetiological significance in a population of 268 cases of first episode schizophrenia. *Psychol. Med.* **17**:371–379.

Kaprio, J., and M. Koskenvuo. 1989. Twins, smoking and mortality: A twelve-year prospective study of smoking-discordant adult pairs. *Soc. Sci. Med.* **29**:1083–1089.

Kaprio J., and M. Koskenvuo. 1990. Cigarette smoking as a cause of lung cancer and coronary heart disease: A study of smoking-discordant twin pairs. *Acta Genet. Med. Gem.* **39**:25–34.

Kendler, K.S., M.C. Neale, A.C. Heath, R.C. Kessler, and L.J. Eaves. 1991. Life events and depressive symptoms: A twin study perspective. In: The New Genetics of Mental Illness, ed. P. McGuffin and R. Murray, pp. 144–162. Oxford: Butterworth Heineman.

Kendler, K.S., M.C. Neale, R.C. Kessler, A.C. Heath, and L.J. Eaves. 1992a. A population-based twin study of major depression in women: The impact of varying definitions of illness. *Arch. G. Psychiat.* **49**:257–266.

Kendler, K.S., M.C. Neale, R.C. Kessler, A.C. Heath, and L.J. Eaves. 1992b. Generalized anxiety disorder in women: A population-based twin study. *Arch. G. Psychiat.* **49**:267–272.

Kendler, K.S., M.C. Neale, R.C. Kessler, A.C. Heath, and L.J. Eaves. 1992c. The genetic epidemiology of phobias in women: The interrelationship of agoraphobia, social phobia, situational phobia, and simple phobia. *Arch. G. Psychiat.* **49**:273–281.

Kendler, K.S., M.T. Tsuang, and P. Hays. 1987. Age at onset in schizophrenia: A familial perspective. *Arch. G. Psychiat.* **44**:881–890.

Kringlen, E. 1967. Heredity and Environment in the Functional Psychoses: An Epidemiological-Clinical Twin Study. Oslo: Oslo Univ. Press.

Kringlen, E. 1986. Status of twin research in functional psychosis. *Psychopathology* **19**:85–92.

Kringlen, E. 1987. Contributions of genetic studies on schizophrenia. In: Search for the Causes of Schizophrenia, ed. H. Häfner et al. Berlin: Springer.

Kringlen, E. 1991. Genetic aspects of schizophrenia with special emphasis on twin research. In: Etiology of Mental Disorder, ed. E. Kringlen, N.J. Lavik, and S. Torgersen. Oslo: Oslo Univ. Press.

Kringlen, E., and G. Cramer. 1989. Offspring of monozygotic twins discordant for schizophrenia. *Arch. G. Psychiat.* **46**:873–877.

McGue, M., and I.I. Gottesman. 1989. Genetic linkage in schizophrenia: Perspectives from genetic epidemiology. *Schizo. Bull.* **15**:453–464.

McGuffin, P., R. Katz, and P. Bebbington. 1988. The Camberwell Collaborative Depression Study: III. Depression and adversity in the relatives of depressed probands. *Br. J. Psychiat.* **152**:775–782.

Miller, J.Z., W.E. Nance, J.A. Norton, R.L. Wolen, R.S. Griffith, and R.J. Rose. 1977. Therapeutic effect of vitamin C: A co-twin control study. *JAMA* **237**:248–251.

Pérusse, L., P.P. Moll, and C.F. Sing. 1991. Evidence that a single gene with gender- and age-dependent effects influences systolic blood pressure determination in a population-based sample. *Am. J. Hum. Genet.* **49**:94–105.

Plomin, R., P. Lichtenstein, N.L. Pedersen, G.E. McClearn, and J.R. Nesselroade. 1990. Genetic influence on life events during the last half of the life span. *Psychol. Aging* **5**:25–30.

Reilly, S.L., R.E. Ferrell, B.A. Kottke, and C.F. Sing. 1992. The gender specific apolipoprotein E genotype influence on the distribution of lipids and apolipoproteins in the population of Rochester, MN: II. Regression relationship with concomitants. Am. *J. Hum. Genet.*, **51**:1311–1324.

Reveley, M.A., A.M. Reveley, and R. Baldy. 1987. Left cerebral hemisphere hypodensity in discordant schizophrenic twins: A controlled study. *Arch. G. Psychiat.* **44**:625–632.

Sing, C.F., and P.P. Moll. 1990. Genetics of atherosclerosis. *Ann. Rev. Genet.* **24**:171–187.

Stassen, H.H. 1985. The similarity approach to EEG analysis. *Meth. Inform. Med.* **24**:200–212.

Stassen, H.H., D.T. Lykken, P. Propping, and G. Bomben. 1988a. Genetic determination of the human EEG: Survey of recent results on twins reared together and apart. *Hum. Genet.* **80**:165–176.

Stassen, H.H., C. Scharfetter, G. Winokur, and J. Angst. 1988b. Familial syndrome patterns in schizophrenia, schizoaffective disorder, mania and depression. *Eur. Arch. Psychiat. Neur.* **237**:115–123.

Tienari, P. 1963. Psychiatric illnesses in identical twins. *Acta Psychiat. Scand.* Suppl. No. 171.

Author Index

Subject Index

Note: Page numbers in *italics* refer to illustrations; those in **bold** refer to tables

g 5–7, 37
  critics of 19
  existence and importance of 90–1
  levels 17, 22
  multiple view on meaning and/or causes
      of 93–4
  see also intelligence; IQ
Galton, Francis, history of twins and
      1–4
Genain quadruplets 258–66, 269, 290
  clinical and psychosocial assessment
      258–9
  EEG, resting 260, *261*
  EEG tomography 258, *261, 262–3,
      264,* 266
  positron emission tomography (PET)
      258, 259, 265–6, *265, 266*
  ventricular size 267
gene mapping 200
genetic covariance structure 151, 166
genetic dominance 25, 94
genetic–environmental (GE) correlations
      7, 86, 89–90, 116–17, 171–2, 209,
      276–7
  in alcohol use and abuse 277
  behavioral strategies 128–30
genetic–environmental covariation
      (G,E,COV) 24
genetic–environmental interaction (G ×
      E) 7, 23–4, 25, 89–90, 102, 116–17,
      171–2, 189
  in alcohol use and abuse 277–8
  in personality 147–60
  measure of 117
genetic linkage analysis *see* linkage gene
      approach
genetic polymorphisms 62
genetic simplex 150
genetic variation 35
genotype 167
  measured approach 293
Gesell infant development scales 77
gestational age 228
Gilles de la Tourette (GTS) syndrome
      196, 232
glucose uptake in brain 73
growth 218–20
  long-term 219–20
  retardation 219
growth curve models 73
Guilford 20

h^2 35
handedness 62, 73, 291
Hanna, Davoren 26
Hawaii Family Study 43
Headstart programs 23
height 60–1, 99
heritability
  age-specific 69–73
  coefficients 11, 19
  differential 149–50, 151–3
  estimation of 11–12
heterotypic continuity 188
hierarchical factor model 43
hierarchical path model 43, *44*
hierarchy negotiation 96
home environment 53–6, 77
Howe, Michael 18
Huntington's disease 175
hydrops fetalis 219
hyperactivity 188
hyperactivity–conduct disorder
      comorbidity 188
hypofrontality and functional imaging 268

idiot savants 27, 94
illegitimacy 57
immune disorders 73
inbreeding studies 94–5
infancy and childhood, personality studies
      in 152
infant health 101
infant–infant relationships 223
infidelity detection 96
inspection time 100
intelligence
  age changes 26–7
  alternative definitions 91–3
  definition and characterization of 90–4
  directional dominance 94–5
  distinction between creativity and 20
  effect of nutrition 28
  fluid (*gf*) 30
  general accounts of 22
  genetic and environmental architecture
      87–8
  heritability of 18, 86–7
  role of twin studies 97–8
  see also g; IQ
intelligence–creativity comparison 20
intelligence tests 69
interactionism 17, 22–5, 29–30

Index compiled by Annette J. Musker